Dental Applications of Nanotechnology

Ramesh S. Chaughule
Editor

Dental Applications of Nanotechnology

Springer

Editor
Ramesh S. Chaughule
Ramnarain Ruia College
Mumbai, India

ISBN 978-3-030-07381-7 ISBN 978-3-319-97634-1 (eBook)
https://doi.org/10.1007/978-3-319-97634-1

Softcover re-print of the Hardcover 1st edition 2018

This Springer imprint is published by the registered company Springer Nature Switzerland AG
The registered company address is: Gewerbestrasse 11, 6330 Cham, Switzerland

This book is dedicated to
Professor R. Vijayaraghavan
Ex Dean, Tata Institute of Fundamental
Research, Mumbai, India
A Mentor, Teacher, Advisor, Inspirator and
Everything

And my wife
Kshama
A constant spirit, Supporting and Delightful
partner through thick or thin in life

Foreword I

Imagine a day when a drop of medicine could be placed on a cavity to kill bacteria and then regenerate the parts of the tooth that were destroyed by microorganisms, or when an injection of stem cells could be placed into the jaw of a car driver just after a car accident to rebuild broken bones, or when an injection of a unique nanomedicine could be placed near wisdom teeth to degrade them so that invasive surgery to remove them would not be necessary. What a change in dental health care these advances would be! These are true revolutions in dental medicine, and are the advances we need to help millions of people around the globe have better dental health, and to promote proper nutrition, self-esteem, and life expectancy.

These and so many more ideas are brought to life in this exciting new book by Dr. Ramesh Chaughule entitled Dental Applications of Nanotechnology. Dr. Chaughule brilliantly intertwines material science with medicine to highlight unprecedented growth areas across all of dental medicine. The focus of such advances relies on nanotechnology, not the unrealistic vision of nanorobots in the body surveying and healing diseases at will, but the more realistic design and use of materials in medicine with dimensions less than 10^{-9}m. For those of you having trouble understanding this dimension, consider that the diameter of a single strand of hair is about 80,000–100,000 nm and we cannot even discern nanometer resolution with your unaided eye. This is small and powerful!

Nanomaterials are excellent candidates for dental medicine, since our teeth, jaw bones, and all tissues in the body are composed of nanomaterials, like proteins and calcium phosphate. Cells in our bodies make nanomaterials every second and live in nanomaterials every day. Dr. Chaughule acknowledges this and emphasizes in this book how to leverage this simple fact to improve all aspects of dental health. It is because of these reasons that nanomedicine is experiencing a boom in research and activity across all of medicine. As just one of many examples, nanomedicine is projected to be a global market worth $528 billion by 2019, which is almost double that from 2014.[1] We have not even reached the tip of the iceberg in the capabilities what nanomedicine can bring. Moreover, with over a hundred nanomedicine products approved by the FDA, it is clear that nanomedicine is here to stay and will continue to revolutionize medicine. This pioneering book highlights just that.

This book covers fundamental research, applied clinical studies, and commercialization potential across all of dental medicine. It is comprehensive and presents ideas we need to improve dental care for patients and, most importantly, stimulates new ideas rarely discussed in other books. It is an excellent resource for any educator, medical device industry person, entrepreneur, clinician, and simply any person interested in science, engineering, and medicine. After reading this book, it is hard to imagine anyone not seeing the promise nanomedicine will have in dentistry!

So, do not put away that toothpaste just yet, but a nanomedicine revolution in dental care is right around the corner—this book shows it! Dr. Chaughule's pioneering book in this area gets us all thinking how dental care will significantly change in the coming years, and we all need to pay close attention!

Boston, MA USA

Thomas J. Webster, Ph.D.
Art Zafiropolou Chair; Department Chair, Chemical Engineering, Past-President, US Society For Biomaterials; Fellow, AANM, AIMBE, BMES, IUSBE, and NAI
Northeastern University

[1] Commercialization of New Technologies Driving Big Market Growth in Nanomedicine, BCC Research LLC; accessed May 25, 2018, at https://www.bccresearch.com/pressroom/hlc/commercialization-of-new-technologies-driving-big-market-growth-in-nanomedicine.

Foreword II

It gives me great pleasure to write the foreword for this book which presents a collection of topics from eminent authors on the role of nanotechnology and nanobiomaterials in dental health.

From its humble beginnings in the 1970s, nanotechnology has today become mainstream in almost every aspect of science and engineering. Nanotechnology involves the manipulation of individual atoms and molecules in the 1–100 nm range to produce unique and interesting structures with myriad applications in physical, chemical, and biological systems. To date, it has had a huge impact in fields as diverse as semiconductors and electronic devices, production of fuels, electrochemical energy conversion and storage, advanced composite structures for the aerospace industry, improving air and water quality, as well as medical applications including diagnostics and therapeutics. In particular, nanotechnology has arguably had the greatest impact in medicine, for example, with the use of nanoparticles to deliver chemotherapy drugs or vaccines to specifically targeted cells, insulin release via nanocapsules, the removal of toxins from the bloodstream via nanosponges, use of nanoparticles as free radical scavengers, as well as diagnostic tools such as the use of carbon nanotubes coated with antibodies to detect cancer cells, monitor the level of blood-borne gases, and many others. This book now extends the field of nanotechnology and nanobiomaterials to new and exciting applications in dental health.

Written by a group of leading experts, this book presents several important chapters that address dental applications of nanotechnology. Periodontal disease and tooth decay represent two of the biggest threats to dental health. Apart from local infections of the structures around the teeth, periodontal disease has also been linked to other health problems such as heart disease, diabetes, and respiratory disease. Therefore, improving dental health can benefit not just one's teeth and gums, but also the entire body. Two chapters contained herein describe the therapeutic applications of nanomaterials to reduce the inflammation of the dental tissues and promote bone regeneration. Prosthodontics is also a major focus; three chapters relate the application of nanobiomaterials to the improvement of oral function and appearance of patients with deficient teeth, oral, and maxillofacial tissues. Similarly, orthodontics is covered by two chapters which address the use of nanotechnology-tailored agents to combat biofilms, for example. Other chapters delve into the application of nanomaterials to the soft inner tissue of the teeth or pulp, the properties of advanced dental nanocomposites, and the role of nanomedicine in the assessment and treatment of oral biofilms, as well as diagnosis and therapeutic drug delivery in dentistry.

The editor, Dr. Chaughule, should be commended for assembling a group of experts from across the globe to present a comprehensive and timely book on the dental applications of nanotechnology. The book should appeal equally to scientists and researchers, students, and practitioners in the field. I fully expect that the readers will find the material useful and enjoyable.

Newark, DE, USA

Ajay K. Prasad
Engineering Alumni Distinguished
Professor and ChairDepartment of
Mechanical Engineering University of
Delaware
prasad@udel.edu

Preface

Nanotechnology has immense applications in almost all the fields of science and human life. Nanoparticles constitute a crucial and technology-intensive area of research and development in the burgeoning field of nanotechnology and nanoscience. Scientists are exploring many research areas to understand bulk materials at the nanoscale. Engineered nanoparticles with promising properties have been tailored and produced on a technical scale. Nanotechnology is one of the most popular areas of current research and has developed in multiple disciplines, including dentistry. The foremost goal of dentistry is the rehabilitation of the stomatognathic system. Nanotechnology-based treatment modalities like nanomaterials and nanorobots are finding their way in routine dental health care. Unlike bulk materials, nano-sized particles are quite unique in nature because of the increase in surface-to-volume ratio which alters their physical, chemical, and biological properties which trigger chemical activity with distinct crystallography. Nanoparticles comprise a size range from 10 to 100 nm in diameter. Various methods have been employed for the synthesis of the nanoparticles. The two approaches mainly used are bottom-up and top-down (more details are given in the book). The primary aim of restorative dentistry is to restore the form and function of the tooth. The extensive range of restorative materials being manufactured should combine innovation with long-lasting clinical success. The physical properties and handling characteristics of these restorative materials should constantly improve with time, enabling dental professionals to meet the varying demands of dental patients and the different requirements of practice. Nanotechnology has made significant inroads into the fields of preventive, reconstructive, regenerative, restorative, rehabilitative, and diagnostic domains.

I am pleased to introduce this book, “Dental Applications of Nanotechnology,” to aspiring and working scientists, dental practitioners, and as a ready reference for the dental students to understand the principles of nanotechnology, its applications, and latest techniques. This book covers important topics such as pulp/periodontal regeneration, tissue engineering, restorative dentistry, endodontics, prosthodontics, orthodontics, and therapeutics.

Implantable nanomaterials can be applied in various fields, such as tissue healing and substitution, coatings for implants, tissue regeneration scaffolds, implant materials, osseous repair, bioresorbable materials, smart materials, and diagnostic and therapeutic devices. The chapter by Porenczuk discusses tissue engineering processes that require cell lines, bioactive molecules, and supporting matices, such as synthetic polymers, including bioglass, to be used as regenerative treatment of the pulps. The use of nano-bioglass and hydroxyapatite nanocomposites in the field of regenerative endodontics is also introduced. Periodontal regeneration leads to the formation of new bone, cementum, and periodontal ligament on a previously diseased root surface. Deepa and Arunkumar in their chapter discuss this issue using various sizes of nanoparticle graft materials in the treatment of intrabony defects, bone regeneration around implants, and its role in tissue engineering. Recent developments in nanomaterials and nanotechnology have provided a promising insight into the commercial applications of nanomaterials in the management of periodontal diseases. The Chapter by Arjunkumar discusses the various ways in which nanotechnology has influenced the field of periodontics, in the form of nanodentifrices, dental hypersensitivity cure, drug delivery systems, antibiofilm approaches and resolution of inflammation. Dental restorative resins are explored to further enhance their physical and mechanical properties, as the traditional dental materials usually show weak mechanical properties, elastic modulus, and poor abrasion resistance. Nanomaterials are used in the preparation of nanocomposites. These are resin-based composites with inorganic filler particles, a coupling agent, and polymerization initiator. Chaughule et al. have synthesized the composites using titania to show enhanced mechanical, chemical, and biological properties than that of materials available in the market. The Chapter by Hend Mahmoud Abou El Nasr and Makbule Bilge Akbulut shows how nanotechnology has invaded every aspect in endodontics. This includes improvement in radiography, local anesthesia, dentin hypersensitivity, root canal disinfection, endodontic filling materials, and functionalization/conjugation. Regenerative endodontics and endodontic surgery are also improved by using nanobiomaterials. Nanomaterials play an important role in innovation and clinical technological changes in the field of prosthodontics. It is an important branch of oral health care and rehabilitation. The chapter by Jadhav mainly focuses on the various applications of novel nanomaterials in the field of dentistry and the advances in nanotechnology, with a focus on promising applications in prosthodontics. Using silver nanoparticles, she has explored suitable applications in acrylic resin, tissue conditioner, dental adhesives, dental porcelain, dental composites, dental cements, implants, and maxillofacial prosthesis. The chapter by Aeran and Seth illustrates that nanotechnology applied to implants also increased osseointegration by 150%, which is quite a significant result. In their chapter, they have discussed the applications of nanometals, nanoceramics, nanoresin, and other nanomaterials in prosthodontics. The performance of composites can also be enhanced by adding appropriate nanomaterials. Nanomaterials have been playing a significant role in basic scientific innovation and clinical technological change of prosthodontics. In another chapter on prosthodontics, Praveena and her coauthors have elucidated how bulk materials, when reduced to the nanoscale, change the physicochemical properties of materials. They also present comprehensive information regarding the recent advancements of nanobiomaterials with respect to removable, fixed, and maxillofacial prosthodontics and their advantages and limitations. In addition, they have discussed the significance of nanotechnology in the field of implant prosthodontics. Orthodontics is a field of dentistry that deals primarily with malpositioned teeth and jaws. The chapter by Lekhadia enlightens the application of nanotechnology in biomaterials and biomechanics in orthodontics and its use to improve and speed up orthodontic treatment. Another chapter on orthodontics by Batra focuses on various materials whose properties can be modified by the application of nanoparticles, and describes tests that can be performed to detect the physical and biological properties of the new materials. She discusses the advantages and disadvantages of using nanoparticles, and the precautions that one needs to take while researching with nanoparticles. Biofilms form when bacteria adhere to surfaces in some form of watery environment and begin to excrete a slimy, glue-like substance that can stick to implant materials, biological tissues, etc. One promising approach to combating these biofilms is based on nanotechnology-tailored agents. Shetty and Gupta have explored this technique by conventional approaches that could be augmented by interference with the factors that enable the cariogenic bacteria to escape from the normal homeostatic mechanisms to restrict their growth in plaque and outcompete the organisms associated with health. Further, they focus on recent research on the creation, characterization, and evaluation of nanoparticles for the prevention or treatment of biofilms in the oral cavity.

Nanotechnology presently faces many technical, ethical, and biological challenges. Particularly, due to their small size, nanoparticles can cause various adverse health effects. Hence, there is a critical requirement to standardize nanotechnology-based products and devices and improve our understanding of how to exploit the benefits while diminishing the risks. Bhardwaj et al. have enlightened this situation in their chapter by describing nanotechnology in depth and explain the importance of nanoencapsulation and nanotherapeutics used in dental drug delivery systems. Besides technology improvements, there are also risk safety assessments of crucial interest for further developments in this field. Dental materials should be harmless to all oral tissues, and should not contain leachable and diffusible toxic substances, which could pass into circulatory system and contribute to systemic toxicity responses. The chapter by Dragana et al. deals with the construction and physical characteristics, biocompatibility, bioactivity, and biofunctionality of new materials based on active silicate systems and hydroxyapatite. They have suggested the use of endodontic cement based on dicalcium and tricalcium silicate and hydroxyapatite for further clinical trials.

The editor wishes to thank all the distinguished and expert contributors for their enthusiastic participation in this endeavor. I am confident that the book will serve as a valuable guide for researchers and students of dentistry, materials engineering, bioengineering, and medicine. Support from Dr. Suhas Pednekar, Principal, Ramnarain Ruia College, Mumbai, and my family members is gratefully acknowledged. Last but not least, the editor sincerely thanks the Springer staff for showing faith in bringing out this book to their expectations.

Mumbai, India Ramesh S. Chaughule

Contents

1 **Nano-materials in Regenerative Pulp Treatment** 1
Alicja Porenczuk

2 **Nanobiomaterials and Their Role in Periodontal Rehabilitation** 23
D. Deepa and K. V. Arunkumar

3 **Advanced Nanomaterials and Their Functionalization in Clinical Endodontics** 37
Hend Mahmoud Abou El Nasr and Makbule Bilge Akbulut

4 **Nanocomposites and Their Use in Dentistry** 59
Ramesh Chaughule, Dipika Raorane, Suhas Pednekar and Rajesh Dashaputra

5 **Applications of Nanoparticles in Orthodontics** 81
Panchali Batra

6 **Nanomaterials: A Boon to Prosthodontics** 107
Rajashree Dhananjay Jadhav

7 **Nanomaterials: The Changing Phase of Prosthodontics** 121
Himanshu Aeran and Jyotsna Seth

8 **Nanostructures in Dentistry: In Diagnosis, Drug Delivery and Oral Cancer Therapy, and their Biocompatibility** 133
Archana Bhardwaj, Abhishek Bhardwaj and R. Nageswar Rao

9 **Nanotechnology in Orthodontics—Futuristic Approach** 155
Dhaval Ranjitbhai Lekhadia

10 **Nanobiomaterials and Their Application in Prosthodontics** 177
Channamsetty Praveena, Prakash Manne, Lohitha Kalluri and Ravikanth Anne

11 Nanomaterials for the Management of Periodontal Diseases 203
Radhika Arjunkumar

12 Oral Biofilms: From Development to Assessment and Treatment 217
Heeresh Shetty and Pankaj Gupta

13 Physical Properties and Biocompatibility of Nanostructural Biomaterials Based on Active Calcium Silicate Systems and Hydroxyapatite 247
Opačić-Galić Vanja, Petrović Violeta, Popović-Bajić Marijana, Jokanović Vukoman, Živković Slavoljub, Nikolić Biljana and Mitić-Ćulafić Dragana

Correction to: Advanced Nanomaterials and Their Functionalization in Clinical Endodontics E1
Hend Mahmoud Abou El Nasr and Makbule Bilge Akbulut

Index 273

Chapter 1
Nano-materials in Regenerative Pulp Treatment

Alicja Porenczuk

1.1 The Construction of Tooth's Tissues

Tooth's hard tissues, such as enamel and dentin, are built of both the organic and inorganic compounds, the latter made of a basic unit called a hydroxyapatite (HPA; general chemical structure $Ca_{10}(OH)_2(PO_4)_6$). HPA and its derivatives, such as fluorohydroxyapatite (FHA; general chemical structure $[Ca_5(PO_4)_3OH_{1-x}F_x]$ and carbonate hydroxyapatite (CHA)), make the most of the enamel's structure. Although the chemical composition of the enamel's HPA is like the one forming bone, it contains less carbonates, sodium and magnesium, and forms needle-like crystal network [1]. The enamel is unique, as it is the only epithelial-derived calcified tissue in vertebrates. Its hardness, resulting from a high mineral content, sets between iron and carbon steel, yet its elasticity is higher than that [1]. Dentin is a complex, bone-like structure forming the bulk of the tooth. It is much softer and more elastic than the enamel due to a high protein content (20–30 wt%), most of which is collagen type I followed by non-collagenous proteins and proteoglycans [1, 2]. The non-collagenous proteins (phosphorylated and non-phosphorylated matrix proteins, proteoglycans, metalloproteinases) and growth factors (transforming growth factor beta 1 (TGF β1), fibroblast growth factor (FGF-2), insulin-like growth factor (IGF-I), IGF-II, platelet-derived growth factor (PDGF), vascular endothelial growth factor (VEGF), LIM mineralization protein 1 (LMP-1)) are considered to be important in dentin's both mineralization and remineralization processes [3–5]. Dentin and enamel are bound at the dentin–enamel junction (DEJ) [2]. The enamel protects the tooth from the occlusal forces, whereas dentin's elasticity protects hard, yet brittle, enamel from crashing under them. Dentin's inner structure is formed by closely packed 18,000 and 21,000 tubules/mm^2 dentinal tubules, the diameter of which varies between 2

A. Porenczuk (✉)
Restorative Dentistry Department, Warsaw Medical University,
Miodowa 18 St., 00-246 Warsaw, Poland
e-mail: alicja.porenczuk@gmail.com

R. S. Chaughule (ed.), *Dental Applications of Nanotechnology*,
https://doi.org/10.1007/978-3-319-97634-1_1

and 4 μm [2]. The dentinal tubules are more numerous in the inner third layer than the outer third layer of the tissue. Dentin's layer close to DEJ is called mantle dentin, and though it is less mineralized than the rest of the tissue, its elasticity protects the overlying enamel from detaching from DEJ [2, 6]. The inner, or circumpulpal, dentin makes the bulk of the tissue and is composed of intertubular and peritubular dentin. The thickness of this layer continuously increases (4 mm/day) at the expense of the space initially occupied by the pulp. When demineralized, the intertubular dentin reveals a dense collagenous network, whereas the peritubular dentin shows a thin network of non-collagenous proteins and phospholipids [2].

Dentin's formation is proceeded by the odontoblasts and depends on the stage of tooth's development and its response to various stimuli [2]. A tight linkage between the pulpal cells and the dentin makes it clear why in the literature they are referred to as "the dentin–pulp complex." The secondary dentin, formed throughout the tooth's lifetime, can be physiological, related to the normal aging process, and/or pathological, deposed as a mechanical barrier during carious attack or non-carious injuries. In the last two cases, the deposed barrier (it is also known as the tertiary or reactionary dentin) lacks the phosphorylated proteins [2]. The presence of reactionary dentin may be related to both the virulence of the bacteria, speed of carious process and/or to chemical irritation caused by dental materials. The interactions between the dentin–pulp complex and dental materials may be influenced by both the materials' features, such as chemical structure, composition, and concentration of any eluted components or degradation products, and the cells' response to them. Nowadays, the attention is paid to the understanding of how dental materials may contribute to the regenerative processes of the caries-affected tissues. There are numerous studies focused on their cytotoxicity, and so, for example, zinc oxide/eugenol (ZOE) may cause a severe pulp irritation [7]. Also, approximately 40% of teeth reconstructed with the amalgam fillings manifest moderate to severe pulp inflammation, compared to 24–48% filled with composite polymer resins [8]. Even the materials regarded as biocompatible, such as glass-ionomer cements (GICs) or calcium hydroxide (CH), can affect the pulpal cells' vitality. Accorinte et al. proved that even the most widely used pulp-capping materials, such as CH and/or mineral trioxide aggregate (MTA), could induce chronic pulp inflammation attributable to their high pH, even though the tissue bridge was eventually created [9]. The cytotoxicity of GICs is still a subject of a dispute, as they may decrease the number of pulpal cells by approximately 32.5% and their metabolism by 42.5% [10].

1.2 Caries

Tooth decay is an infectious disease caused by acid-producing bacteria and their metabolism products. Its occurrence and development were once attributed to low socioeconomic status in developing countries, resulting from the lack of effective medical care and prophylaxis programs, yet at present this has been redeployed to industrial, wealthy communities, such as the USA or Norway, where tooth decay is rapidly evolving, especially in children [11]. It is a well-known fact that a coexistence of three factors, such as cariogenic diet rich in fermentable sugars, bacteria biofilm on the tooth's surfaces, and the individual susceptibility, favors caries incidence [11]. Recent findings from the Human Microbiome Project showed that dental caries is related to bacteria shift in the biofilm composition [12]. Along with frequent sugars uptake, the ecology of the dental plaque moves towards more acidic-tolerant species. A novel approach states that the biofilm accumulated on tooth's surfaces contains both harmless and cariogenic species. The occurrence of specific conditions, such as dramatic fluctuations in nutrient availability, pH, oxygen tension, and osmolality, would dictate the transition between the species and their virulence [13, 14]. The environmental stress provokes bacteria responses to biofilm formation, competence development, and acid tolerance. The co-regulation processes are regulated by a set of two components—a kinase able to sense the bacterial environment, which is activated by an auto-phosphorylation, and the response regulator that modulates gene expression at one or more promoter sites [14]. Although little is still known about the etiology and role of each species in caries development, three major values decide upon its incidence—adhesion, or effective competitiveness in the biofilm, ability to produce acids, and acid tolerance [14]. It is a well-known fact that *Streptococcus* (*S.*) spp. is responsible for the initiation of the carious process, with *S. mutans* being the pioneer bacteria associated with its onset. *S. sobrinus* plays an important role in caries development, as it produces more acids than *S. mutans* [13]. Further progression and maturation of the carious lesions are caused by *Lactobacillus* spp., whereas root caries development is promoted by acidophilic and aciduric *Actinomyces* spp., *Atopobium* spp., *Olsenella* spp., *Pseudoramibacter* spp., *Propionibacterium* spp., and *Selenomonas* spp. [15].

Remaining dentin thickness (RDT) is defined as the amount of healthy dentin separating the bottom of the carious lesion from the pulp. With the decrease in this distance, the permeability of the remaining dentin increases [16, 17]. The dentin's permeability plays an important role, as the bacteria and chemical agents may penetrate through the opened dentinal tubuli causing the cells irritation. Dentin's tubular structure decides upon its permeability to bacteria and their toxic products of metabolism. As the dentinal tubuli number increases toward the pulp chamber, the acidic irritation may evoke the pulp inflammation or necrosis. Pulp's biological response to various stimuli or injury is dependent on the existing cell population. Dentin matrix contains a wide range of bioactive molecules with potent cell signaling properties, which may be released into the pulp during injury [18]. At the end of the twentieth century, Roberts-Clark and Smith speculated that the presence of angiogenic growth factors, such as the platelet-derived growth factor (PDGF-AB) and vascular endothelial growth factor (VEGF), may stimulate new capillary formation in the injured sites [19]. This has been confirmed by other studies indicating that the pulpal cells would secrete large amounts of PDGF-AB, VEGF, and fibroblast growth factor 2 (FGF-2), and this process could be stimulated by HEMA present in polymer restorative resins [20]. However, the role and depth of RDT during restorative treatment, and the influence of applied materials in terms of their distance from the pulp, have been a controversy. A layer of RDT < 3 mm is believed to be the pulp's critical barrier when the total-etch bonding agents are applied [10]. For example, the usage of the resin-modified liners, such as Vitrebond™ (3M ESPE, St. Paul, MN, USA) or Ultra-Blend™ plus (Ultradent Products, Inc., South Jordan, UT, USA), caused no inflammation or tissue disorganization, even with the RDT = 272 μm [10]. In other study, the cells' injury was observed in 33.7% when the RDT = 0.5–0.01 mm [17].

1.3 Fluoride and Remineralization Process

The term "remineralization" refers to a process of carious tissue repair induced by a continuous flow of fluoride, calcium, and phosphates ions between the oral fluids and the tissues. The HPA's solubility depends on the level of calcium and phosphate ions, present in both the tooth's structures and the saliva. The mutual ionic flow in the environment, regulated by pH fluctuations during the acidic attack, may contribute to the onset of the enamel decay. When the pH value returns to neutral, the ions incorporation into the tooth's tissues promotes a process called recrystallization, which can repair the damaged site at the very early stage [11, 12, 21]. Fluoride is one of the most electronegative elements and, when ionized, is highly attractive to the hydroxyl ions building the HPA. Calcium fluoride serves as a reservoir of fluoride and is only decomposed in very low pH. Since the electrostatic forces are higher for combined calcium and fluoride ions than for calcium and the hydroxyl ions, the first combination provides better physical–chemical stability of the crystalline network. Therefore, the acidic solubility of FHA is lower than the HPA's [11]. Although high amounts of fluoride within the carious-affected tissues seem to be crucial for the remineralization process, it was proved that frequent exposures to small amounts of fluoride protect dental tissues even better. Even a minor (0.03–0.11 ppm) increase of fluoride ions in the oral cavity through fluoride slow-release devices would reduce adult caries imminence by 64% [21]. The same result can be obtained through periodic topical fluoride gel applications or providing bioactive restorations freeing small amounts of fluoride ions at a constant level [22].

1.4 Minimally Invasive Dentistry

Minimally invasive dentistry (MID) involves ultraconservative, atraumatic carious tissue removal. The direct cavity excavation, meaning complete caries removal, bears the risk of the pulp exposure and need for the endodontic treatment. The discolored dentin left at the bottom of the cavity, demineralized yet hard in clinical inspection, may undergo remineralization if the cavity is completely sealed off the outer environment (Fig. 1.1). Thus, the bacteria remnants in the infected tissue have no conditions for further development, and the progression of the lesion is stopped [23, 24]. Deep cavities preparation bears the risk of implementing of the endodontic treatment, in which the inflamed pulp is removed and the system of the root canals is cleaned,

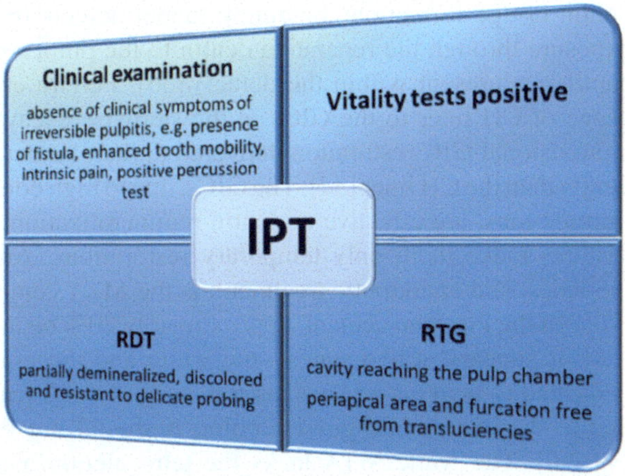

Fig. 1.1 Selection criteria for the IPT procedures

shaped, and filled with biologically inert material. The idea of leaving partially demineralized tissue in the cavity is supported by the fact that the bacteria persisting in the cavity are sealed off the nutrients delivery. The minimally invasive therapies of deep carious lesions involve an indirect pulp treatment (IPT), in which the bulk of the caries-affected dentin is removed and the demineralized dentin is left at the bottom. The IPT is divided into two other procedures—a one-sitting partial removal and stepwise excavation. In the stepwise excavation, the carious tissue is removed in one sitting and the cavity is temporarily restored with a biologically active material. At the interval, the temporary restoration and the remnants of the carious tissue are removed and the cavity is sealed with a permanent filling.

Studies on the efficiency of the IPT procedures show success rates varying from 69 to 97% [24–28]. The MID strategy involves also the usage of remineralizing materials, which would promote the repair process within the cavity. The CH and/or MTA, thanks to their chelating ability and high pH, can extract bioactive proteins and the metalloproteinases metallo proteinases (MMPs) from dentin. Acidic agents, such as GICs and their derivatives, dentin adhesive systems, can also induce this process. The extracted biomolecules are involved in cell signaling and differentiation, which lead to the extracellular matrix deposition and mineralization. Therefore, the remineralization process of the demineralized dentin may be reached through the biomaterials application in the cavity. Protection of the pulp–dentin complex against irritants is done with dental materials called liners or sealers. One of the most studied and recommended for application in deep carious lesions is the CH liner [29, 30]. Chisini et al. have shown that the majority of Brazilian dentists used the CH as first-choice material for direct (86.3%) and indirect (80.3%) pulp protection [31]. The stepwise excavation using CH cement as liner showed clinical and radiological success, even after 4 years of observation [32]. However, the CH cement cannot adhere to dentin and is hydrolytically degradable over time, leaving empty spaces underneath the restoration [33]. Moreover, it may induce mild to severe inflammation of the cells adjacent to the application site [34, 35] and may lead to an incompetent dentin bridge formation containing tunnel defects (patency sites leading from the exposure through the reparative dentin to the pulp, sometimes with fibroblasts and capillaries present within the defect) [36]. Pereira et al. compared the short-term usage of CH liner to the GICs in the stepwise excavation and have observed that a provisional GIC restoration delivered darker, harder, drier, and less contaminated dentin than the CH liner [37]. The GICs, due to high content of released and recharged fluoride ions, are effective in dentin remineralization, yet their limited mechanical features make them only temporary restorations. Another material used in MID therapies, and endodontic treatment, is the MTA cement, which is made of powder (70% of the Portland cement, approximately 20% bismuth oxide and 5% of gypsum) [38]. It appears in two states—the white and the gray cement. The gray MTA is dedicated to heavy load sites, as it is built of 1–10 μm powder particles. However, due to the unaesthetic grayish color, it should not be used in the aesthetic zone [39, 40]. The white MTA lacks the tetracalcium aluminoferrite, and the powder particles are smaller than the gray MTA (1–30 μm). Therefore, it is more aesthetically acceptable, yet at the cost of the physical properties [41, 42]. Generally, all types of the MTA are hydrophilic in nature and need water for setting [43]. The mean setting time is 165 min, which means that the restoration must be divided into two separate appointments [44]. Also, the application is quite problem-making for the clinicians [40]. The morphology of joint of the MTA cement to dentin is close to the hybrid layer, with tag-like structures of the cement infiltrating deep into the tissue. Therefore, it can be rated as the biofixation [43]. The comparison in the efficiency of clinical usage of the white MTA and the CH liner showed that MTA produced significantly thicker dentin bridge and caused less inflammation in a 90-days observation period than the CH [34, 35]. The direct pulp capping with the MTA was found to be more reliable, with less inflammation signs and predictable dentin bridge formation [30]. Kundzina et al. have observed the 85% success rate in direct pulp capping with the MTA, yet no significance between the MTA and the CH liner application and the postoperative pain's occurrence [45]. Elshamy et al. proved that the antibacterial activity of MTA was significantly higher than the CH liner, especially against *L. acidophilus* [46]. In the overall evaluation, the MTA was found to be less costly, as the further retreatments were avoided [47].

Recent advances in dental materials sciences aim at introducing bioactive materials ready to induce biomineralization, bear heavy occlusal loads, and behave differently depending on the situation. Biodentine (Septodont, Saint Maur des Fossées, France) is a restorative material based on calcium silicates. It can be used as a dentin substitute, due to high physical properties and bioactivity [48, 49]. It comprises an encapsulated powder and a liquid. The powder is tricalcium and dicalcium silicates ($3CaO{\cdot}SiO_2$; $2CaO{\cdot}SiO_2$) and calcium carbonate ($CaCO_3$), whereas the liquid consists of calcium chloride ($CaCl_2 * 2H_2O$) and a hydro-soluble polymer [50]. The application of the material into the cavity is done with a spatula, without any additional preparation. Biodentine's application is particularly needed in the IPT techniques. Serving as the dentin's replacement, it should be partially removed after the control time, with the rest of the material left as a liner. Koubi et al. observed that Biodentine was suitable for posterior teeth restoration for up to 6 months [51]. The operating time is longer for Biodentine than for GICs [51]. In their studies, Nowicka et al. and Tran et al. compared Biodentine and the MTA cements, and have proved that they both were similarly effective in the cell proliferation induction. Moreover, the clinical application of both materials resulted in a homogenous dentin bridge formation over directly exposed pulp [29, 48]. However, it was revealed that Biodentine should not be used as a dentin replacement under a polymer restoration, as leakage was detected [50]. Also, when acid etched, significant changes to its structure and a lower calcium to silicon ratio would occur [48]. It may be freely used as a baseline under GIC restorations, though. Biodentine showed a high washout and resorption values, and the addition of admixtures seemed to affect the physical properties of the material [52].

1.5 Characteristics of Dental Biomaterials

Biomineralization is the process of minerals deposition within or outside of the cells [53]. Dental materials capable of bonding to the tooth's tissues and inducing repair are called the biomaterials. Bioactivity is defined as the material's ability to adhere to the living tissue [54]. After placement, the chemical interaction between the biomaterial's surface and the tissue's compounds takes place—this process is called a "bioactive fixation" [3]. Biomaterials in dentistry include bioceramics divided into a large area of other subjects, e.g., glass ceramics, silica-based glass (popularly called the bioglass), zirconia, alumina, titania [55]. Generally, they are described as bioactive, biocompatible, and resorbable. The bioglass, which is the most widely used biomaterial, can form FHA crystals, thus inducing biomineralization. Moreover, this process proved to be thermodynamically stable at the physiological pH, which, in case of pulpal cells irritation caused by caries, may be decisive upon their survival. The ratio of calcium to phosphorus in bioglass is stable in body fluids, which decides on its prolonged bioactivity and resistance to being chemically interfered [55].

The involvement of both the life sciences and the engineering in medicine to provide solutions for the damaged tissues' repair or improvement is called tissue engineering [56]. This area of medicine covers usage of isolated cells or their substitutes, growth factors, and/or scaffolds (membranes, bases), on which the new tissue could be developed [57]. The idea of creating a biologically active material that would contain all substituents indispensable for the cells adhesion and cultivation on its surface arose. The tissue engineering together with the regenerative medicine (TERM) determines the involvement of material science research aimed at obtaining a material that would be biocompatible and easily degradable, without leaving any toxic remnants [58]. The regenerative endodontics is one of the TERM's areas of interest and is aimed at reparation of the pulp–dentin complex [59]. Even though most of the synthesis methods are still on the verge of science fiction, some of the created materials were stable enough to find their place in a clinical practice. For instance, barrier membranes are used in periodontal surgery, particularly to support the periodontal tissues regeneration by blocking the epithelial cells migration to

Table 1.1 Comparison of polymer scaffolds' characteristics

Advantages of different types of scaffolds	
Natural polymers	Synthetic polymers
Very high biocompatibility	High biocompatibility
Support cell attachment, migration, and proliferation	Support cell attachment, migration, and proliferation Can be functionalized with bioactive molecules, e.g., growth factors
Naturally possess cells-to-scaffold binding ligands (Arg-Gly-Asp (ROD) binding sequences)	Require incorporation of cells-to-scaffold ligands
Mimic the natural substrates needed for cell functioning	Allows cells to attach, grow, and differentiate
Complete bioresorbability	Some, but not all, are biodegradable Various degradation kinetics
Graft acceptance	High risk of rejection due to reduced bioactivity
Limited application in load-bearing sites	Can be used in load-bearing sites
High elasticity	High stiffness (brittleness)
Easy to shape for implantation	Difficult to shape for implantation
Difficult to provide homogenous structure	Fabricated in predetermined shapes and composition Fabrication ability to provide mesoporous scaffold

the operated site. Smart bioactive materials are used in the restorative dentistry and enhance the remineralization process through stimulating of the ionic flow. Since the chemical structure of HPA is close to the bone, the studies on its exploitation in TERM evolved, leading to the usage of bioactive scaffolds in bone regeneration, especially in orthopedics [11, 57].

TERM provides tissue regeneration via cells application into the damaged sites. The foundation (scaffold) acts as a template for the cells to adhere and proliferate, thus providing the healing effect. The bioactive particles are dispersed in the scaffold, which, in this case, acts as a 3D network for all the biological actions. The scaffolds may be either natural or synthetic polymers. Their properties are depicted in Table 1.1 [60–64].

The scaffolds need to be porous so as to facilitate the water flow through them and enable the ions liberation, cells migration and proliferation as well as its vascularization in the tissue [62]. The porosity should also facilitate the flow of the nutrients to the cells and their waste products out of the scaffold site. The density and size of the pores are determined by the specific surface area, in which the cells adhere. Large pores are crucial for the cells to adhere to the cells-to-scaffold binding ligands and, simultaneously, small enough to establish a high specific surface, with a minimal ligand density to allow efficient binding of a critical number of cells [64]. The scaffolds should also possess properties that determine their application in TERM, among which the biocompatibility, biodegradability, and mechanical strength adequate to usage are required [58]. There are various types of scaffolds used in TERM. Biomaterials, ceramics, natural and synthetic polymers may be distinguished in this area. All of them possess different properties, which allow them to be used in specific situations. Due to high mechanical properties, ceramic scaffolds such as HAP and tricalcium phosphate (TCP) are mostly used for bone regeneration procedures, whereas the biological materials are useful in all situations where an excellent cell adhesion and growth are required.

1.6 Nano-materials

The term "nano-materials" refers to materials built of nano-sized (1–100 nm) particles, thus possessing different properties from their normal-size equivalents. In restorative dentistry, these include nano-hydroxyapatite (nano-HPA) composites, nano-bioactive glass (nano-BAG) composites.

1.6.1 Nano-hydroxyapatite (Nano-HPA)

The mechanical properties of HPA make it inapplicable for bearing heavy mechanical loads, e.g., orthopedic appliances [65]. Nano-hydroxyapatite (nano-HPA) powders possess improved sinterability and enhanced densification, which may improve their physical properties [57]. Along with the biocompatibility, their bioactivity level was described as being higher than their coarser equivalents [66]. The synthesis of multiform nano-HPA is technically challenging and may require using various methods, which are depicted in Fig. 1.2.

To mediate and control the nucleation, growth, and the stability of the nano-crystals, various precipitants are used, among which citrates (calcium/citrate/phosphate solutions), acidic amino acids, and ethylenediamine tetraacetic acid (EDTA) stand out [67, 68]. The nano-HPA powder particles or nano-HPA nano-fibers (60-600 nm), in most cases obtained through electrospinning, had a positive impact on the polymer composites' properties. Their addition enhanced the polymers' mechanical stability, collagen-supported cell proliferation and provided a substrate (nano-HPA's surface) for the mineralization process [55, 69]. The nano-HPA particles dispersed on the surfaces of various scaffolds, e.g., poly(3-hydroxybutyrate) (PHB), showed interesting morphology, with the bioglass sticking out of the nano-HPA nano-fibers (spindle-like morphology) [55, 70]. The morphology of the nano-HPA-PHB complex was rough enough to induce the cell attachment and their further mineralizing activity [55]. In other studies, where the nano-HPA particles were attributed on biodegradable scaffolds such as the collagen, poly(lactic-co-glycolic acid) (PLGA) or collagen–polycaprolactone (PCL), their addition was

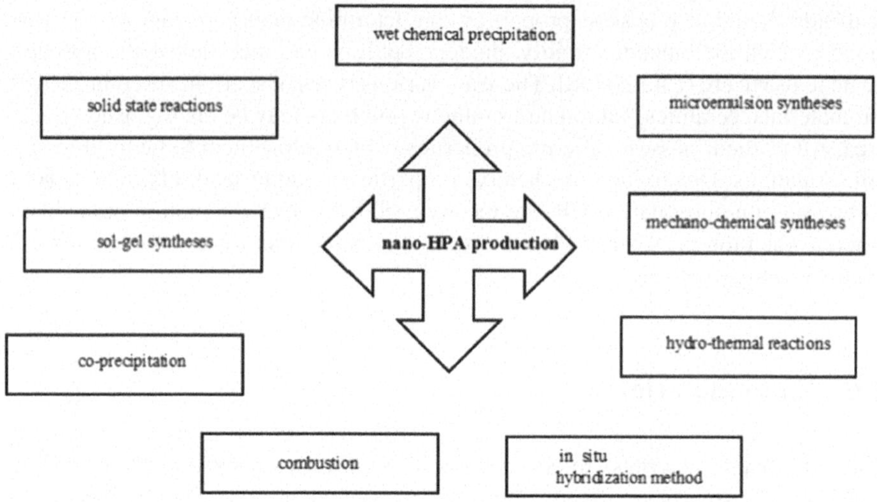

Fig. 1.2 Methods of synthesis of nano-HPA

beneficial, as the mineralization taking place on their surfaces increased the bioactivity of the whole scaffold [55, 71, 72].

1.6.2 Nano-bioactive glass (Nano-BAG)

Bioactive glass was developed as an inhibitor for the interfacial mobility of the implant inserted into the vivid tissue [73]. The first material containing bioglass, introduced by Hench in 1969 and further ameliorated in 1991, is called Bioglass 45S5 (chemical composition: 45 wt% SiO_2, 24.4 wt% CaO, 24.5 wt% Na_2O, 6 wt% P_2O_5) [60, 73]. Its synthesis had a revolutionary response, as it was easily interacting with the living tissue. Founded by the United States Army Medical Research and Development Command, the novel material's usefulness in quick rehabilitation of the wounded soldiers was undisputable. Hench proved that a bioglass of a chemical characteristic of $SiO_2 \cdot CaO \cdot Na_2O \cdot P_2O_5$ could biologically bond to the living tissue. At first, the requirements placed before the bioactive materials were the biocompatibility and non-toxicity of their derivatives or degradation products. Shortly, making usage of the remineralizing properties of the bioglass, resulting from the calcium and phosphorous content, became a primary goal for many scientists. Novel mesoporous bioglass particles possess similar chemical characteristics as their predecessor, yet due to the mesoporous structure, large surface area, and high porosity, they are far more bioactive [74].

The bioactivity can be measured with a bioactivity index (BI), which divides the bioglass materials into four separate categories (Fig. 1.3). At first, it was believed that

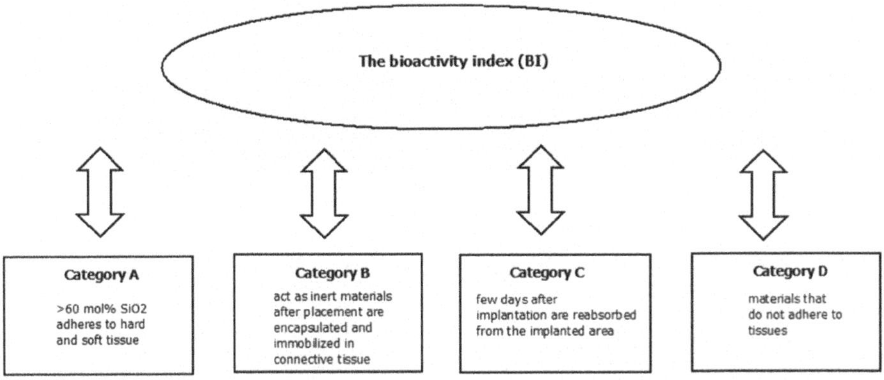

Fig. 1.3 A schematic division of the nano-BAG materials

the bioglass material should release phosphorous for the mineralization process to occur. Lately, it is been stated that the bioglass composed of only the $SiO_2 \cdot CaO \cdot Na_2O$ would exhibit spectacular remineralizing potential, even with SiO_2 to up to 85 mol% [75]. Generally, the bioglass' bioactivity remains unchanged, even if the CaO would be replaced by MgO or CaF_2 and/or Na_2O would be replaced by K_2O. However, the addition of fluoride, which is indispensable in the remineralization process, decreases solubility and shifts the material's position in "A–C" categories. The addition of multi-covalent cations shrinks the "A" category and may completely inhibit the material's bioactivity.

In order to obtain the desired adhesion of the bioglass material to the bone, a layer of CHA must be formed on the surface of the implanted material. The ability to form this layer results from the bioglass reactivity, which may be divided into phases (Fig. 1.4). The amorphous layer built of calcium phosphates and later crystallized forms the apatite layer able to induce the biofixation [76]. The kinetics of the glass filler dissolution is strictly dependent on its chemical structure and individual elements content [77]. The research conducted on the bioglass dissolution and its ability to release fluoride ions has implied that materials containing up to 15% of fluoride would be more effective in HPA formation than the ones with higher fluoride content [77, 78]. High concentrations of calcium and phosphates ions may lead to the amorphous calcium phosphate formation. The remineralizing potential of bioglass is decreased in contact with serum proteins [77].

The apatite formation on the scaffolds' surfaces may both improve the cell-to-cell, scaffold–cell, and scaffold–tissue interactions. The adhesive zone created on the scaffold's surface helps to withstand most of the mechanical forces acting on it. Recent reports have indicated that the development of nano-bioactive glass (nano-BAG) particles can improve the absorption of the remineralizing ions and enhance the mechanical properties of all bioglass materials [79]. The nano-BAG (20–30 nm) can be obtained in a sol-gel method [61]. Foroughi et al., using the electrospinning technique, synthesized a nano-composite comprising of 9 wt% polyhydroxybutyrate, 10,

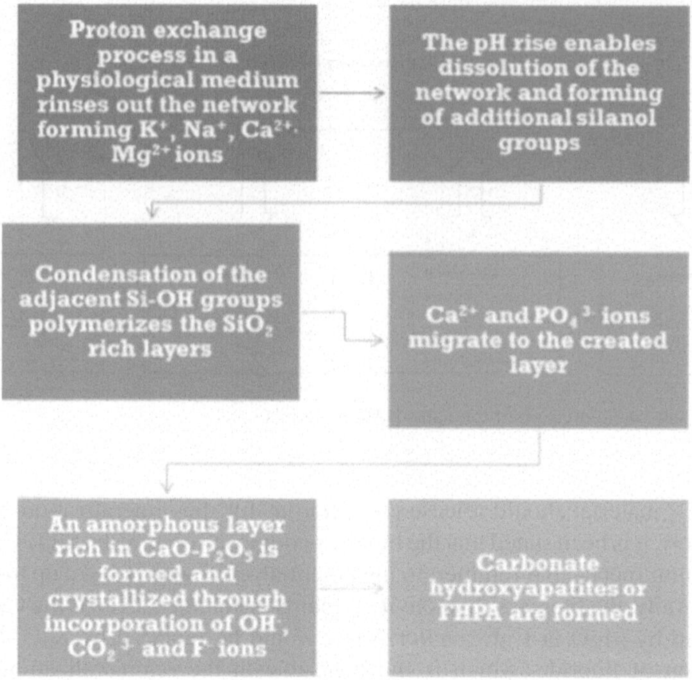

Fig. 1.4 A schematic presentation of the CHA layer formation on the bioactive material's surface

15, and 20 wt% chitosan, and 7.5, 10, and 15 wt% nano-BAG (35–55 nm). The uniform morphology of the obtained nano-BAG fibers has led to a significant enhancement of the tensile strength (3.42 MPa), which was even four times greater than the control sample [79]. Due to their high surface energy and ability to move toward the direction of stretching, the nano-BAG particles would act as temporary cross-links and, thus, enhance the scaffold's mechanical features. It may seem that along with the increasing level of the nano-BAG in the material, the physical endurance would be higher; thus, higher levels of the added nanoparticles should be favorable. However, high amount of the nano-BAG particles (15 wt%) has led to their agglomeration on the scaffold's surface and formation of the stress points, which weakened the material [79]. The most suitable amount of the nano-BAG added to the scaffold was set to be 10 wt%, as the rising nanoparticles content had a significant impact on Young's modulus, leading to a decrease in the scaffold's flexibility [79].

The porosity of the scaffold plays a crucial role in the cells migration and proliferation on its surface. The pores enable vascularization of the scaffold and its integration with the tissue [61, 79]. Establishing of how much nano-BAG should be added to the scaffold so as not to have a great impact on the porosity is thus essential. Foroughi et al. showed that 15 wt% of the nano-BAG enrichment resulted in larger pores. However, the overall porosity was low, as the agglomeration of the nano-BAG occurred [79]. Maji et al. have determined that along with the increase in the wt%

of the nano-BAG in the scaffold, the pore sizes diminished and their shape became irregular [61].

As the nanoparticles possess different properties from their normal-sized analogues, both the chemical and physical properties can be boosted up to the desired level. The chosen BAG biomaterials' characteristics and application are presented in Table 1.2. Maji et al. have added 30 wt% of nano-BAG particles into a natural biopolymer-based composite scaffold and have obtained a maximum compressive strength of 2.2 ± 0.1 MPa, hydrophilicity and biodegradability. Their material was also non-toxic and supported the mesenchymal cell attachment, proliferation, and differentiation, which were confirmed in various cytotoxicity assays (MTT, RUNX-2 expression). The authors speculated that the material's stability in aqueous environment could have been obtained through the chemical interactions between the gelatin, chitosan, and silica phases compounds [61].

As the nano-BAG is known for its proliferation enhancement ability, it may be necessary to specify its properties before application. Moorthi et al. put an effort to investigate the nano-BAG particles with various amounts of CaO and SiO_2 ratio and their role on the osteoblast proliferation. They found out that the nano-BAG material ($SiO_2{\cdot}CaO{\cdot}P_2O_5$; mol% ~70:25:5) stimulated the osteoblast proliferation and promoted more cells to enter G2/M cell cycle phase than the other nano-BAG material ($SiO_2{\cdot}CaO{\cdot}P_2O_5$; mol% ~64:31:5) [80]. Thanks to that, the nano-BAG proved to be suitable for orthopedic and periodontal tissue engineering applications [81].

The nano-BAG incorporation into the conventional GIC structure was introduced, so as to combine the remineralizing potential of both the glass filler from GICs and the nano-BAG. Moreover, it was suspected that an enhancement in both the biological and mechanical properties could be obtained through the nano-BAG addition. Only 10–30 wt% of the nano-BAG particles are recommended to be added to a GIC powder, as it was observed that the higher nano-BAG particles content, the lower was the compressive strength. This could have resulted from the reduction of Al^+ ions content in the GIC powder when replaced with the nano-BAG particles [82]. Kim et al. speculated that nano-BAG addition to a conventional GIC structure would increase the surface area and, thus, enhance the biomineralization capability. The investigation of mechanical and biological properties of nano-BAG particles (amorphous and spherical in shape, 42 nm in diameter, 5 wt%) incorporated into the conventional GIC revealed that the enhanced material manifested similar curing time as the conventional GIC (6 min.) and a minimal weight loss over 28 days of the observation period. The cytotoxicity of this material was tested on the immortalized human dental pulp stem cell line (ihDPSC) using the MTS assay and the alizarin red staining. Both tests confirmed no significant differences in cytotoxicity of this material and revealed the presence of the mineralized nodules. The material has also shown good biological surface activity, which resulted in high biomineralization capacity, mechanical properties, such as compressive (200.1 ± 15.9 MPa), diametral tensile (11.3 ± 1.9 MPa), and flexural strengths (24.2 ± 2.2 MPa) [82]. To compare the nano-BAG-enhanced GIC cements, the material comprising of a conventional GIC powder enriched with synthesized nano-ceramic particles (nano-HPA combined with nano-fluoroapatite and N-vinylpyrrolidone) was investigated. In

Table 1.2 Chosen biomaterials characteristics and application

Investigator	Active particle	Active particle's characterization	Method of synthesis	Scaffold	Scaffold's characterization	Mechanical properties	Application
Maji et al. (2016) [61]	Nano-BAG	57.44% SiO_2, 35.42% CaO, 7.15% P_2O_5 Spherical 20-30 nm	Sol-gel	Gelatin/chitosan nanocomposite	Nano-BAG addition 10-30 wt% 3D interconnected network due to chitosan-gelatin cross-linking Average pore sizes 100-250 μm Mean porosity 81-89% Pore size distribution in the scaffold 100–400 μm Hydrophilic Degradability rate (27.5% / 16 days)	Young's modulus with 10 wt% of nano-BAG (55 ± 7.12 MPa) Compressive strength with 10 wt% of nano-BAG (1.2 ± 0.01 MPa)	Bone regeneration

(continued)

Table 1.2 (continued)

Investigator	Active particle	Active particle's characterization	Method of synthesis	Scaffold	Scaffold's characterization	Mechanical properties	Application
Foroughi et al. (2017) [79]	Nano-BAG	46.13% SiO_2, 48.03% $CaCO_3$, 41.64% Na_2CO_3, 3.59% H_3PO_4 Spherical 35-55 nm	Electro-spinning	Polyhydroxybutyrate/chitosan nanocomposite	Nano-BAG addition 7.5, 10, 15 wt% Fully porous, uniform, bead-free Average pore size 12 μm Average fiber diameter 354 ± 72 nm Hydrophilic Less prone to agglomeration Degradability rate (~65% / 60 days)	Young's modulus with 10 wt% of nano-BAG (188.67±0.02 MPa)	Bone regeneration Dentistry
Kim et al. (2017) [82]	Nano-BAG	Atomic ratio Si:Ca 85:15 Spherical, amorphous Size 42 nm Density 2.51±0.02 g/cm^3	Sol-gel	GIC cement enhanced with chitosan	Nano-BAG addition 5 wt%	Compressive strength with 5 wt% of nano-BAG (200.1±15.9 MPa) Tensile strength with 5 wt% of nano-BAG (11.3±1.9 MPa)	Dentistry Pulp tissue regeneration

(continued)

Table 1.2 (continued)

Investigator	Active particle	Active particle's characterization	Method of synthesis	Scaffold	Scaffold's characterization	Mechanical properties	Application
	GIC glass filler	Density 3.06±0.01 g/cm^3				Flexural strength with 5 wt% of nano-BAG (24.2±2.2 MPa)	
Caridade et al. (2013) [86]	Nano-BAG	46.02% SiO_2, 27.18% CaO, 22.96% Na_2O, 3.77% P_2O_5 Spherical 30-50 nm	Flame spray synthesis (nano-BAG) Solvent casting (scaffold)	Chitosan composite	Hydrophilic	Young's modulus with nano-BAG (20 MPa) Young's modulus with micro-BAG (17 MPa)	Dentistry Pulp tissue regeneration
	Micro-BAG	45% SiO_2, 24.5% CaO, 24.5% Na_2O, 0.6% P_2O_5 Irregular 5 μm					
Xu et al. (2015) [89]	Nano-BAG	Approx. 30 nm	Electro-spinning with pulsed laser deposition	Composite scaffold with a micro-pattern, nano-sized fiber matrix	Hydrophilic	Not defined	Re-epithelialization of skin wounds
	Nano-fibers	500 – 1000 nm					

SiO_2 – silicon dioxide; CaO – calcium oxide; P_2O_5 – phosphorus (V) oxide; BAG – bioactive glass; wt% – weight percent; $CaCO_3$ – calcium carbonate; Na_2CO_3 – sodium carbonate; H_3PO_4 – phosphorus acid; Na_2O – sodium oxide; GIC – glass-ionomer cement; approx. - approximately their study, Moshaverinia et al. proved that after 24 h of setting, the diametral tensile and biaxial flexural strengths of the modified glass-ionomer cements exhibited higher compressive (184 MPa), diametral tensile (22 MPa), and flexural strengths (33 MPa), as compared to the control group [83]. In a similar study where the nano-HPA/nano-fluoroapatite were added to the conventional GIC, the improved cements exhibited relevantly higher compressive (177–179 MPa), diametral tensile (19–20 MPa), and biaxial flexural strengths (26–28 MPa) as compared to the control group (160 MPa in compressive strength, 14 MPa in diametral tensile strength, and 18 MPa in biaxial flexural strength). What's more, their bond strengths to dentin were higher after 7 and 30 days of storage in distilled water [84]. The composition of the nano-BAG (chemical composition close to Bioglass 45S5: 46.08 wt% SiO_2, 22.96 wt% Na_2O, 27.18 wt% CaO, 3.77 wt% P_2O_5, particle size 30–50 nm) and chitosan resulted in a more intensive calcium and phosphorous release with subsequent formation of the a HPA layer (thickness of ~1.2–0.2 μm). The nano-BAG incorporation into chitosan has also inclined its hydrophobic character to hydrophilic [85, 86]. The conclusion from these reports clearly states that the nano-BAG addition to different materials would enhance their mechanical properties. The calcium ions concentration on nano-BAG-enhanced membranes in the presence of the human periodontal ligament cells was relevantly higher than bare chitosan composite membranes, suggesting that inorganic particles would let out high levels of the remineralizing ions [87]. In comparison with the nano-HPA particles set in chitosan membranes, the addition of nano-HPA would decrease both the tensile strength and elongation at failure, but increased the elastic modulus. These membranes were much more flexible under wet conditions, compared to chitosan/nano-BAG membranes [88].

Xu et al. have successfully prepared a composite scaffold with a controlled micropattern, nano-sized fiber matrix (500–1000 nm), and surface-modified nano-BAG component (approximately 30 nm) for wound healing. The composite scaffold consisted of akermanite (Ca, Mg, and Si-containing bioceramic stimulating angiogenesis), and the nano-bioactive glass particles were deposited on the surface of the fibers through pulsed laser deposition technique [89, 90]. The authors claimed that the hydrophilicity of dense scaffolds could be increased by placing hydrophilic inorganic nano-sized particles on their surfaces. The process of coating of the scaffold with nano-BAG through pulsed laser deposition technique led to a hydrophobic-to-hydrophilic transition of the scaffold, which then released most of the Ca, Mg, and Si ions during the first 24 h of the immersion period. Their material manifested a wound healing ratio of 95%, significantly higher than the control (76%) within 11 days. Their results indicated that not only was the material useful at accelerating wound healing, but also improved the quality of the produced tissue through enhanced angiogenesis [89].

1.7 Summary

The acidic activity of the carious bacteria may cause a severe damage or death of the primary pulp cells—the odontoblasts. The severity of this damage depends on the bacteria shift in the biofilm and their virulence. The remaining pulpal cells are able to migrate to the damaged site and differentiate into the odontoblast-like cells, possessing the ability of the dentin components secretion. Thus, reparation of the damaged carious lesion can be gained. A minimally invasive dentistry aims at preserving as much tissues as possible through the carious-affected tissue partial removal and application of bioactive materials ready to induce the remineralization process within the cavity. The tissue engineering and regenerative medicine require the presence of main components—an appropriate cell line, bioactive molecules, and supporting matrices. The latter, called scaffolds, may be either natural or synthetic polymers or bioceramics, including bioglass. At present, the nano-bioactive glass and nano-hydroxyapatite are in the spectrum of scientific research. They were proven to be biocompatible to different cell lines. Moreover, their addition to scaffolds can enhance their mechanical properties. Their usage in regenerative dentistry is thus promising.

References

1. Bartlett JD (2013) Dental enamel development: proteinases and their enamel matrix substrates. ISRN Dent 2013(684607):1–24
2. Goldberg M, Kulkarni AB, Young M, Boskey A (2012) Dentin: structure, composition and mineralization: the role of dentin ECM in dentin formation and mineralization. Front Biosci (Elite Ed) 3:711–735
3. Ferracane JL, Cooper PR, Smith AJ (2010) Can interaction of materials with the dentin–pulp complex contribute to dentin regeneration? Odontology 98:2–14
4. Wang X-Y, Zhang Q, Chen Z (2007) A possible role of LIM mineralization protein 1 in tertiary dentinogenesis of dental caries treatment. Med Hypotheses 69:584–586
5. Tziafas D (2004) The future role of a molecular approach to pulp-dentinal regeneration. Caries Res 38:314–320
6. Wang RZ, Weiner S (1998) Strain-structure relations in human teeth using Moiré fringes. J Biomech 31(135–41):31
7. Brännström M, Nyborg H (1976) Pulp reaction to a temporary zinc oxide/eugenol cement. J Prosthet Dent 35(2):185–191
8. Chandwani ND, Pawar MG, Tupkari JV, Yuwanati M (2014) Histological evaluation to study the effects of dental amalgam and composite restoration on human dental pulp: an in vivo study. Med Princ Pract 23:40–44 [PubMed: 24217468]
9. Accorinte MLR, Loguercio AD, Reis A, Carneiro E, Grande RHM, Murata SS, Holland R (2008) Response of human dental pulp capped with MTA and calcium hydroxide powder. Oper Dent 33(5):488–495
10. de Souza Costa CA, Hebling J, Garcia-Godoy F, Hanks CT (2003) In vitro cytotoxicity of five glass-ionomer cements. Biomaterials 24:3853–3858
11. ten Cate JM (2013) Contemporary perspective on the use of fluoride products in caries prevention. Brit Dent J 214(3):161–167
12. Strużycka I (2014) The oral microbiome in dental caries. Polish J Microbiol 63(2):127–135

13. Peterson SN, Snesrud E, Schork NJ, Bretz WA (2011) Dental caries pathogenicity: a genomic and metagenomics perspective. Int Dent J 61(01):11–22
14. Lemos JA, Abranches J, Burne RA (2005) Responses of cariogenic streptococci to environmental stresses. Curr Issues Mol Biol 7:95–107
15. Preza D, Olsen I, Aas JA, Willumsen T, Grinde B, Paster BJ (2008) Bacterial profiles of root caries in elderly patients. J Clin Microbiol 46(6):2015–2021
16. Camps J, Dejou M, Remusat I, Remusat M, About I (2000) Factors influencing pulpal response to cavity restorations. Dent Mater 16:432–440
17. Murray PE, Smith AJ, Windsor LJ, Mjör IA (2003) Remaining dentine thickness and human pulp responses. Int Endod J 36:33–43
18. Smith AJ (2003) Vitality of the dentin-pulp complex in health and disease: growth factors as key mediators. J Dent Edu 67(6):678–689
19. Roberts-Clark D, Smith AJ (2000) Angiogenic growth factors in human dentine matrix. Arch Oral Biol 45:1013–1016
20. Tran-Hung L, Laurent P, Camps J, About I (2008) Quantification of angiogenic growth factors released by human dental cells after injury. Arch Oral Biol 53(1):9–13
21. Marinho VC, Chong LY, Worthington HV, Walsh T (2016) Fluoride mouth rinses for preventing dental caries in children and adolescents. Cochrane Database Syst Rev 29(7) (Article No. CD002284)
22. Toumba KJ, Curzon ME (2005) A clinical trial of a slow-releasing fluoride device in children. Caries Res 39(3):195–200
23. Al-Abdi A, Paris S, Schwendicke F (2017) Glass hybrid, but not calcium hydroxide, remineralized artificial residual caries lesions in vitro. Clin Oral Investig 21:389–396
24. Maltz M, Garcia R, Jardim JJ, de Paula LM, Yamaguti PM, Moura MS, Garcia F, Nascimento C, Oliveira A, Mestrinho HD (2012) Randomized trial of partial vs. stepwise caries removal. J Dent Res 91(11):1026–1031
25. Petrou MA, Alhamoui FA, Welk A, Altarabulsi MB, Alkilzy M, Splieth CH (2014) A randomized clinical trial on the use of medical Portland cement, MTA and calcium hydroxide in indirect pulp treatment. Clin Oral Investig 48(5):1383–1389
26. Al-Zayer MA, Straffon LH, Feigal RJ, Welch KB (2003) Indirect pulp treatment of primary posterior teeth: a retrospective study. Pediatr Dent 25:29–36
27. Bjørndal L, Reit C, Bruun G, Markvart M, Kjaeldgaard M, Näsman P, Thordrup M, Dige I, Nyvad B, Fransson H, Lager A, Ericson D, Petersson K, Olsson J, Santimano EM, Wennström A, Winkel P, Gluud C (2010) Treatment of deep caries lesions in adults: randomized clinical trials comparing stepwise vs. direct complete excavation, and direct pulp capping vs. partial pulpotomy. Eur J Oral Sci 3:290–297
28. Trairatvorakul C, Sastararuji T (2014) Indirect pulp treatment vs antibiotic sterilization of deep caries in mandibular primary molars. Int J Paediatr Dent 24(1):23–31
29. Nowicka A, Lipski D, Parafiniuk M, Sporniak-Tutak K, Lichota D, Kosierkiewicz A, Kaczmarek W, Radlińska-Buczkowska J (2013) Response of human dental pulp capped with biodentine and mineral trioxide aggregate. J Endod 39(6):743–747
30. Zhaofei L, Lihua C, Mingwen F, Qingan X (2015) Direct pulp capping with calcium hydroxide or mineral trioxide aggregate: a meta-analysis. J Endod 41(9):1412–1417
31. Chisini LA, Conde MCM, Correa MB, Dantas RVF, Silva AF, Pappen FG, Demarco FF (2015) Vital pulp therapies in clinical practice: findings from a survey with dentist in southern Brazil. Braz Dent J 26(6):566–571
32. de Andrade Massara MDL, Tavares WLF, Sobrinho APR (2016) Maintenance of pulpal vitality in a tooth with deep caries: a case report. Gen Dent 64(4):30–32
33. Buyukgural B, Cehreli ZC (2008) Effect of different adhesive protocols vs calcium hydroxide on primary tooth pulp with different remaining dentin thickness: 24-month results. Clin Oral Investig 12(1):91–96
34. Eskandarizadeh A, Shahpasandzadeh MH, Shahpasandzadeh M, Torabi M, Parirokh M (2011) A comparative study on dental pulp response to calcium hydroxide, white and grey mineral trioxide aggregate as pulp capping agents. J Conserv Dent 14(4):351–355

35. Asgary S, Eghbal MJ, Parirokh M, Ghanavati F, Rahimi H (2008) A comparative study of histologic response to different pulp capping materials and a novel endodontic cement. Oral Surg Oral Med Oral Pathol Oral Radiol Endod 106:609–614
36. Cox C, Subay R, Ostro E, Suzuki S, Suzuki SH (1996) Tunnel defects in dentin bridges: their formation following direct pulp capping. Oper Dent 21(1):4–11
37. Mestrener S, Holland M, Dezan R Jr (2003) Influence of age on the behavior of dental pulp of dog teeth after capping with an adhesive system or calcium hydroxide. Dent Traumatol 19:255–261
38. Pereira MA, Santos-Júnior RBD, Tavares JA, Oliveira AH, Leal PC, Takeshita WM, Barbosa-Júnior AM, Bertassoni LEB, Faria-e-Silva AL (2017) No additional benefit of using a calcium hydroxide liner during stepwise caries removal. A randomized clinical trial. J Am Dent Assoc 148(6):369–376
39. Camilleri J, Pitt Ford TR (2006) Mineral trioxide aggregate: a review of the constituents and biological properties of the material. Int Endod J 39:747–754
40. Parirokh M, Torabinejad M (2010) Mineral trioxide aggregate: a comprehensive literature review—part I: chemical, physical, and antibacterial properties. J Endod 36:16–27
41. Roberts HW, Toth JM, Berzins DW, Charlton DG (2008) Mineral trioxide aggregate material use in endodontic treatment: a review of the literature. Dent Mater 24:149–164
42. Camilleri J, Montesin FE, Di Silvio L, Pitt Ford TR (2005) The chemical constitution and biocompatibility of accelerated Portland cement for endodontic use. Int Endod J 30(12):834–842
43. Chang S-W (2012) Chemical characteristics of mineral trioxide aggregate and its hydration reaction. Restor Dent Endod 37(4):188–193
44. Czarnecka B, Coleman NJ, Shaw H, Nicholson JW (2008) The use of mineral trioxide aggregate in endodontics—a status report. Dent Med Probl 45(1):5–11
45. Kundzina R, Stangvaltaite L, Eriksen HM, Kerosuo E (2017) Capping carious exposures in adults: a randomized controlled trial investigating mineral trioxide aggregate versus calcium hydroxide. Int. Endod. J. 50:924–32
46. Elshamy FMM, Singh GB, Elraih H, Gupta I, Idris FAI (2016) Antibacterial effect of new bioceramic pulp capping material on the main cariogenic bacteria. J. Contemp. Dent. Pract. 17(5):349–53
47. Schwendicke F, Brouwer F, Stolpe M (2015) Calcium hydroxide versus mineral trioxide aggregate for direct pulp capping: a cost-effectiveness analysis. J Endod 41(12):1969–1974
48. Tran XV, Gorin C, Willig C, Baroukh B, Pellat B, Decup F, Opsahl Vital S, Chaussain C, Boukpessi T (2012) Effect of a calcium-silicate-based restorative cement on pulp repair. J Dent Res 91:1166–1171
49. Peng W, Liu W, Zhai W, Jiang L, Li L, Chang J, Zhu Y (2011) Effect of tricalcium silicate on the proliferation and odontogenic differentiation of human dental pulp cells. J Endod 37:1240–1246
50. Camilleri J (2013) Investigation of Biodentine as dentine replacement material. J Dent 41:600–610
51. Koubi G, Colon P, Franquin JC, Hartmaan A, Richard G, Faure MO, Lambert G (2013) Clinical evaluation of the performance and safety of a new dentine substitute, Biodentine, in the restoration of posterior teeth – a prospective study. Clin Oral Investig 17(1):243–249
52. Grech L, Mallia B, Camilleri J (2013) Investigation of the physical properties of tricalcium silicate cement based root end filling materials. Dent Mater 29(2):20–28
53. Orimo H (2010) The mechanism of mineralization and the role of alkaline phosphatase in health and disease. J Nippon Med Sch 77(1):4–12
54. Wiegand A, Buchalla W, Wittin T (2007) Review on fluoride-releasing restorative materials—fluoride release and uptake characteristics, antibacterial activity and influence on caries formation. Dent Mater 23:343–362
55. Balu R, Singaravelu S, Nagiah N (2014) Bioceramic nanofibers by electrospinning. Fibers 2:221–239
56. Langer R, Vacanti JP (1993) Tissue engineering. Science 260:920–926
57. Zhou H, Lee J (2011) Nanoscale hydroxyapatite particles for bone tissue engineering. Acta Biomater 7:2769–2781

58. Yamada Y, Ito K, Nakamura S, Ueda M, Nagasaka T (2011) Promising cell-based therapy for bone regeneration using stem cells from deciduous teeth, dental pulp, and bone marrow. Cell Transplant 20:1003–1013
59. Prescott RS, Alsanea R, Fayad MI, Johnson BR, Wenckus CS (2008) In vivo generation of dental pulp-like tissue by using dental pulp stem cells, a collagen scaffold, and dentin matrix protein 1 after subcutaneous transplantation in mice. J Endod 34(4):421–426
60. Montazerian M, Zanotto ED (2016) History and trends of bioactive glass-ceramics. J Biomed Mater Res A 104A(5):1231–1249
61. Maji K, Dasgupta S, Pramanik K, Bissoyi A (2016) Preparation and evaluation of gelatin-chitosan-nanobioglass 3D porous scaffold for bone tissue engineering. Int J Biomater 2016:1–14 (Article No. 9825659)
62. Demarco EF, Conde MCM, Cavalcanti BN, Casagrande L, Sakai VT, Nör JE (2011) Dental pulp tissue engineering. Braz Dent J 22(1):3–14
63. Baino F, Fiorilli S, Vitale-Brovarone C (2016) Bioactive glass-based materials with hierarchical porosity for medical applications: review of recent advances. Acta Biomater 42:18–32
64. O'Brien F (2011) Biomaterials and scaffolds for tissue engineering. Mater Today 14(3):88–95
65. Chen F, Tang QL, Zhu YJ, Wang KW, Zhang ML, Zhai WY, Chang J (2010) Hydroxyapatite nanorods/poly(vinylpyrolidone) composite nanofibres, arrays and three-dimensional fabrics: electrospun preparation and transformation to hydroxyapatite nanostructures. Acta Biomater 6:3013–3020
66. Stupp SI, Ciegler GW (1992) Organoapatites: materials for artificial bone. I. Synthesis and microstructure. J Biomed Mater Res 26:169–183
67. Martins MA, Santos C, Almeida MM, Costa MEV (2008) Hydroxyapatite micro-and nanoparticles: nucleation and growth mechanisms in the presence of citrate species. J Colloid Interface Sci 318:210–216
68. Boanini E, Fini M, Gazzano M, Bigi A (2006) Hydroxyapatite nanocrystals modified with acidic amino acids. Eur J Inorg Chem 2006:4821–4826
69. Venugopal J, Vadgama P, Kumar TSS, Ramakrishna S (2007) Biocomposite nanofibres and osteoblasts for bone tissue engineering. Nanotechnology 18(5):1–8
70. Zhang Y, Venugopal JR, El-Turki A, Ramakrishna S, Su B, Lim CT (2008) Electrospun biomimetic nanocomposite nanofibers of hydroxyapatite/chitosan for bone tissue engineering. Biomaterials 29:4314–4322
71. Liao S, Murugan R, Chan CK, Ramakrishna S (2008) Processing nanoengineered scaffolds through electrospinning and mineralization suitable for biomimetic bonetissueengineering. J Mech Behav Biomed Mater 1:252–260
72. Yang F, Wolkeand JGC, Jansen JA (2008) Biomimetic calcium phosphate coating onelectrospun poly(ε-caprolactone) scaffolds for bone tissue engineering. Chem Eng J 137:154–161
73. de Aza PN, de Aza AH, Pena A, de Aza S (2007) Bioactive glasses and glass-ceramics. Bol Soc Esp Ceram 46(2):45–55
74. Zhang X, Zeng D, Li N, Wen J, Jiang X, Liu C, Li Y (2016) Functionalized mesoporous bioactive glass scaffolds for enhanced bone tissue regeneration. Sci Rep 6 (Article No. 19361)
75. de Aza PN, Guitian F, de Aza S (1994) Bioactivityofwollastonite ceramics: in vitro evaluation. Scripta Metall Mater 31:1001–1005
76. Hench LL, Andersson Ö (1993) Bioactive glasses. In: Hench LL, Wilson J (eds) An introduction to bioceramics. World Scientific Publishing, Singapore, pp 41–62
77. Shah FA, Brauer DS, Hill RG, Hing KA (2015) Apatite formation of bioactive glasses is enhanced by low additions of fluoride but delayed in the presence of serum proteins. Mater Lett 153:143–147
78. Khvostenko D, Hilton TJ, Ferracane JL, Mitchell JC, Kruzic JJ (2016) Bioactive glass fillers reduce bacterial penetration into marginal gaps for composite restorations. Dent Mater 32(1):73–81
79. Foroughi MR, Karbasi SB, Khoroushi M, Khademi AA (2017) Polyhydroxybutyrate/chitosan/bioglass nanocomposite as a novel electrospun scaffold: fabrication and characterization. J Porous Mater 1–14

80. Moorthi A, Parihar PR, Saravanan S, Vairamani M, Selvamurugan N (2014) Effects of silica and calcium levels in nanobioglass ceramic particles on osteoblast proliferation. Mater Sci Eng, C 43:458–464
81. Shalumon KT, Sowmya S, Sathish D, Chennazhi KP, Nair SV, Jayakumar R (2013) Effect of incorporation of nanoscale bioactive glass and hydroxyapatite in PCL/chitosan nanofibers for bone and periodontal tissue engineering. J Biomed Nanotech 9(3):430–440
82. Kim D-A, Lee J-H, Jun S-K, Kim H-W, Eltohamy M, Lee H-H (2017) Sol-gel-derived bioactive glass nanoparticle-incorporated glass ionomer cement with or without chitosan for enhanced mechanical and biomineralization properties. Dent Mater 33:805–817
83. Moshaverinia A, Ansari S, Movasaghi Z, Billington RW, Darr JW, Rehman IU (2008) Modification of conventional glass-ionomer cements with N-vinylpyrrolidone containing polyacids, nano-hydroxy and fluoroapatite to improve mechanical properties. Dent Mater 24:1381–1390
84. Moshaverinia A, Ansari S, Moshaverinia M, Roohpour N, Darr JA, Rehman I (2008) Effects of incorporation of hydroxyapatite and fluoroapatite nanobioceramics into conventional glass ionomer cements (GIC). Acta Biomater 4(2):432–440
85. Caridade SG, Merino EG, Alves NM, de Zea Bermudez V, Boccaccini AR, Mano JF (2013) Chitosan membranes containing micro or nano-size bioactive glass particles: evolution of biomineralization followed by in situ dynamic mechanical analysis. J Mech Behav Biomed Mater 20:173–183
86. Caridade SG, Merino EG, Alves NM, Mano JF (2013) Biomineralization in chitosan/Bioglass® composite membranes under different dynamic mechanical conditions. Mater Sci Eng, C 33:4480–4483
87. Mota J, Yu N, Caridade SG, Luz GM, Gomes ME, Reis RL, Jensen JA, Walboomers XF, Mano JF (2012) Chitosan/bioactive glass nanoparticle composite membranes for periodontal regeneration. Acta Biomater 8:4173–4180
88. Teng SH, Lee EJ, Yoon BH, Shin DS, Kim HE, Oh JS (2009) Chitosan/nanohydroxyapatite composite membranes via dynamic filtration for guided bone regeneration. J Biomed Mater Res A 88:569–580
89. Xu H, Lv F, Yi Z, Ke Q, Wu C, Liu M, Chang J (2015) Hierarchically micro-patterned nanofibrous scaffolds with a nanosized bio-glass surface for accelerating wound healing. Nanoscale 7(44):18446–18452
90. Zhai W, Lu H, Chen L, Lin X, Huang Y, Dai K, Naoki K, Chen G, Chang J (2012) Silicate bioceramics induce angiogenesis during bone regeneration. Acta Biomater 8:341–349

Chapter 2
Nanobiomaterials and Their Role in Periodontal Rehabilitation

D. Deepa and K. V. Arunkumar

2.1 Introduction

2.1.1 Brief Introduction of Nanotechnology in Periodontics

Scientists in the field of regenerative medicine and tissue engineering are continually in quest of new ways to apply the principles of cell transplantation, material science, and bioengineering to construct biological substitutes that will restore and maintain normal function in diseased and injured tissues. The regenerative treatment of periodontal defects with grafts, or procedure, has attracted enormous attention from material scientists and also from both private and government organizations because of its considerable financial potential and scientific significance. One of the emerging areas is tissue engineering that seeks to develop techniques and materials to aid in the formation of new tissues to replace damaged tissues. The definitive goal in periodontal therapy is creation of an environment that is conducive to maintaining the patient dentition in health, comfort, and function [1]. Periodontal surgery as a part of treatment of periodontal disease is mainly performed to gain access to diseased areas for adequate cleaning, achieve pocket reduction or elimination and to restore periodontal tissues loss as a consequence of disease process. The shift in therapeutic concepts from resection to regeneration has significantly impacted the practice of periodontology in the recent times. Periodontal regeneration leads to the formation of new bone, cementum, and periodontal ligament on a previously diseased root sur-

D. Deepa (✉)
Department of Periodontology, Swami Vivekanand Subharti University,
Meerut 250005, India
e-mail: deepa_arun@rediffmail.com

K. V. Arunkumar
Department of Oral & Maxillofacial Surgery, Swami Vivekanand Subharti University,
Meerut 250005, India

R. S. Chaughule (ed.), *Dental Applications of Nanotechnology*,
https://doi.org/10.1007/978-3-319-97634-1_2

face. It requires a sequence of biological events including cell adhesion, mitogenesis, chemotaxis, differentiation, and metabolism. Bone grafts provide structural framework for clot development, maturation, and remodeling that support bone formation in osseous defects.

According to **Ashman**, an ideal synthetic bone material should be [2]:

(a) Biocompatible.
(b) Able to serve as a framework for new bone formation.
(c) Resorbable in the long term and have potential for replacement by host bone.
(d) Osteogenic, or at least facilitate new bone formation.
(e) Radiopaque.
(f) Easy to manipulate clinically.
(g) Not supporting the growth of oral pathogens.
(h) Hydrophilic.
(i) Available in particulate and molded forms.
(j) Having surface electrical activity (i.e., be charged negatively).
(k) Microporous and provide added strength to the regenerating, host bone matrix, and permit biological fixation.
(l) Readily available.
(m) Non-allergenic.
(n) Effective in a broad range of medical situations (e.g., cancer, trauma, and infective bone destroying diseases).
(o) Having a surface that is amenable to grafting.
(p) Acting as matrix or vehicle for other materials (e.g., bone protein inducers, antibiotics, and steroids).
(q) Having high compressive strength.

2.2 Nanocrystalline Bone Grafts in Bone Regeneration

Nanotechnology is a multidisciplinary field that covers a vast and diverse array of devices derived from engineering, physics, chemistry, and biology. The prefix "nano" has found in last decade an ever-increasing application to different fields of the knowledge. Within the convention of International System of Units (SI), it is used to indicate a reduction factor by 109 times. So, the nanosized world is typically measured in nanometers (1 nm corresponding to 10^{-9} m) and it encompasses systems whose size is above molecular dimensions and below macroscopic ones (generally >1 nm and <100 nm). At this size, atoms and molecules work differently and provide a variety of surprising and interesting uses. The term "nanotechnology" was first used in 1974 by Norio Taniguchi, a researcher at the University of Tokyo. Also, nanoparticles have special properties, including chemical, optical, magnetic, and electro-optical, which differ from those of either individual molecules or bulk species [3]. The important feature of nanostructured materials is the development of self-assembly. Self-assembly is the essential part of nanotechnology. It is common to many dynamics, multi-component systems, from smart materials and self-healing structures to netted sensors and computer networks. When choosing a material for self-assembly, the materials should have a critical number of charged groups, below which the assembling procedure does not work at all. To form a well-defined stable multilayer, the appropriate opposite charge density is required for the matched materials. To facilitate analytical studies, the materials should bear some functional groups, which can be detected by analytical instruments [3]. Nanosized hydroxyapatite (NHA) is the main component of mineral bone. Living bone constantly undergoes a coupled resorptive–formative process known as bone remodeling. NHA possesses exceptional biocompatibility and bioactivity properties with respect to bone cells and tissues, probably due to its similarity with the hard tissues of the body. Calcium phosphate biomaterials have been extensively used clinically in the form of powders, granules, dense and porous blocks and various composites. These materials form the main mineral part of calcified tissues. However, calcium phosphate presence in bone is in the form of nanometer-sized needle-like crystals of approximately 5–20 nm width by 60 nm length, with a poorly crystallized non-stoichiometric apatite phase containing CO_3^{2-} Na^+, F−, and other ions in a collagen fiber matrix. Current research deals with new hydroxyapatite (HA) formulations aiming at better and more effective biomedical applications, producing this material with properties closer to those of living bone, such as nanosized and monolithic structures. NHA exhibits greater surface area and is expected to have better bioactivity than coarser crystals [4].

Bone grafts in nanosized particles mimic the natural bone mineral in its size, composition, and morphology is said to have better osteoconductive property than the conventional bone grafts. Therefore, nanosized ceramics may represent a promising class of bone graft substitute in intrabony defects due to their improved osseointegrative properties. For example, advantages of nanocrystalline hydroxyapatite (NCHA) materials are their osteoconductivity, bioresorbability, and close contact than the conventional HA. A special feature of nanostructured materials is an extremely high number of molecules on the surface of materials. It can promote proliferation, migration, and survival of periodontal ligament cells.

Webstar et al. [5] evaluated the effect of osteoblast function on NHA. Crystalline HA was prepared via established wet chemical synthesis. It was concluded that osteoblast adhesion as well as proliferation were significantly ($p < 0.01$) greater on nanocrystalline (50 nm grain size) than on conventional (250 nm grain size) HA at all time periods tested. Therefore, NHA binds to the bone and stimulates bone healing by stimulation of osteoblast activity.

2.3 Preparation Methods of HA Nanoparticles

(a) Dry methods: Dry methods, which could be identified in contrast to wet methods where a solvent is always used, could be performed in two main ways: solid-state synthesis and mechanochemical process. These methods have the convenience of producing highly crystalline HA from relatively inexpensive raw materials. The main disadvantage is the large size of particles in case of solid-state synthesis and the low phase purity of HA in case of mechanochemical process. In the recent years, progress in preparing HA using dry methods, especially solid-state method, has been very slow [6].

(b) Wet methods: Aqueous solutions of various sources for phosphate and calcium ions are employed, and HA crystals are normally produced by precipitation. Wet processes could be performed by a number of technical routes classified into six groups: conventional chemical precipitation, hydrolysis method, sol–gel method, hydrothermal method, emulsion method, and sonochemical method. Wet chemical methods have the advantages in precise control over the morphology and size of particles, and based on the statistical analysis, they are the most promising methods for the synthesis of nanosized HA. However, difficulties in controlling the crystallinity and phase purity of nanoparticles, and some technically intricate and time-consuming details make some wet procedures unsuitable for scaling up to produce large quantities of HA. Most commonly used methods are wet chemical method and modified sol–gel method.

(c) High-temperature processes: These methods, which have the convenience of avoiding undesirable calcium phosphate phases, are used to produce HA with high crystallinity and good chemical homogeneity. Two possible routes for high-temperature synthesis of HA are combustion method and pyrolysis process, of which the former has received more attention. Poor control over the processing variables and generation of the secondary aggregates, especially during the pyrolysis, are the main disadvantages.

(d) Synthesis from biogenic sources: To produce HA ceramics, various natural materials, mainly bone waste, eggshells, exoskeleton of marine organisms, naturally derived biomolecules, and biomembranes, have been employed over the past decade. This field is expected to attract more attention in the near future, owing to the better physicochemical properties of the HA generated from biogenic sources. However, some of these materials could be exclusively used to produce HA blocks or at most HA particles of large size. This stresses the need for further work in the area.

(e) Combination procedures: These methods, as relatively new strategies, employ two or more distinct procedures to synthesize HA nanoparticles. Among several possibilities, combinations of hydrothermal–mechanochemical, hydrothermal–hydrolysis, and hydrothermal–microemulsion have received more attention. Reports on various hybrids employed microwave method have also been published recently. In general, combination procedures open exciting possibilities to improve the characteristics of HA nanoparticles [6].

NHA powder exhibits greater surface area and is expected to have better bioactivity than coarser crystals [4]. Osteoconductivity, solubility, sinterability, and mechanical reliability of HA could be promoted by controlling its particle size and structural morphology in order of a nanoscale [7]. In the evaluation of bioactivity of NHA as compared to conventional HA, it was also found that the pH of the conventional HA was of unvarying trend, as it was not resorbed in the medium, which indicates its physiological stability. The NHA showed drastic changes in the pH suggesting that it dissolves much faster than conventional HA which may remain as a permanent implant thus inhibiting the healing process [8]. The pH value depends on solubility or resorbability of HA. As pH decreases, solubility increases. This is in accordance with the study of Faithi et al. who quantitatively estimated the release of calcium ions from NHA to support its *in vitro* bioresorbability. The results showed that more calcium ions were released from the NHA in comparison with conventional HA [4]. The ionic dissolution rate of NHA is much similar to that of natural bone mineral *in vitro* in comparison with conventional HA. In this manner, the resorbability of HA could be promoted by engineering the crystallite size to nano-submicron level. Accordingly, it is clear that the rate of bioresorbability of NHA is higher than that in conventional HA and is similar to that in biological apatite of bone. NHA possesses exceptional biocompatibility and bioactivity properties with respect to bone cells and tissues, probably due to its similarity with the hard tissues of the body [9]. Besides a good biocompatibility, a synthetic bone substitute should also ensure the formation of new bone after their implantation. In fact, an optimal synthetic bone substitute is resorbed by hydrolytic and cellular degradation process involving the action of macrophages and further is replaced by vital bone over time. A major prerequisite for this process is angiogenesis because newly formed blood vessels transport oxygen and nutrients into the implanted bone substitute, a physiological milieu is created, which supports the ingrowth of bone cells and also differentiation of pluripotential stem cells from the surrounding tissue to an osteoblastic phenotype. Nanostructured HA promotes up-regulation of FGF-2 and primes endothelial cells to VEGF action. FGF-2 plays a biological pleiotropic role, including cell migration, angiogenesis, bone development, and repair. Also, a synthetic bone substitute must not only support the growth and foster the phenotype of the cell type for the tissue it is to replace (i.e., osteoblasts), but also support the cells that are responsible for maintaining the bone cells (i.e. fibroblasts). The up-regulation of FGF-2 through gene transcription is an important event which explains the role of HA nanocrystals in inducing a positive and controlled angiogenic phenotype in endothelium. The robust increase of FGF-2mRNA (3–6 folds) translates into a significant production of the soluble FGF-2 isoform [10].

To take advantage of NHA in treating periodontal defects, which are open defects with a constant flow of gingival crevicular fluid that causes such minute and potentially easily phagocytosed particles to be exfoliated, their local retention must be improved.

In one of the studies, NHA was prepared by adding 0.5 M Ca (NO_3) 24 H_2O (Merck) in ethanol with 0.5 M $(NH_4)_2$ PO_4 (Merck). The formation and quality of the synthesized compound were studied using X-ray diffractometer and SEM. Elemental analysis of the synthesized powder was determined by energy dispersive X-ray analysis (EDAX) attached with the SEM (particle size ~20 nm). A similar method for synthesis of NHA was used by Loo et al. who synthesized hydroxyapatite nanoparticles through chemical precipitation, and hydrothermal was performed as a post-synthesis treatment to achieve particles of well-defined size and morphology

[11]. Another form of nanocrystalline hydroxyapatite used in past studies is the paste form (**OSTIM**, Heraeus Kulzer, Hanau, Germany) containing 65% water and 35% nanostructured apatite particles. According to Kasaj et al., the paste form of NHA has low consistency and limited space making potential. The low consistency of NHA might result in poor resistance to collapse the mucoperiosteal flap into the intrabony defect, allowing undesirable cell types to enter the secluded wound area. This collapse may be prevented by means of implantation of additional bone grafts or bone graft substitutes into the defect to support NHA preserving its original position [12]. However, in the present study powder form of NHA was used and no such limitation was observed during or in follow period for 6 months. In contrast, Huber et al. reported that the paste does not harden off when mixed with blood or spongiosa and therefore favoring cell migration into the implantation area coinciding with revascularization. Further, its viscosity enables it to be applied to form fit in close contact with the bone [13]. The NHA paste has been used for the treatment of metaphyseal fractures in orthopaedic surgery, tooth perforations, jaw cysts, and peri-implantitis lesions. However, as stated above the present study opted for the powder form of NHA graft which was a more economical alternative.

The main feature of bioactive bone graft materials such as BCP ceramics is their ability to form a strong direct bond with the host bone, resulting in a strong interface, compared to bioinert or biotolerant materials which form a fibrous interface.

Also, NHA and microparticles of β-tricalcium phosphate (β-TCP): (1) reduce the gross amount of NHA required to obturate large defects and (2) make the NHA particles more resistant to direct dissemination or phagocytosis by possible adsorption of the nanosized particles on the microsized particulate surface, improving its local availability [14].

Our study was undertaken to evaluate and compare clinically and radiographically the effectiveness of NHA and NHA with β-TCP in the treatment of human intrabony defects. This was the first study to our knowledge in which NHA with β-TCP was used in ratio of 70:30 for the intrabony defects [15]. Twenty intrabony defects in 10 systemically healthy patients were selected for the study. Ten defects were treated with NHA (Sybograf™) (control group) and ten defects were treated with NHA with β-TCP graft (Sybograf™ Plus) (test group).

The clinical and radiographic parameters were evaluated at baseline, 3 months and 6 months. Gingival index, plaque index, and gingival bleeding index were recorded. The probing depth and relative attachment level were recorded using a custom made acrylic stent with a UNC-15 probe. Radiographically, the area of the defect was assessed. The results were evaluated and subjected to statistical analysis. The results demonstrated that there was significant reduction in the gingival index, plaque index, and gingival bleeding index scores. However, there was no statistically significant difference in the scores when compared between the test and the control groups. This suggested that strict plaque regimens were reinforced, and meticulous oral hygiene was followed by the subjects during the study. There was significant reduction of probing pocket depth (PPD) from baseline to 6 months in control group and test group. Also, there was significant reduction in relative attachment level (RAL) from baseline to 6 months in control group and test groups. On intergroup comparison, there was no statistically significant difference.

In adjunct to the probing depth and the attachment level, the radiographic area of the defect showed a similar trend within the groups and no significant difference was seen when compared between the two groups. Hence, it was concluded that the results of the study demonstrated significant improvement in all the parameters with good radiographic defect fill; however, there was no statistical significant difference on intergroup comparison (Figs. 2.1, 2.2, 2.3, 2.4, 2.5, 2.6, and 2.7).

Klawitter et al. stated that trabeculae of bone vary in size from 20 to over 100 μm. When trabeculae reach about 100 μm, it carries its own blood vessels; much in the

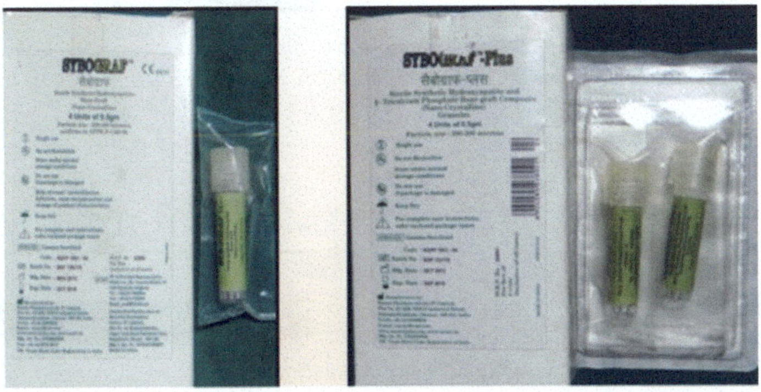

Fig. 2.1 Sybograf^{TM} and Sybograf plus^{TM}

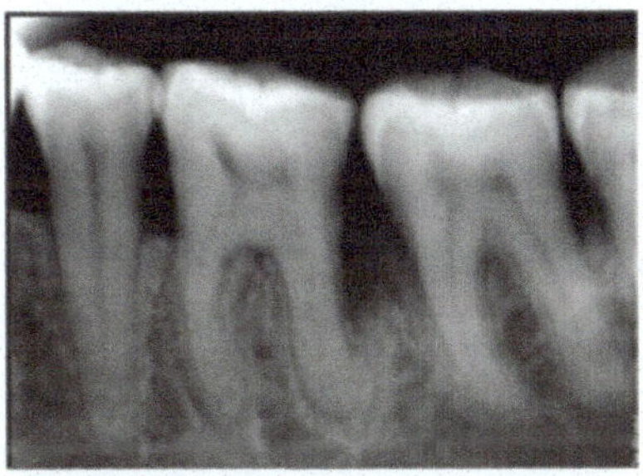

Fig. 2.2 Preoperative radiograph at baseline

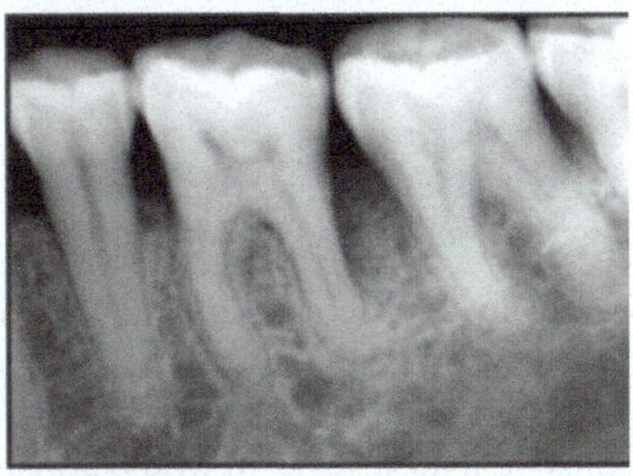

Fig. 2.3 Three months after treatment with Sybograf^{TM}

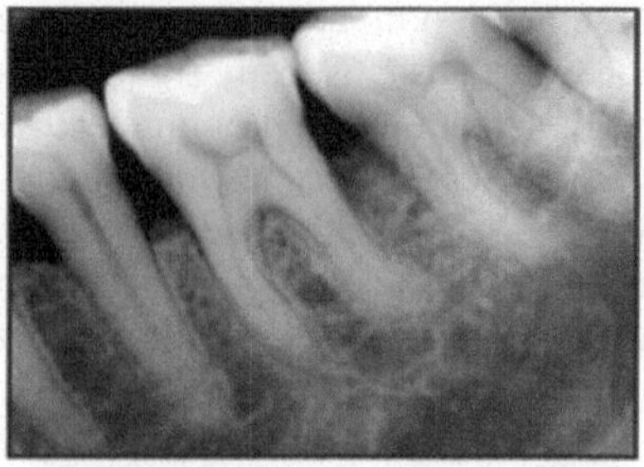

Fig. 2.4 Six months after treatment with Sybograf™

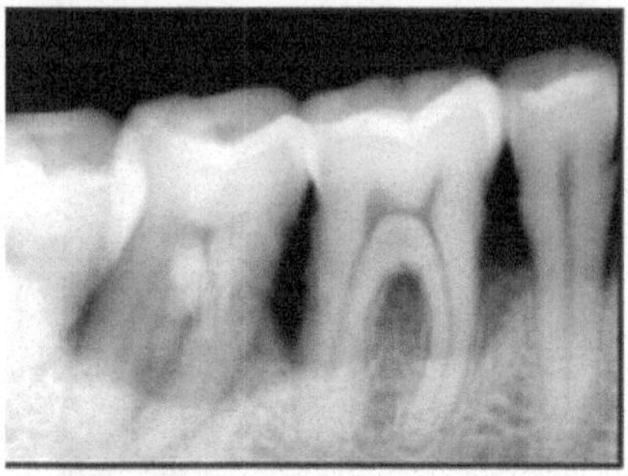

Fig. 2.5 Preoperative radiograph at baseline

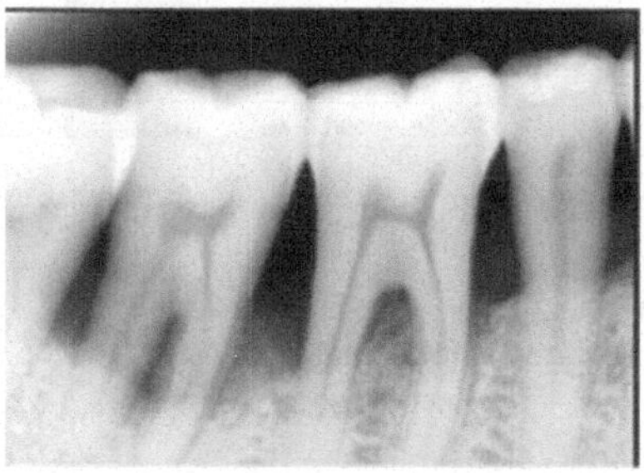

Fig. 2.6 Three months after treatment with Sybograf plus™

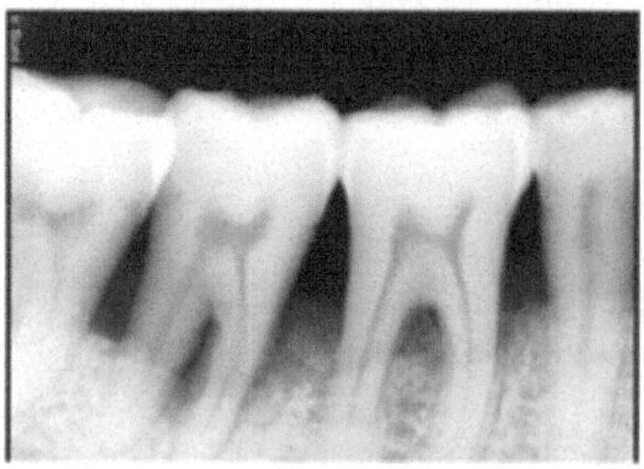

Fig. 2.7 Six months after treatment with Sybograf plus™

same way an osteon does via Haversian canal. Compact bone has Haversian systems or osteons of between 50 and 250 μm. Thus, to support trabecular bone ingrowth, the pores would need to be at least 40–100 μm, and to support osteonal bone ingrowth, pores of at least 100 μm would appear necessary [16]. Later Hirschorn et al. reported that particle size of about 380 μm in diameter would yield this minimal dimension and particles yielding under 100 μm space dimensions may possess less mineralization potential [17].

Another vital property of nanomaterials which play a major role in regeneration in the current study is development of self-assembly which helps to integrate different functions into synthetic extracellular matrices. These synthetic extracellular matrices will need to perform functions such as to sustain cell viability and proliferation, allow the establishment of a blood vessel network formation, and provide sufficient support to prevent tissue collapse. In recent times, developments in this field have seen the use of pH-induced self-assembly of a peptide amphiphile to artificially construct a nanostructured fibrous scaffold with the structural features of extracellular matrices. After cross-linking, the newly produced fibers are able to direct mineralization of HA to form a composite material in which the crystallographic axes of HA are aligned with the long axes of the fibers. This alignment is similar as that observed in vivo between collagen fibrils and HA crystals in bone [3].

Material properties including porosity, surface geometry, and surface chemistry play significant role in determining the osteoconductive capacities of a graft which must be analyzed histologically.

2.4 Nanotechnology and Dental Implants

Nanotechnology is one of the current fields of interest to adjust the properties of implants and tissue engineering scaffolds. It deals with systems and structures that result in novel properties due to their small size (1–100 nm). The two main approaches: (i) bottom-up (from molecular level to organized molecular films, supramolecular structures, or nanoparticles) and (ii) top-down technologies (bits of bulk material are removed resulting in nanoscale features; e.g., surface patterning) are used. One of the challenges in implantology is to achieve and maintain the osseointegration as well as the epithelial junction of the gingiva with implants. An intimate junction of the gingival tissue with the neck of dental implants may prevent bacterial colorizations leading to peri-implantitis while direct bone bonding may ensure a biomechanical anchoring of the artificial dental root.

Nanotechnology could be utilized to create surfaces with controlled topography and chemistry that would assist understanding biological interactions and developing novel implant surfaces with predictable tissue-integrative properties.

Nanoscale alteration of titanium implant surfaces can alter cellular and tissue responses that may promote osseointegration. Three nanostructured implant coatings have been developed:

- **Nanostructured Diamond**: It has ultrahigh rigidity, improved toughness, low friction, and good adhesion to titanium alloys.

- **Nanostructured Hydroxyapatite Coatings**: This is used to achieve the desired mechanical characteristics and enhanced surface reactivity and has been found to increase osteoblast adhesion, proliferation, and mineralization [18].
- **Nanostructured Metallo-Ceramic Coatings**: These offer continuous variation from a nano crystalline metallic bond at the interface to the hard ceramic bond on the surface [19].

Besides surface contact area and surface topography, bone bonding, and stability play a major role in implant success and osseointegration. Bone growth and implant success could also be accelerated by the use of nanotechnology. Osteoblast configuration on a more complex implant surface is produced by the addition of nanoscale deposits of hydroxyapatite and calcium phosphate particles [20]. Material engineering, and hence implant dentistry, has advanced extensively on the basis of researches conducted on the effects and subsequent optimization of micro-topography and surface chemistry. These new implants constructed on the basis of this technology are more acceptable as they enhance the integration of nano-coatings resembling biological materials to the periodontal tissues [21]. In addition, implant surfaces coated with titanium oxide nanotubes and laced with silver nanoparticles serve the purpose of fighting infection, thus increases the shelf life of the implants.

2.5 Nanotechnology in the Management of Peri-implantitis

Peri-implantitis affected parts become covered by an infected smear layer of instrumentation debris after routine implant preparation, which seems to compromise fibrin clot adhesion to such altered surfaces. In addition, continuous flap mobility and early clot retraction could draw the clot-blended graft complex away from the implant surface with subsequent creation of a micro-gap graft, epithelialization, and eventual implant surface recontamination. For that reason, treating a peri-implantitis affected implant surface should include complete removal of the infected biofilm and the smear-like layer, resulting in complete exposure of the roughened titanium implant surface. This should be followed by maximizing clot-blended graft adhesion to the implant surface through application of a graft material with a particle size smaller than that of the implant surface pores. Such mechanical integration is considered to reduce the possibility of the implant bone micro-gap, a factor that could ensure complete protection of the underlying defect-filled regenerative materials or blood clot. This, along with subsequent enhanced bone regeneration, could be interpreted as a true re-osseointegration.

Particle size seemed to be an important factor that optimizes adhesion of particles to the exposed implant surface. Such improved adhesion is likely to retard the apical migration of the epithelial attachment, a factor that could enhance reosseointegration. Further peri-implantitis affected surface conditioning with citric acid improves nanoparticle-sized hydroxyapatite-blended clot adhesion to titanium implant surfaces (Gamal et al. 2013 [22]). Growth factors such as platelet-derived growth factor

(PDGF) have significantly enhanced periodontal therapy outcomes with a high degree of variability, mostly due to the lack of continual supply for a required period of time. One method to overcome this barrier is gene therapy. Platelet-derived growth factor, PDGF-B gene delivery in fibroblasts using nanosized calcium phosphate particles (NCaPP) as vectors, has found to significantly enhance fibroblast proliferation.

2.6 Nanomaterials for Periodontal Tissue Engineering

Tissue engineering is an evolving interdisciplinary field integrating biology, engineering, materials science, and medicine that focuses on the development of biological substitutes to restore, replace, maintain, or enhance tissue and organ function. Tissue engineering concepts for periodontal regeneration are focused on the utilization of synthetic scaffolds for cell delivery purposes [23]. Although the employment of such systems offers promise, it is very likely that the next generation of materials will rely heavily on nanotechnology and its potential to produce non-biologic self-assembling systems for tissue engineering purposes [24]. As detailed above, self-assembling systems for biologic systems are those which automatically undergo prespecified assemblies much in line with known biologic systems associated with cells and tissues. Using these principles, it is possible to construct systems on a nano-, micro- or even macroscale.

Nanotechnology has got the potential to produce non-biologic self-assembling systems for tissue engineering purposes. It is possible to create polymer scaffolds in the future for cell seeding, growth factor delivery and tissue engineering via nanodevices implanted to sites of tissue damage. As native tissues or organs are composed of proteins within nanoscale and cells directly interact with nanostructured extracellular matrices (ECM), nanobiomaterials such as nanofibers, nanotubes, nanoparticles, and other nanofabricated devices with smaller than 100 nm in at least one dimension are able to contribute cell growth and tissue regeneration. Recent advances in nanotechnology, however, have enabled the design and fabrication of biomimetic microenvironment at the nanoscale, providing an analog to native ECM.

Presently, materials available for such constructs are metals, ceramics, polymers, and even composite materials, the like of which have not yet been developed. The clinical utility of these nano-constructed self-assembling materials is their capacity to be developed into nanodomains or nanophases, leading to unique nanobuilding blocks with inbuilt nanocontrol and nanodelivery capabilities. Our present capacity to create polymer scaffolds for cell seeding, growth factor delivery and tissue engineering purposes is well recognized. In the future, these processes may well be manipulated via nanodevices implanted to sites of tissue damage. Example that defines the benefits of nanotopographic patterning is the addition of chemically etched, nanoscale roughness to the surface of porous polylactic-co-glycolic acid, PLGA scaffolds—a method that has been shown to enhance cell adhesion and growth, as well as the expression of matrix components [25]. Moreover, several advanced techniques like multiphoton polymerization and layer-by-layer assembly have shown that nanotopographic cues hold the potential to exponentially improve scaffold design.

Apart from the benefits discussed above, nanotechnology is also expected to play an important role in creating novel tissue regeneration strategies (e.g., cell sheet engineering), and in surmounting other important obstacles in tissue engineering—such as the development and characterization of new biomaterials and stem cell engineering. For example, high-throughput assays based on micro-/nanotechnologies are emerging for the cell-based screening of biomaterials [26].

One exciting opportunity that lies ahead is the idea of "organ-on-a-chip" which, in the foreseeable future, will substitute the expensive and life-costing animal testing used for drug development and for evaluation/optimization of nanoparticulate systems for drug delivery. For tissue engineering, nanotechnology has also opened the door to new approaches that could stimulate the reconstruction of complex tissue architectures.

Therefore, further studies using more subjects and histologic analysis of newly formed bone could clarify the benefits of the new nanocrystalline bone grafting material using tissue engineering principles. In addition, a longer post-treatment observation interval may be needed to confirm the stability of clinical outcomes using these materials.

2.7 Conclusion

Advances in nanomaterials and nanotechnology have provided a hopeful insight into the commercial applications of nanomaterials in the management of periodontal diseases. While current work is focused on the recent development particularly of nanoparticles and nanotubes for periodontal management, the materials developed from them such as the hollow nanospheres, core–shell structures, nanocomposites, nanoporous materials, and nanomembranes will play an escalating role in materials development for the dental industry. With the clearing up of nanotechnology-specific medical regulations and a continued influx of investments and time, we believe that nanomedicine will not only improve conventional therapies, but also bridge the shortcomings of conventional medicine to help people on both global and individual levels.

References

1. Nasr HF, Reidy ME, Yukna RA (2000) Bone and bone substitutes. Perio 1999(19):74–86
2. Ashman A (1992) The use of synthetic bone materials in dentistry. Compendium 13(11):1020, 1022, 1024–1026
3. Kong LX, Peng Z, Li SD, Bartold PM (2000) Nanotechnology and its role in the management of periodontal diseases. Perio 2006(40):184–196
4. Fathi MH, Mortazav V, Esfahani SIR (2008) Bioactivity evaluation of synthetic nanocrystalline hydroxyapatite. Dent Res J 5(2):81–87

5. Webster TJ, Ergun C, Doremus RH, Siegel RW, Bizios R (1999) Nanocrystalline hydroxyapatite enhances osteoblast function. In: Engineering in medicine and biology, proceedings of the first joint BMES/EMBS conference Atlanta, USA, vol 2, pp 744–750
6. Shojai MS, Khorasani MT, Khoshdargi ED, Jamshidi A (2013) Synthesis methods for nanosized hydroxyapatite with diverse structures. Acta Biomater 9:7591–7621
7. Rauschmann MA, Wichelhaus TA, Stirnal V, Dingeldein E, Zichner L, Schnettler R et al (2005) Nanocrystalline hydroxyapatite and calcium sulphate as biodegradable composite carrier material for local delivery of antibiotics in bone infections. Biomaterials 26(15):2677–2684
8. Figueroa IA, Peralta ON, Morales CF, Tenorio RG, Pina-Barba MC (2012) Synthesis and characterization of biocompatible-nanohydroxyapatite crystals obtained by a modified sol-gel processing. Biomatter 2(2):71–76
9. Ferraz MP, Monteiro FJ, Manuel CM (2004) Hydroxyapatite nanoparticles: a review of preparation methodologies. J App Biomater Biomech 2:74–80
10. Pezzatini S, Solito R, Morbidelli L, Bigi A, Ziche M (2007) Nanocrystalline hydrxyapatite promotes angiogenesis in vitro by up-regulation of FGF-2. Eur Cells Mater 14(3):107
11. Loo SCJ, Siew YE, Ho S, Boey FYC, Ma J (2008) Synthesis and hydrothermal treatment of nanostructured hydroxyapatite of controllable sizes. J Mater Sci Mater Med 19:1389–1397
12. Kasaj A, Rohrig B, Zafiropoulos GG, Willershausen B (2008) Clinical evaluation of nanocrystalline hydroxyapatite paste in the treatment of human periodontal bony defects-A randomized controlled clinical trial: 6 month results. J Periodontol 79(3):394–400
13. Huber FX, Berger I, McArthur N, Huber C, Kock HP, Hillmeier J et al (2008) Evaluation of a novel nanocrystalline hydroxyapatite paste and a solid hydroxyapatite ceramic for the treatment of critical size bone defects(CSD) in rabbits. J Mater Sci Mater Med 19(1):33–38
14. Gamal AY, Iacono VJ (2013) Mixed nano/micro-sized calcium phosphate composite and EDTA root surface etching improve availability of graft material in intrabony defects: an in vivo scanning electron microscopy evaluation. J Periodontol 84(12):1730–1739
15. Singhal A, Deepa D, Tomar N (2018) Comparative evaluation of synthetic nanocrystalline hydroxyapatite (sybograf^{tm}) and nanocrystalline hydroxyapatite with β-tricalcium phosphate (sybograf^{tm}plus) in the treatment of intrabony defects—a clinical and radiographic study (Under publication process)
16. Klawitter JJ, Hulbert SF (1971) Application of porous ceramics for the attachment of load bearing internal orthopedic applications. J Biomed Mater Res Symposium 5(6):161–229
17. Hirschorn JS, McBeath AA, Dustoor MR (1971) Porous titanium surgical implant materials. J Biomed Mater Res 5(6):49–67
18. Miller R, Guo Z, Vogler EA, Siedlecki CA (2006) Plasma coagulation response to surfaces with nanoscale chemical heterogeneity. Biomaterials 27:208–215
19. Sawase T, Jimbo R, Baba K, Shibata Y, Ikeda T, Atsuta M (2008) Photo-induced hydrophilicity enhances initial cell behavior and early bone apposition. Clin Oral Implants Res 19:491–496
20. Albrektsson T, Sennerby L, Wennerberg A (2008) State of the art of oral implants. Perio 2000; 47:15–26
21. Saiz E, Zimmermann EA, Lee JS, Wegst UG, Tomsia AP (2013) Perspective on the role of nanotechnology in bone tissue engineering. Dent Mater 29:103–115
22. Ay Gamal, Abdel-Ghaffar KA, Iacono VJ (2013) A novel approach for enhanced nanoparticle-sized bone substitute adhesion to chemically treated periimplantitis-affected implant surfaces: an invitro proof-of-principle study. J Periodontol 84(2):239–247
23. Bartold PM, McCulloch CAG, Narayanan AS, Pitaru S (2000) Tissue engineering: a new paradigm for periodontal regeneration based on molecular and cell biology. Periodontology 2000(24):253–269
24. Bayne SC (2005) Dental biomaterials: where are we and where are we going? J Dent Educ 69:571–585
25. Pattison MA, Wurster S, Webster TJ, Haberstroh KM (2005) Biomaterials 26:2491–2500 [PubMed: 15585251]
26. Anderson DG, Levenberg S, Langer R (2004) Nat Biotechnol 22:863–866 [PubMed: 15195101]

Chapter 3
Advanced Nanomaterials and Their Functionalization in Clinical Endodontics

Hend Mahmoud Abou El Nasr and Makbule Bilge Akbulut

3.1 Introduction

The term nanotechnology, also known as molecular nanotechnology or molecular engineering, was first introduced by Richard Feynman in 1959 to the scientific approach of creating functional materials, devices, and systems through the control of atoms on a nanometer scale (i.e., in range of 0.1–100 nm) [1, 2]. It involves the tailoring of materials at atomic level (one nm is one billionth (10^{-9}) of a meter) to attain unique properties, which can be suitably manipulated for the desired applications [3].

3.1.1 Types of Nanotechnologies

Broadly, nanotechnology consists of three mutually overlapping molecular technologies that are highly interdependent [4]:

(a) Wet nanotechnology: It is the study of biological systems that exist primarily in a water environment. The functional nanometer-scale structures of interest here are genetic materials, membranes, enzymes, and other cellular components.

The original version of this chapter was revised: Incorrect co-author name has been corrected. The correction to this chapter is available at https://doi.org/10.1007/978-3-319-97634-1_14

H. M. Abou El Nasr (✉)
Department of Endodontics, Faculty of Oral and Dental Medicine,
Cairo University, Cairo, Egypt
e-mail: Hend.abouelnasr@dentistry.cu.edu.eg

M. B. Akbulut
Department of Endodontics, Faculty of Dentistry Necmettin,
Erbakan University, Konya, Turkey

R. S. Chaughule (ed.), *Dental Applications of Nanotechnology*,
https://doi.org/10.1007/978-3-319-97634-1_3

(b) Dry nanotechnology: It is derived from surface science and physical chemistry and focuses on the fabrication of structures in carbon, silicon, and other inorganic materials.
(c) Computational nanotechnology: This permits the modeling and simulation of complex nanometer-scale structures.

3.1.2 *Types of Nanoparticles (NPs)*

Nanoparticles can be organic polymers (organic NPs) and/or inorganic elements (inorganic NPs) [5].

Liposomes, dendrimers, and carbon nanomaterials are examples of organic NPs, while inorganic NPs, such as polystyrene, magnetic, ceramic, and metallic NPs, have a central core composed of inorganic material that define their fluorescent, magnetic, electronic, and optical properties [5].

3.1.3 *Properties of Nanoparticles*

At the nanometer-length scale, materials behave differently due to the increased number of atoms near the surface compared to the bulk structure.

- Two principal factors cause the properties of nanomaterials to differ from other materials: the increase in relative surface area and the physical effects [4].
- The smaller a particle, the greater is its surface-area-to-volume ratio and the higher its chemical reactivity and biological activity.
- The greater the chemical reactivity of nanomaterials results in increased production of reactive oxygen species (ROS) that have potent antimicrobial effect. However, the same properties that make nanoparticles so unique—that is, primarily, their small size, large surface area, chemical composition, solubility, and geometry—could also be responsible for their potential hazard to human health.

3.2 Nanotechnology in Medicine

Nanometer-sized particles are in the same range of dimension as antibodies, membrane receptors, nucleic acids and proteins, among other biomolecules. These biomimetic features, together with their high surface-to-volume ratio, make NPs powerful tools for imaging, diagnosis, and therapy [5].

Therefore, the integration of nanomaterials with biology led to the development of diagnostic devices, contrast agents, analytical tools, physical therapy application, and drug delivery vehicles.

Moreover, based on enhanced effectiveness, the new-age drugs are nanoparticles of polymers, metals, or ceramics, which can combat conditions like cancer and fight human pathogens like bacteria.

Thus, applying nanotechnology for diagnosis, treatment, monitoring, and control of disease has been referred to as "nanomedicine" [6].

3.3 Nanotechnology and Nanobiomaterials in Dentistry

New potential treatment opportunities in dentistry may include different approaches:

(a) Building up particles by combining atomic elements: i.e., bottom-up approach. This is applied in local anesthetics, hypersensitivity cure, dentifrices, orthodontic treatment, diagnosis and treatment of oral cancer.
(b) Using equipment to create mechanically nanoscale objects: i.e., top-down approach [1]. Examples include nanocomposites, impression materials, nanosolutions, nanoneedles, nanotweezers, and bone replacement materials.
(c) In addition to bottom-up and top-down approaches, there is the functional approach. This approach disregards the method of production of a NP; the objective is to produce a NP with a specific functionality [2].

Nanotechnology has been applied to many specialties in the dental field. For instance, effective drug delivery systems were generated for the treatment of periodontal diseases. NPs impregnated with antimicrobial agents were produced in order to maintain therapeutic concentration of the agent in the periodontal pocket for a sufficient length of time to ensure eradication of the bacteria present [6, 7]. Moreover, toothpastes containing nanosized calcium carbonate enabled remineralization of early enamel lesions as a prophylactic measure to prevent dental caries [6]. In the same context, NPs of amorphous calcium phosphate ($CaPO_4$) dental composite restorative materials were developed and were described as "smart" because they release calcium and phosphate ions at a cariogenic pH [8] and remineralize tooth lesions [9].

3.4 Nanotechnology in Endodontics

Almost every aspect in endodontics has gained advantage of the newly emerging nanotechnology. Different types of nanoparticles used in endodontics and other applications are summarized in Table 3.1.

Table 3.1 Summary of different types of nanoparticles used in endodontics and their applications

Type of nanoparticle	Hypersensitivity	Root canal disinfection	Obturation		Dentin stabilization	Vital pulp therapy	Regenerative	Surgery
			Filling material	Sealer				
I. Inorganic								
I.A. Metallic								
Silver nanoparticles		✔	✔	✔				✔
I.B. Metal oxides								
Zinc oxide		✔		✔				✔
Calcium phosphate				✔				✔
Bioactive glass	✔	✔	✔			✔		
Bioceramic			✔	✔				
Nano-hydroxyapatite	✔			✔		✔		
Zirconium oxide								✔
Titanium dioxide								
Nanosilica	✔							
II. Organic								
Chitosan		✔		✔	✔		✔	
Quaternary ammonium polyethyleneimine		✔		✔				

3.4.1 Radiography

By using dental digital radiographies obtained with nanophosphor scintillators, the radiation dose is diminished and high-quality images are obtained [6].

3.4.2 Local Anesthesia

To induce oral anesthesia in the era of nanodentistry, a colloidal suspension containing millions of active analgesic micron-sized dental NPs is instilled on patient's gingivae. After contacting the surface crown or mucosa, ambulating nanorobots reach the pulp. Once installed there, analgesic dental NPs may be commanded by dentist to shut down all sensitivity in any particular tooth that requires treatment [1].

3.4.3 Dentin Hypersensitivity

Dentin hypersensitivity is defined as sharp pain arising from exposed dentin as a result of various stimuli such as heat, cold, chemical, or osmotic and that cannot be ascribed to any other pathology.

Occlusion of the exposed and patent dentinal tubules is an effective method to provide relief from dentinal pain [10].

Nanoparticles of carbonated apatite, silica, bioglass, and functionalized hydroxyapatite have been used for the management of dentin hypersensitivity [10–12]. They occlude the dentinal tubules within minutes offering patients a quick and permanent cure [1].

The main advantage of using nanoparticles is their size, which allows them to be pushed deeper into dentinal tubules. Furthermore, these nanoparticles, once modified, could promote remineralization of the demineralized dentin [13].

3.4.4 Cleaning of Endodontic Instruments

Saghiri et al. [14] found that nanostructured foam was more efficient in removing debris from endodontic files, especially from those with complex cross-section than ProFile file cleaner.

3.4.5 Root Canal Disinfection: Nanoparticle-Based Disinfection

Bacterial infection is the primary cause of endodontic disease. Biofilm is the preferable mode for bacterial growth in oral disease conditions such as apical periodontitis [15]. Irrigants, alone or in combination, failed to achieve "microbe-free" root canals prior to obturation. Therefore, newer antimicrobials were warranted for root canal disinfection. Nanoparticles exhibit antibacterial activity as a result of their polycationic/polyanionic nature with higher surface area and charge density, resulting in greater degree of interaction with the bacterial cell [16]. The size of NPs plays an important role in their antibacterial activity with smaller particles showing higher antibacterial activity than the macroscaled ones. Moreover, bacteria are less likely to acquire resistance against nanoparticles than other conventional antibiotics. However, direct or close contact between the NPs and the bacterial membrane is essential for effective destruction of bacteria [16].

3.4.6 Endodontic Filling Materials

Different types of nanoparticles were investigated for their antimicrobial potential against endodontic pathogens:

A. Silver Nanoparticles (Ag NPs)

The antimicrobial properties of Ag NPs have been generally exploited in nanofiber materials, bandages, wound dressings, and ointments. Ag NPs also prevented bacterial colonization on various surfaces such as catheters, prostheses, and clothing [1].

Mechanism of action: Silver (Ag) works in a number of ways to disrupt critical functions in a microorganism. It attacks multiple sites within the cell to inactivate critical physiological functions such as cell wall synthesis, membrane transport, nucleic acids (DNA and RNA) synthesis, protein folding and with transport [1]. The mode of action of Ag NPs has been proposed to be identical to one of the Ag ions. The positively charged Ag NPs interact strongly with the cell membrane modifying its permeability [17, 18]. Moreover, the Ag NPs interact with multiple targets in the microbial cell, such as enzymes and plasmids, providing the bacteria least capacity to gain resistance [18].

Ag NPs with size in the range of 10–100 nm showed powerful bactericidal potential against gram-positive and gram-negative bacteria. A root canal irrigating solution of 0.1% Ag NPs imparted limited antibacterial effect against *Enterococcus faecalis* biofilm, while an intracanal medicament of 0.2% Ag NPs gel applied for 7 days led to the destruction of *E. faecalis* biofilm and the number of viable bacterial cells was significantly decreased [18]. Further, a mixture of calcium hydroxide [$Ca(OH)_2$] and Ag NPs had a potential advantage in eliminating *E. faecalis* from root canals [19].

The surface charge on Ag NPs was important in bactericidal efficacy against *E. faecalis*. When NPs with positive, negative, or neutral surface charges were compared; the positively charged Ag NPs were effective against *E. faecalis* and exhibited a high level of cytocompatibility [20]. Ag NPs dispersion was found to be biocompatible, especially in lower concentrations [21].

B. Chitosan Nanoparticles (CS NPs)

Chitosan is a polysaccharide derived from chitin, one of the most abundant natural polymers in the biosphere. Chitin is the main component of the exoskeleton of marine crustaceans (e.g., shrimps, crabs). It is currently receiving a great deal of interest for medical, pharmaceutical, and agricultural applications because of its biodegradability, biocompatibility, nontoxicity, and antiviral properties [22]. It can be synthesized in various forms such as powder, capsules, films, scaffolds, hydrogels, and beads [22].

Mechanism of action: Chitosan acts mainly on the outer surface of the bacteria. One of the proposed mechanisms of its antibacterial properties is that—similar to Ag NPs—the interaction between positively charged chitosan and negatively charged cell would alter the bacterial cell wall permeability resulting in the leakage of intracellular components and cell death [22].

It was also hypothesized that, due to its chelating property, it sequesters essential nutrients and inhibits enzymatic activities essential for bacterial cell survival [22]. Its antifungal property was referred to its penetration of the cells, reaching the nuclei where they bind with DNA and inhibit RNA and protein synthesis. Although the antibacterial effect of chitosan is not affected by dentin, dentin matrix, or lipopolysaccharide [23], it shows its antibacterial activity only in an acidic medium because of its poor solubility above pH 6.5 [22]. Thus, its use is limited because of its insolubility in water, high viscosity, and tendency to coagulate with proteins at high pH [22]. Incorporating chitosan NPs into pastes of calcium hydroxide was effective as intracanal medicament at 7 and 14 days [24]. Figure 3.1 depicts the antibacterial mechanism of nanoparticles with positive charge (e.g., chitosan). Mature bacterial biofilm consisting of abundant EPS and bacteria enclosed. When cationic nanoparticles are introduced for the treatment of biofilms, it can interact with both EPS and bacterial cells. The initial electrostatic interaction takes place between positively charged nanoparticles and negatively charged bacterial surface. Bacterial killing occurs upon contact-mediated lipid peroxidation via production of reactive oxygen species (ROS). The membrane damage and increased permeability of unstable membrane eventually lead to the ingress of nanoparticles into the cytoplasm and release of cytoplasmic constituents. EPS secreted by bacteria in biofilm may interact with the nanoparticles and prevent from interacting with bacteria and thus reducing the antibacterial efficacy.

Nanoparticles of calcium hydroxide (CaOH NPs) itself were prepared and were compared to chitosan NPs as intracanal medicaments for 1 week and 1 month. The highest depth of penetration into dentinal tubules was recorded by the CaOH NPs, while the greatest fracture resistance was found with the CS NPs. On the other hand,

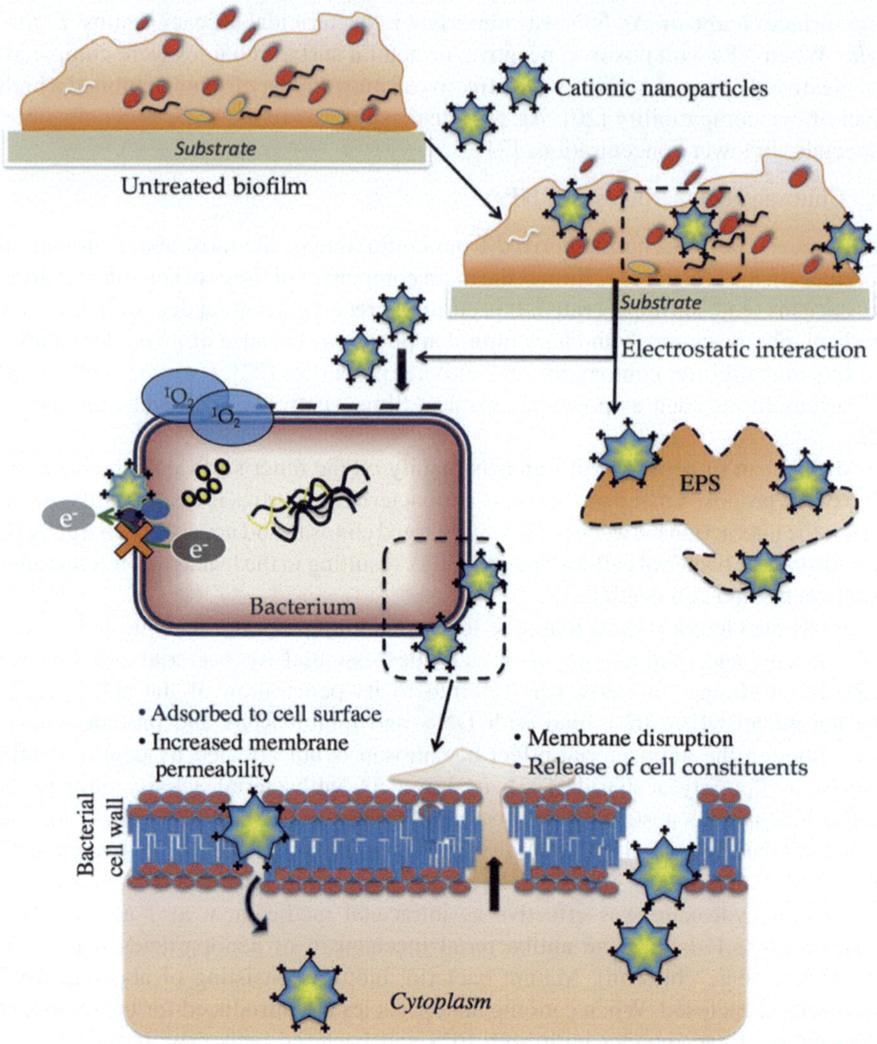

Fig. 3.1 A schematic diagram illustrating the antibacterial mechanism of nanoparticles with positive charge (e.g., chitosan)

both types of nanoparticles showed larger zones of inhibition of *E. faecalis* compared to their regular counterparts [25].

C. Zinc Oxide NPs (ZnO NPs)

ZnO NPs and a mixture of ZnO/CS NPs produced a significant reduction in the number of *E. faecalis* adhering to dentin [26] and were able to disrupt the multilayered biofilm architecture [16].

Mechanism of action: The antibacterial activity of the polycationic ZnO NPs was attributed to the electrostatic attraction with the negatively charged bacterial cell, which might lead to the altered cell wall permeability, resulting in leakage of the proteinaceous and other intracellular components and death of the cell [16]. ZnO nanocrystals were biocompatible and allowed new bone formation and remodeling [27].

D. Bioactive Glass (BG)

Bioactive glass received considerable interest in root canal disinfection due to its antibacterial properties.

Mechanism of action: The antibacterial mechanism of BG has been attributed to several factors acting together: the high pH due to release of ions in aqueous environments [28], the increase in osmotic pressure more than 1%, which is inhibitory for many bacteria, and the induced mineralization on the bacterial surface due to calcium/phosphorus precipitation [29]. However, it was less effective than calcium hydroxide in preventing residual bacterial growth [28]; it was also less effective in eliminating biofilms compared to the planktonic counterparts [30, 31].

3.4.7 *Root Canal Disinfection*

A. Photodynamic therapy (PDT)/photosensitizers

The recent advances toward achieving predictable endodontic disinfection focused on newer alternatives such as PDT. PDT is based on the injection, ingestion, or topical application of a photosensitizer dyes followed by visible light activation [7, 23, 32]. There are two classical PDT mechanisms: production of radical ions of oxygen (type I) or singlet oxygen (type II) [33].

Mechanism of action: PDT uses a specific wavelength of light to activate a non-toxic dye (photosensitizer), leading to the formation of ROS and singlet oxygen.

These reactive oxygen or singlet oxygen molecules are highly reactive and damage bacterial cell wall, membrane lipids and proteins, and nucleic acids, which eventually promote bacterial cell death [23, 34]. This antibacterial mechanism confers PDT with the advantage of broad-spectrum antibacterial activity and a lower risk of developing resistance. However, a reduction in the uptake of the photosensitizers into the bacterial cells could be one of the clinical factors compromising the antibacterial effect of PDT [23]. Moreover, the antibacterial activity of PDT was compromised in the presence of root canal constituents such as pulpal tissue, serum, dentin matrix, and bacterial remnants [35].

- Photosensitizers: Phenothiazines and xanthenes are two classes of photosensitizers commonly tested for antibacterial efficacy [23].

B. Rose bengal (RB) is an anionic xanthene dye [36]. It is water soluble, nontoxic and absorbs light in the visible spectral region [37]. However, its application is strongly limited by the fact that it tends to aggregate in aqueous solutions.

C. Methylene blue (MB) is a well-established photosensitizer that falls under the category of cationic phenothiazines [23] and has been used in PDT for targeting various gram-positive and gram-negative oral bacteria. MB possesses a greater antibacterial effect than RB because of its hydrophilicity, low molecular weight, and cationicity [23]. Nonetheless, the reduced susceptibility of biofilms to PDT was attributed to reduced penetration of the photosensitizer.
D. Indocyanine green (ICG) has gained special attention as a new photosensitizer due to its low host tissue toxicity, high absorption in the near infrared spectrum, fungal and bacterial elimination efficacy, and powerful photosensitized cellular damage. It affects the target cell mainly through a photothermal effect rather than a photochemical effect and performs its bactericidal effect through oxidative stress [38]. The use of ICG with PDT caused a significant reduction in *E. faecalis* colony units. Moreover, when Ag NPs were added to the ICG, the eradication of the microorganism was intensified [39].

3.4.8 Functionalization/Conjugation

Functionalization could alter the surface composition, charge, and structure of the material wherein the original bulk material properties are left intact. Coating or surface attachment of photosensitizers to NPs significantly improves their antibacterial properties [35]. Functionalized NPs with photosensitizers offer unique physicochemical advantages such as allowing higher concentration of photosensitizer uptake per cell, reduce the efflux of photosensitizer from target cell, permit greater interaction with cells because of the surface charge, provide greater stability to photosensitizer molecules after conjugation, and allow controlled release of ROS [32, 36].

Chitosan offered an attractive material for conjugation with other reactive molecules. This was attributed to the membrane destabilizing/permeabilizing effect of chitosan that could subsequently enhance the effect of singlet oxygen on bacterial cells [35]. In fact, the antibacterial effect of PDT was significantly improved when chitosan was conjugated with rose bengal, even in the presence of tissue inhibitors (dentin, dentin matrix, lipopolysaccharides [35].

3.4.9 Dentin Stabilization

Collagen is the main structural protein of human dentin [40] and contributes to its viscoelasticity, ultrastructural stability, and tensile strength [41]. Degradation of collagen by bacterial enzymes can facilitate microbial penetration, compromise structural integrity, and lower the resistance of dentin to fracture [24]. Moreover, matrix metalloproteinases (MMP-1, MMP-2, MMP-8, and MMP-9) are found in dentin collagen matrix and play an important role in its physiological and pathological degradation [42]. Activation of these MMPs can occur by some root canal irrigants during endodontic treatment. Among the materials tested to inhibit the action of these MMPs were chitosan and its derivatives, due to its ability to neutralize MMPs [42].

PDT enhances cross-linking of collagen fibrils in the dentin matrix, thereby increases the tensile strength and elastic modulus of collagen, and improves dentin stability [23]. It is a rapid process in which the proteins and collagen molecules are covalently cross-linked when illuminated in the presence of appropriate photosensitizers [43]. However, the cross-linked collagen may become stiff and brittle. Thus, the incorporation of polymeric fillers would reinforce the collagen structure and serve as spacers in between the collagen fibrils [43, 44]. One promising method of improving collagen cross-linking and reducing degradation is the use of PDT in combination with polycationic chitosan nanoparticles. Chitosan nanoparticles (CS NPs) and PDT inhibit the activity of bacterial collagenases. Thus, combining PDT and binding to CS NPs inhibit bacterial-mediated collagenolytic activity and enhance the mechanical properties of collagen matrix [34].

3.4.10 Endodontic Filling Materials

Root canal obturation is one of the essential steps for successful endodontic treatment aiming at preventing the reinfection of root canals that have been chemo-mechanically cleaned, shaped, and disinfected. Adequate root canal obturation is achieved through entirely filling the root canal system and providing a tight seal. Three-dimensional root canal obturation requires the use of a core material (gutta-percha, resilon, or coated cones) and root canal sealer.

3.4.10.1 Obturating Core Material

Gutta-percha has always been and remains the most widely used root canal obturating material. It presented a slight antibacterial activity due to its zinc oxide component [45]. However, studies have shown that endodontic microorganisms have high affinity for gutta-percha [46, 47]. This could lead to persistence of infection in the root canals. Therefore, antimicrobial agents were integrated into gutta-percha points to improve their antibacterial properties. Calcium hydroxide gutta-percha points (Roeko Gmbh + Co., Langenau, Baden-Württemberg, Germany), Active Point (chlorhexidine containing gutta-percha), MGP (iodoform containing gutta-percha) (Medidenta), and TGP (tetracycline containing gutta-percha) (Medidenta) could be given as examples for commercially available preparations. On the other hand, gutta-percha was coated with nanosilver particles in an attempt to enhance its antibacterial effect [48]. This new material has displayed significant antibacterial and antifungal properties against *E. faecalis*, *Staphylococcus aureus*, *Candida albicans*, and *Escherichia coli* [48]. Moreover, nanosilver-coated gutta-percha was cytotoxic to mouse fibroblasts after 1 hour and decreased with time to become similar to standard gutta-percha after 24 h and 1 week [49].

EndoSequence® BC Points™ has been introduced with a patented formula of coating each cone with bioceramic nanoparticles. Bioceramic nanoparticle-impregnated cones were claimed to provide gap-free seal when used in conjunction with BC Sealer™ through the interaction and bonding of bioceramic particles between sealer and gutta-percha cones [50].

Bioactive glass nanoparticles (BG NPs) integrated into polyisoprene or polycaprolactone-based root canal filling materials were proposed as promising single obturation materials without the need of a sealer [51]. This is because hydroxyapatite formation was observed on their surface [51].

3.4.10.2 Root Canal Sealers

One of the desired properties of root canal sealers is their antimicrobial properties. The common sealers possess antibacterial activity for a maximum period of 1 week, with most of them showing a significant reduction in antibacterial properties immediately after it sets [52, 53]. With recent advances in material science and nanotechnology, nanoparticles were incorporated into novel root canal sealers to improve their physical, chemical, biological, and antibacterial properties. They highly interact with microbial cells due to their higher surface area and charge density.

Types of nanoparticles added to root canal sealers: The development and characterization of root canal sealers were performed with nanoparticles of quaternary ammonium polyethyleneimine nanoparticles (QPEI NPs), zinc oxide nanoparticles (ZnO NPs), chitosan nanoparticles (CS NPs), calcium phosphate nanoparticles (CP NPs), and bioactive glass nanoparticles (BG NPs).

A. Antibacterial Properties

Incorporation of cationic nanoparticles such as CS NPs and ZnO NPs in zinc oxide-based and resin-based root canal sealers improved their antibacterial effect [54, 55]. Del Carpio-Perochena et al. [56] added CS NPs in the epoxy resin-based and calcium silicate-based root canal sealers and found that this addition enhanced the antibacterial activity of epoxy resin-based sealer whereas no extra benefit was shown in calcium silicate-based sealer. Epoxy resin-based and polymer-based sealers with added QPEI NPs eliminated the growth of *E. faecalis* for 4 weeks [57, 58]. Barros et al. [59] revealed that incorporation of QPEI NPs did not change the antimicrobial activities of ZnO-based and resin-based endodontic sealers in fresh state, while addition of QPEI NPs resulted in great increase in the antibacterial effect of ZnO-based sealer after aging for 7 days.

An experimental nano-hydroxyapatite epoxy resin-based sealer was compared with commercial sealers regarding antibacterial activity and exhibited similar antimicrobial inhibitory effect between this experimental sealer and AH Plus [60].

B. Biological Properties
1. Biocompatibility

The biocompatibility of nano-ZnO sealer was comparable to unmodified ZnO-based sealer and less cytotoxic to murine fibroblasts than epoxy resin-based sealer [61]. In an animal study, nano-ZnO sealer induced similar tissue reaction to epoxy resin-based and commercial zinc oxide and eugenol-based sealer. It exhibited satisfactory biocompatibility on subcutaneous tissue of rats after 15, 30, and 60 days of implantation [62]. Additionally, nano-ZnO sealer was more biocompatible with human gingival fibroblasts than commercial micro-sized ZnO- and resin-based sealers [63]. Abramovitz et al. showed that the addition of QPEI NPs to epoxy resin-based, polymer-based, and methacrylate resin-based sealers did not impair their biocompatibility, since QPEI NPs did not leach into surrounding tissues [64].

2. Bioactivity

An experimental bioactive methacrylate endodontic sealer containing amorphous calcium phosphate nanoparticles (CP NPs) was developed. It aimed at helping the remineralization of root dentin and thus strengthening the roots [65]. This bioactive sealer released calcium and phosphate ions for a long period of time and has been proposed as a promising root canal filling material because of its anti-biofilm activity and remineralization capacity [66].

C. Physical Properties

Current publications investigated the physical properties of nanomodified sealers. Addition of CS NPs did not alter the flow characteristics of ZnO-based root canal sealer [54]. Incorporation of QPEI NPs increased the setting time, hydrophilicity, wettability, and surface charge of ZnO-based and epoxy resin-based sealers [59]. Loading with nano-zirconium oxide particles as radiopacifier enabled the penetration of portland-based and epoxy resin-based sealers in dentinal tubules at the apical third, thus improving their retention and sealing ability [66]. It was observed that replacing conventional ZnO powder in Grossman sealer with ZnO NPs decreased the setting time and dimensional changes of the sealer [67]. Moreover, loading with 25% ZnO-NP improved the physicochemical properties such as setting time, flow solubility, dimensional stability, and radiopacity of Grossman sealer [67]. Addition of nano-hydroxyapatite to methacrylate-based root canal sealers caused no alteration in their radiopacity and film thickness [68].

D. Commercial Sealers Produced with Nanotechnology

Guttaflow is a silicon polymer-based root canal sealer that is composed of finely ground gutta-percha, poly-dimethylsiloxane and nanosilver. Nanosilver is biocompatible, prevents the spread of bacteria, and does not cause corrosion or color changes in the sealer [69]. Nanoceramic Sealer (NCS) (B&L Biotech, Fairfax, VA, USA) is another calcium silicate-based sealer that has been compared with other commercial calcium silicate-based sealers and was found to support periodontal ligament stem cell survival [70].

3.4.11 Temporization

Quaternary ammonium polyethyleneimine (QPEI) NPs are long-lasting, stable, biocompatible, and nonvolatile antibacterial polymers. Incorporation of QPEI NPs into standard temporary restorative material improved significantly their sealing ability and antibacterial properties [71].

3.5 Nanobiomaterials in Endodontics

Biomaterials are materials that stay in close contact with vital tissues such as pulp, periradicular tissues, or periodontal ligament for extended periods of time. They have an extensive usage in the field of endodontics including pulp capping, pulpotomy, perforation repair, root-end filling, apexogenesis, and apexification procedures. The majority of endodontic biomaterials are mainly composed of calcium hydroxide or calcium silicate.

Nanotechnology application in endodontic biomaterials emerges in two ways. One of them is nanomodification of the content of the material, and the other way is to add nanoparticles into the composition of the material. Both nanomodification and incorporation of nanoparticles are expected to provide better biological and physicochemical properties.

3.5.1 Vital Pulp Therapy

Vital pulp therapy is aiming at treating pulp injuries by capping the pulp, to stimulate the formation of tertiary dentin and maintain pulp vitality. Nanosized bioactive glass (BG) particles possess a higher specific surface area, a more regular size, and better bioactivity compared with traditional BG particles [72, 73]. Nanocrystalline hydroxyapatite (HA) paste was found biocompatible and superior to formocresol when tested as pulpotomy and pulp capping agent [74]. It was also recommended as a substitute for MTA and Dycal in direct pulp capping [75]. Moreover, incorporation of 3%wt. nano-hydroxyapatite increased the compressive strength of calcium hydroxide cement and enhanced the release of calcium ions which indicates mineralization capacity that is desired for inducing hard tissue formation in pulp capping procedures [76].

3.5.2 Regenerative Endodontics

A well-controlled release system of specific growth factors is a pivotal strategy in dentin-pulp engineering. The different delivery systems share the same goal, which is the release of bioactive molecules in a time-controlled manner. Numerous studies focused on creating appropriate delivery systems such as NPs, microspheres, hydrogels, and scaffolds for tissue regeneration [77]. An antibiotic-containing nano-/microfibrous scaffold has been synthesized to provide a drug delivery system to disinfect necrotic immature permanent teeth through a controllable release of low, yet effective antibiotic doses, while serving as a matrix for the growth and differentiation of stem cells from the apical papilla after bleeding is induced [78, 79].

3.5.3 Endodontic Surgery

A. Root-End Filling Materials

Chogle et al. found that the addition of NPs to the monomer matrix of root-end filling materials reduced apical leakage significantly [80]. The cytotoxicity of a polymer nanocomposite resin was not significantly different from that of ProRoot MTA and Geristore [81].

B. Mineral Trioxide Aggregate (MTA)

Mineral trioxide aggregate (MTA), introduced into the endodontic literature in 1993 [82], has been widely preferred by clinicians due to its biocompatibility, bioactivity, and good sealing ability [83]. However, MTA has some shortcomings like long setting time, weak handling characteristics, and discoloration potential [83].

Akbari et al. [84] found that the addition of nanosilica to MTA accelerated the hydration process, reduced the setting time, and had no adverse effect on its compressive and flexural strengths. Moreover, the addition of Ag NPs to MTA improved its antimicrobial efficacy [85, 86] and induced similar inflammatory reaction to conventional MTA in the subcutaneous tissue of rats [87].

Nanowhite mineral trioxide aggregate (Nano-WMTA) produced via reducing the size of particles of WMTA and adding tricalcium aluminate, calcium sulfate, zeolite, and strontium into the composition exhibited decreased setting time [88], low surface porosity, and more resistance to acidic environment than WMTA [88, 89]. Furthermore, the greater surface area of nano-WMTA and incorporation of nano-tricalcium aluminate induced faster release of calcium ions and significantly raised the calcium content, as well as the pH of the environment compared to conventional WMTA. This displayed favorable histopathological reactions and contributed to bone regeneration in the mandible of rabbits [90–92]. These nanobiomaterials resemble bone neutral minerals in size and composition promoting desirable osteoconductive activity [93]. In fact, nano-WMTA has been proposed in the situation of periapical inflammation [89].

Zirconium oxide (ZrO) nanoparticles were added into the composition of white portland cement as radiopacifier. This ZrO blended cement revealed accelerated hydration process and decreased solubility. Besides, incorporation of ZrO nanoparticles did not cause alteration in antimicrobial activity against *S. aureus* and *E. coli*, but exhibited slightly more antibacterial activity against *Pseudomonas aeruginosa* [93].

The incorporation of titanium dioxide nanoparticles into white portland cement reduced the setting time and improved flexural and compressive strength of the material [94]. However, addition of other nanoparticulate radiopacifiers such as zirconium oxide, niobium oxide, and bismuth oxide or calcium tungstate did not alter the physicochemical properties of calcium silicate-based cements when compared to microparticulate forms [95].

The addition of nano-hydroxyapatite to the portland cement reduced the final setting time and enhanced the antibacterial activity of the cement against *E. faecalis* biofilm. However, it impaired the compressive strength and solubility of the material [96].

C. Bone Replacement Materials

Nanosized HA is the main component of mineral bone in the form of nanometer-sized needle-like crystals of approximately 5–20 nm width by 60-nm length. Utilizing nanotechnology, calcium and phosphate are manipulated at the molecular level and assembled to produce a strong and osteoconductive material with unique structural and functional properties [1].

3.6 Conclusion

From protecting existing pulp tissue to regenerating new pulp tissue, nanotechnology has invaded every aspect in endodontics. The new properties imparted to the nanomaterials by virtue of their smaller particles and altered physicochemical and electrostatic characteristics seemed to be of potential benefit to the endodontic specialty. However, its validity and clinical relevance are still not established. This, certainly, opens new research horizons and highlights the major questions that need to be addressed in order to optimize its benefits to endodontics.

References

1. Mohan S, Gurtu A, Singhal A, Mehrotra A. (2013) Nanotechnology: its implications in conservative dentistry and endodontics. J Dent Sci Oral Rehabil 2013 July–Sept
2. Sivaramakrishnan SM, Neelakantan P (2014) Nanotechnology in dentistry—what does the future hold in store? Dentistry 4:198
3. Shrivastava S, Bera T, Roy A, Singh G, Ramachandrarao P, Dash D (2007) Characterization of enhanced antibacterial effects of novel silver nanoparticles. Nanotechnology 18:1–9

4. Kanaparthy R, Kanaparthy A (2011) The changing face of dentistry: nanotechnology. Int J Nanomed 6:2799–2804
5. Sanvicens N, Marco MP (2008) Multifunctional nanoparticles-properties and prospects for their use in human medicine. Trends Biotechnol 26:425–433
6. Ozak ST, Ozkan P (2013) Nanotechnology and dentistry. Eur J Dent 7:145–151
7. Séguier S, Souza SL, Sverzut AC, Simioni AR, Primo FL, Bodineau A, Corrêa VM, Coulomb B, Tedesco AC (2010) Impact of photodynamic therapy on inflammatory cells during human chronic periodontitis. J Photochem Photobiol, B 101:348–354
8. Xu HH, Moreau JL, Sun L, Chow LC (2011) Nanocomposite containing amorphous calcium phosphate nanoparticles for caries inhibition. Dent Mater 27:762–769
9. Moreau JL, Weir MD, Giuseppetti AA, Chow LC, Antonucci JM, Xu HH (2012) Long-term mechanical durability of dental nanocomposites containing amorphous calcium phosphate nanoparticles. J Biomed Mater Res B Appl Biomater 100:1264–1273
10. Lee SY, Kwon HK, Kim BI (2008) Effect of dentinal tubule occlusion by dentifrice containing nano-carbonate apatite. J Oral Rehabil 35:847–853
11. Tian L, Peng C, Shi Y, Guo X, Zhong B, Qi J, Wang G, Cai Q, Cui F (2014) Effect of mesoporous silica nanoparticles on dentinal tubule occlusion: an in vitro study using SEM and image analysis. Dent Mater J 33(1):125–132
12. Lin X, Xie F, Ma X, Hao Y, Qin H, Long J (2017) Fabrication and characterization of dendrimer-functionalized nano-hydroxyapatite and its application in dentin tubule occlusion. J Biomater Sci Polym Ed 28(9):846–863
13. Besinis A, van Noort R, Martin N (2012) Infiltration of demineralized dentin with silica and hydroxyapatite nanoparticles. Dent Mater 28:1012–1023
14. Saghiri MA, Karamifar K, Mehrvazfar P, Asgar K, Gutmann JL, Lotfi M, Garcia-Godoy F (2012) The efficacy of foam cleaners in removing debris from two endodontic instruments. Quintessence Int 43:811–817
15. Baumgartner C, Siqueira J, Sedgley C, Kishen A (2008) Microbiology of endodontic disease. In: Ingle's endodontics, 6th edn. BC Decker Inc: Hamilton
16. Shrestha A, Shi Z, Neoh KG, Kishen A (2010) Nanoparticulates for antibiofilm treatment and effect of aging on its antibacterial activity. J Endod 36:1030–1035
17. Simon-Deckers A, Brun E, Gouget B, Carrière M, Sicard-Roselli C (2008) Impact of gold nanoparticles combined to X-Ray irradiation on bacteria. Gold Bull. 41:187–194
18. Wu D, Fan W, Kishen A, Gutmann JL, Fan B (2014) Evaluation of the antibacterial efficacy of silver nanoparticles against *Enterococcus faecalis* biofilm. J Endod 40:285–290
19. Javidi M, Afkhami F, Zarei M, Ghazvini K, Rajabi O (2014) Efficacy of a combined nanoparticulate/calcium hydroxide root canal medication on elimination of *Enterococcus faecalis*. Aust Endod J 40:61–65
20. Abbaszadegan A, Nabavizadeh M, Gholami A, Aleyasin ZS, Dorostkar S, Saliminasab M, Ghasemi Y, Hemmateenejad B, Sharghi H (2015) Positively charged imidazolium-based ionic liquid-protected silver nanoparticles: a promising disinfectant in root canal treatment. Int Endod J 48:790–800
21. Gomes-Filho JE, Silva FO, Watanabe S, Cintra LT, Tendoro KV, Dalto LG, Pacanaro SV, Lodi CS, de Melo FF (2010) Tissue reaction to silver nanoparticles dispersion as an alternative irrigating solution. J Endod 36:1698–1702
22. Rabea EI, Badawy ME, Stevens CV, Smagghe G, Steurbaut W (2003) Chitosan as antimicrobial agent: applications and mode of action. Biomacromolecules 4:1457–1465
23. Shrestha A, Kishen A (2012) The effect of tissue inhibitors on the antibacterial activity of chitosan nanoparticles and photodynamic therapy. J Endod 38:1275–1278
24. Del Carpio-Perochena A, Kishen A, Felitti R, Bhagirath AY, Medapati MR, Lai C, Cunha RS (2017) Antibacterial properties of chitosan nanoparticles and propolis associated with calcium hydroxide against single- and multispecies biofilms: an in vitro and in situ study. J Endod 43(8):1332–1336
25. Sireesha A, Jayasree R, Vidhya S, Mahalaxmi S, Sujatha V, Kumar TSS (2017) Comparative evaluation of micron- and nano-sized intracanal medicaments on penetration and fracture resistance of root dentin—an in vitro study. Int J Biol Macromol 104:1866–1873

26. Kishen A, Shi Z, Shrestha A, Neoh KG (2008) An investigation on the antibacterial and antibiofilm efficacy of cationic nanoparticulates for root canal disinfection. J Endod 34:1515–1520
27. Sousa CJ, Pereira MC, Almeida RJ, Loyola AM, Silva AC, Dantas NO (2014) Synthesis and characterization of zinc oxide nanocrystals and histologic evaluation of their biocompatibility by means of intraosseous implants. Int Endod J 47:416–424
28. Zehnder M, Luder HU, Schatzle M, Kerosuo E, Waltimo T (2006) A comparative study on the disinfection potentials of bioactive glass S53P4 and calcium hydroxide in contra-lateral human premolars ex vivo. Int Endod J 39:952–958
29. Stoor P, Soderling E, Salonen JI (1998) Antibacterial effects of a bioactive glass paste on oral microorganisms. Acta Odontol Scand 56:161–165
30. Waltimo T, Mohn D, Paque F, Brunner TJ, Stark WJ, Imfeld T et al (2009) Fine-tuning of bioactive glass for root canal disinfection. J Dent Res 88:235–238
31. Waltimo T, Brunner TJ, Vollenweider M, Stark WJ, Zehnder M (2007) Antimicrobial effect of nanometric bioactive glass 45S5. J Dent Res 86:754–757
32. Pagonis TC, Chen J, Fontana CR, Devalapally H, Ruggiero K, Song X, Foschi F, Dunham J, Skobe Z, Yamazaki H, Kent R, Tanner AC, Amiji MM, Soukos NS (2010) Nanoparticle-based endodontic antimicrobial photodynamic therapy. J Endod 36:322–328
33. Foote CS (1991) Definition of type I and type II photosensitized oxidation. Photochem Photobiol 54:659
34. Persadmehr A, Torneck CD, Cvitkovitch DG, Pinto V, Talior I, Kazembe M, Shrestha S, McCulloch CA, Kishen A (2014) Bioactive chitosan nanoparticles and photodynamic therapy inhibit collagen degradation in vitro. J Endod 40:703–709
35. Shrestha A, Kishen A (2014) Antibacterial efficacy of photosensitizer functionalized biopolymeric nanoparticles in the presence of tissue inhibitors in root canal. J Endod 40:566–570
36. Shrestha A, Kishen A (2014) Antibiofilm efficacy of photosensitizer-functionalized bioactive nanoparticles on multispecies biofilm. J Endod 40:1604–1610
37. Moczek Ł, Nowakowska M (2007) Novel water-soluble photosensitizers from chitosan. Biomacromolecules 8:433–438
38. Fekrazad R, Karamifar K, Bahador A (2016) Comparison of antibacterial effect of photodynamic therapy using indocyanine green (Emundo) with 2% metronidazole and 2% chlorhexidine gel on *Porphyromonas gingivalis* (an in-vitro study). Photodiagnosis Photodyn Ther 15:28–33
39. Afkhami F, Akbari S, Chiniforush N (2017) *Entrococcus faecalis* elimination in root canals using silver nanoparticles, photodynamic therapy, diode laser, or laser-activated nanoparticles: an in vitro study. J Endod 43(2):279–282
40. Tjäderhane L, Carrilho MR, Breschi L et al (2012) Dentin basic structure and composition: an overview. Endod Top 20:3–39
41. Kishen A (2006) Mechanisms and risk factors for fracture predilection in endodontically treated teeth. Endod Top 13:57–83
42. Kishen A, Shrestha A (2015) Nanotechnology in endodontics. Current and potential clinical applications. Springer International Publishing
43. Shrestha A, Hamblin MR, Kishen A (2014) Photoactivated rose bengal functionalized chitosan nanoparticles produce antibacterial/biofilm activity and stabilize dentin-collagen. Nanomedicine 10(3):491–501
44. Sionkowska A, Wisniewski M, Skopinska J, Kennedy CJ, Wess TJ (2004) Molecular interactions in collagen and chitosan blends. Biomaterials 25(5):795–801
45. Moorer WR, Genet JM (1982) Antibacterial activity of gutta-percha cones attributed to the zinc oxide component. Oral Surg Oral Med Oral Pathol 53(5):508–517
46. Senges C, Wrbas KT, Altenburger M, Follo M, Spitzmuller B, Wittmer A et al (2011) Bacterial and Candida albicans adhesion on different root canal filling materials and sealers. J Endod 37:1247–1252
47. George S, Basrani B, Kishen A (2010) Possibilities of gutta-percha-centered infection in endodontically treated teeth: an in vitro study. J Endod 36:1241–1244

48. Corrêa JM, Mori M, Sanches HL, da Cruz AD, Poiate Jr. E, Poiate IVP (2015) Silver nanoparticles in dental biomaterials. Int J Biomat Article 485275
49. Shantiaee Y, Dianat O, Mohammadkhani H, Akabarzadeh BA (2011) Cytotoxicity comparison of nanosilver coated gutta-percha with Guttaflow and normal gutta-percha on L929 fibroblast with MTT assay. J Dent Sch Shahid Beheshti Univ Med Sci 29:63–69
50. Available from: http://brasselerusadental.com/products/bc-points/
51. Mohn D, Bruhin C, Luechinger N, Stark W, Imfeld T, Zehnder M (2010) Composites made of flame-sprayed bioactive glass 45S5 and polymers: bioactivity and immediate sealing properties. Int Endod J 43(11):1037–1046
52. Kayaoglu G, Erten H, Alacam T, Orstavik D (2005) Short-term antibacterial activity of root canal sealers towards *Enterococcus faecalis*. Int Endod J 38:483–488
53. Siqueira JF Jr, Favieri A, Gahyva SM, Moraes SR, Lima KC, Lopes HP (2000) Antimicrobial activity and flow rate of newer and established root canal sealers. J Endod 26:274–277
54. Kishen A, Shi Z, Shrestha A, Neoh KG (2008) An investigation on the antibacterial and antibiofilm efficacy of cationic nanoparticulates for root canal disinfection. J Endod 34(12):1515–1520
55. DaSilva L, Finer Y, Friedman S, Basrani B, Kishen A (2013) Biofilm formation within the interface of bovine root dentin treated with conjugated chitosan and sealer containing chitosan nanoparticles. J Endod 39(2):249–253
56. Del Carpio-Perochena A, Kishen A, Shrestha A, Bramante CM (2015) Antibacterial properties associated with chitosan nanoparticle treatment on root dentin and 2 types of endodontic sealers. J Endod 41(8):1353–1358
57. Kesler Shvero D, Abramovitz I, Zaltsman N, Perez Davidi M, Weiss EI, Beyth N (2013) Towards antibacterial endodontic sealers using quaternary ammonium nanoparticles. Int Endod J 46(8):747–754
58. Kesler Shvero D, Zaltsman N, Weiss EI, Polak D, Hazan R, Beyth N (2016) Lethal bacterial trap: Cationic surface for endodontic sealing. J Biomed Mater Res A 104(2):427–434
59. Barros J, Silva MG, Rodrigues MA, Alves FR, Lopes MA, Pina-Vaz I et al (2014) Antibacterial, physicochemical and mechanical properties of endodontic sealers containing quaternary ammonium polyethylenimine nanoparticles. Int Endod J 47(8):725–734
60. Aal-Saraj AB, Ariffin Z, Masudi SM (2012) An agar diffusion study comparing the antimicrobial activity of Nanoseal with some other endodontic sealers. Aust Endod J 38(2):60–63
61. Javidi M, Zarei M, Omidi S, Ghorbani A, Gharechahi M, Shayani Rad M (2015) Cytotoxicity of a new nano zinc-oxide eugenol sealer on murine fibroblasts. Iran Endod J 10(4):231–235
62. Omidi S, Javidi M, Zarei M, Mushakhian S, Jafarian A (2017) Subcutaneous connective tissue reaction to a new nano zinc-oxide eugenol sealer in rat model. Iran Endod J 12(1):64–69
63. Javidi M, Dastmalchi P, Zarei M, Shayani Rad M, Ghorbani A (2017) In vitro cytotoxicity of a new nano root canal sealer on human gingival fibroblasts. Iran Endod J 12(2):220–225
64. Abramovitz I, Beyth N, Weinberg G, Borenstein A, Polak D, Kesler-Shvero D et al (2012) In vitro biocompatibility of endodontic sealers incorporating antibacterial nanoparticles. J Nanomater. Article ID 858073
65. Wang L, Xie X, Li C, Liu H, Zhang K, Zhou Y et al (2017) Novel bioactive root canal sealer to inhibit endodontic multispecies biofilms with remineralizing calcium phosphate ions. J Dent 60(5):25–35
66. Viapiana R, Guerreiro-Tanomaru J, Tanomaru-Filho M, Camilleri J (2014) Interface of dentine to root canal sealers. J Dent 42(3):336–350
67. Versiani MA, Abi Rached-Junior FJ, Kishen A, Pecora JD, Silva-Sousa YT, de Sousa-Neto MD (2016) Zinc oxide nanoparticles enhance physicochemical characteristics of Grossman sealer. J Endod 42(12):1804–1810
68. Collares FM, Leitune VC, Rostirolla FV, Trommer RM, Bergmann CP, Samuel SM (2012) Nanostructured hydroxyapatite as filler for methacrylate-based root canal sealers. Int Endod J 45(1):63–67
69. De-Deus G, Brandao MC, Fidel RA, Fidel SR (2007) The sealing ability of GuttaFlow in oval-shaped canals: an ex vivo study using a polymicrobial leakage model. Int Endod J 40(10):794–799

70. Collado-Gonzalez M, Garcia-Bernal D, Onate-Sanchez RE, Ortolani-Seltenerich PS, Lozano A, Forner L et al (2017) Biocompatibility of three new calcium silicate-based endodontic sealers on human periodontal ligament stem cells. Int Endod J 50:875–884
71. Abramovitz I, Beyth N, Paz Y, Weiss EI, Matalon S (2013) Antibacterial temporary restorative materials incorporating polyethyleneimine nanoparticles. Quintessence Int 44:209–216
72. Mohn D, Zehnder M, Imfeld T, Stark WJ (2010) Radio-opaque nanosized bioactive glass for potential root canal application: evaluation of radiopacity, bioactivity and alkaline capacity. Int Endod J 43:210–217
73. Gong W, Huang Z, Dong Y, Gan Y, Li S, Gao X, Chen X (2014) Ionic extraction of a novel nano-sized bioactive glass enhances differentiation and mineralization of human dental pulp cells. J Endod 40:83–88
74. Shayegan A, Atash R, Petein M, Abbeele AV (2010) Nanohydroxyapatite used as a pulpotomy and direct pulp capping agent in primary pig teeth. J Dent Child (Chic). 77:77–83
75. Swarup SJ, Rao A, Boaz K, Srikant N, Shenoy R (2014) Pulpal response to nano hydroxyapatite, mineral trioxide aggregate and calcium hydroxide when used as a direct pulp capping agent: an in vivo study. J Clin Pediatr Dent 38:201–206
76. Yasaei M, Zamanian A, Moztarzadeh F, Ghaffari M, Mozafari M (2013) Characteristics improvement of calcium hydroxide dental cement by hydroxyapatite nanoparticles. Part 1: formulation and microstructure. Biotechnol Appl Biochem 60(5):502–509
77. Shrestha S, Diogenes A, Kishen A (2014) Temporal-controlled release of bovine serum albumin from chitosan nanoparticles: effect on the regulation of alkaline phosphatase activity in stem cells from apical papilla. J Endod 40:1349–1354
78. Bottino MC, Kamocki K, Yassen GH, Platt JA, Vail MM, Ehrlich Y, Spolnik KJ, Gregory RL (2013) Bioactive nanofibrous scaffolds for regenerative endodontics. J Dent Res 92:963–969
79. Palma PJ, Ramos JC, Martins JB, Diogenes A, Figueiredo MH, Ferreira P, Viegas C, Santos JM (2017) Histologic evaluation of regenerative endodontic procedures with the use of chitosan scaffolds in immature dog teeth with apical periodontitis. J Endod 43(8):1279–1287
80. Chogle SM, Duhaime CF, Mickel AK, Shaikh S, Reese R, Bogle JH, Chan DK, Potluri S (2011) Qutubuddin S. Preliminary evaluation of a novel polymer nanocomposite as a root-end filling material. Int Endod J 44:1055–1060
81. Modareszadeh MR, Chogle SA, Mickel AK, Jin G, Kowsar H, Salamat N, Shaikh S, Qutbudin S (2011) Cytotoxicity of set polymer nanocomposite resin root-end filling materials. Int Endod J 44:154–161
82. Lee SJ, Monsef M, Torabinejad M (1993) Sealing ability of a mineral trioxide aggregate for repair of lateral root perforations. J Endod 19(11):541–544
83. Parirokh M, Torabinejad M (2010) Mineral trioxide aggregate: a comprehensive literature review–Part III: Clinical applications, drawbacks, and mechanism of action. J Endod 36(3):400–413
84. Akbari M, Zebarjad SM, Nategh B, Rouhani A (2013) Effect of nano silica on setting time and physical properties of mineral trioxide aggregate. J Endod 39:1448–1451
85. Samiei M, Aghazadeh M, Lotfi M, Shakoei S, Aghazadeh Z, Vahid Pahdel SM (2013) Antimicrobial efficacy of mineral trioxide aggregate with and without silver nanoparticles. Iran Endod J 8:166–170
86. Bahador A, Pourakbari B, Bolhari B, Hashemi FB (2015) In vitro evaluation of the antimicrobial activity of nanosilver-mineral trioxide aggregate against frequent anaerobic oral pathogens by a membrane-enclosed immersion test. Biomed J 38(1):77–83
87. Zand V, Lotfi M, Aghbali A, Mesgariabbasi M, Janani M, Mokhtari H et al (2016) Tissue reaction and biocompatibility of implanted mineral trioxide aggregate with silver nanoparticles in a rat model. Iran Endod J 11(1):13–16
88. Saghiri MA, Asgar K, Lotfi M, Garcia-Godoy F (2012) Nanomodification of mineral trioxide aggregate for enhanced physiochemical properties. Int Endod J 45(11): 979–988
89. Saghiri MA, Godoy FG, Gutmann JL, Lotfi M, Asatourian A, Sheibani N, Elyasi M (2014) The effect of pH on solubility of nano-modified endodontic cements. J Conserv Dent 17:13–17

90. Saghiri MA, Orangi J, Tanideh N, Janghorban K, Sheibani N (2014) Effect of endodontic cement on bone mineral density using serial dual-energy X-ray absorptiometry. J Endod. 40(5):648–651
91. Saghiri MA, Orangi J, Tanideh N, Asatourian A, Janghorban K, Garcia-Godoy F et al (2015) Repair of bone defect by nano-modified white mineral trioxide aggregates in rabbit: a histopathological study. Med Oral Patol Oral Cir Bucal 20(5):e525-31
92. Saghiri MA, Asatourian A, Orangi J, Lotfi M, Soukup JW, Garcia-Godoy F et al (2015) Effect of particle size on calcium release and elevation of pH of endodontic cements. Dent Traumatol 31(3):196–201
93. Li Q, Coleman NJ (2014) Hydration kinetics, ion-release and antimicrobial properties of white Portland cement blended with zirconium oxide nanoparticles. Dent Mater J 33(6):805–810
94. Khataee R, Heydari V, Moradkhannejhad L, Safarpour M, Joo SW (2013) Self-cleaning and mechanical properties of modified white cement with nanostructured TiO_2. J Nanosci Nanotechnol 13(7):5109–5114
95. Bosso-Martelo R, Guerreiro-Tanomaru JM, Viapiana R, Berbert FL, Duarte MA (2016) Tanomaru-Filho M. Physicochemical properties of calcium silicate cements associated with microparticulate and nanoparticulate radiopacifiers. Clin Oral Invest 20(1):83–90
96. Guerreiro-Tanomaru JM, Vazquez-Garcia FA, Bosso-Martelo R, Bernardi MI, Faria G, Tanomaru MF (2016) Effect of addition of nano-hydroxyapatite on physico-chemical and antibiofilm properties of calcium silicate cements. J Appl Oral Sci 24(3):204–210

Chapter 4
Nanocomposites and Their Use in Dentistry

Ramesh Chaughule, Dipika Raorane, Suhas Pednekar and Rajesh Dashaputra

4.1 Introduction

Dental caries (tooth decay) has historically been considered as the common problem in the oral disease burden globally in the individuals of all ages. Dental caries is still a major public health problem as the disease affects 60–90% of school-aged children and the vast majority of adults [1]. It occurs mainly due to acid attack from food and bacteria buildup. Silver amalgam has been used more than 150 years as a restorative material to overcome this decayed tooth structure [2]. Amalgam is a kind of metallic filling material composed of mercury (nearly 50%) and a powdered alloy made mostly from silver, zinc, etc. Continuous use of silver amalgam for tooth fillings or capping of the tooth has the disadvantages like mercury toxicity, its potential health hazard, poor aesthetic nature, odd look and therefore inconvenient for the use in restorative dentistry. It is thus replaced by composite restorative materials. The term 'composite' actually only refers to the fact that the material is composed of several components, i.e., at least two different phases. According to this broad definition, glass ionomers, compomers, resin-based composites, ormocers, etc., are included in this group. They all have something in common that is they cure to form a polymer network with glass, quartz, ceramic filler particles, etc., embedded in it. Among them, glass inonomer cement is primarily used in prevention of dental caries and its treatment as a filling material [3]. It acts as sealants when pits and fissures in the tooth occur and release fluoride to prevent further enamel demineralization and promote remineralization. But due to less polishability, brittle nature, poor wear resistance,

R. Chaughule (✉) · D. Raorane · S. Pednekar
Department of Chemistry, Ramnarain Ruia College,
Mumbai, India
e-mail: rsctifr@yahoo.com

R. Dashaputra
Department of Dentistry, Bhabha Atomic Research Centre Hospital,
Mumbai, India

R. S. Chaughule (ed.), *Dental Applications of Nanotechnology*,
https://doi.org/10.1007/978-3-319-97634-1_4

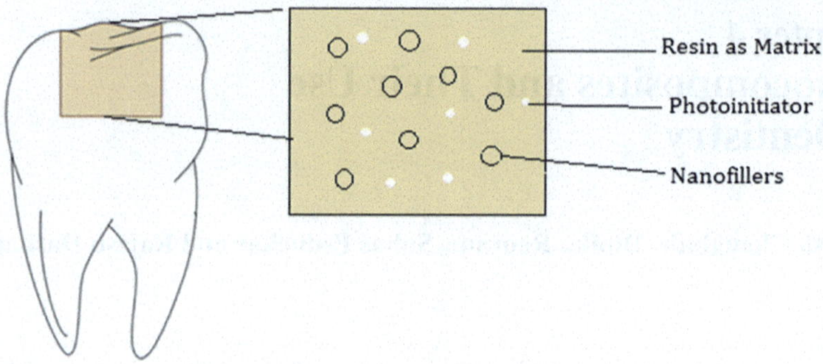

Fig. 4.1 Elucidation of structure of dental composite resin materials

tensile strength, aesthetic, etc., than resin composites make it inconvenient to use by dental practitioners. A general elucidation of dent in widely used resin composite is shown in Fig. 4.1.

Resin-based composites are the most popular restorative materials providing very good aesthetics and a long period of good clinical performance [4]. Dental composite materials are increasingly important in modern dentistry because of their advantages in dental applications [5]. They come up with number of advantages like good aesthetic appearance, acquires tooth-like restoration, no metal content and the shade of a natural tooth.

4.2 History of Composite Materials

Composites are physical mixtures of metals, ceramics, and/or polymers. Normally, dental composites are the classical mixture of ceramic particles and a polymer matrix. A growing number of dental practitioners are now using composite materials in their day-to-day practice because these materials have a capability to replace biological tissue in the function with their appearance. Classification of such dental composites is shown in Table 4.1.

4.3 Dental Composite Materials

The composites typically involve organic phase, i.e., matrix, inorganic phase, i.e., fillers and coupling phase (coupling agent) that binds the organic and inorganic entities together. Apart from matrix and inorganic phase, composite materials need photo initiator to initiate polymerization [4]. Thus, dental composite material

Table 4.1 Classification of composites [6]

Types of composites	Filler
Densified [7] – Midway filled Ultrafine Fine – Compact-filled [8] Ultrafine Fine	<60% by volume Particles <3 μm Particles >3 μm >60% by volume Particles <3 μm Particles >3 μm
Microfine [9] – Homogeneous and heterogeneous	Average particle size = 04 μm
Industrial use	Blends of densified and microfine composites [10]
Traditional	Equivalent to what are termed macrofill composites in other classifications
Fiber-reinforced [11]	Industrial-use composites

mainly consists of resin matrix, different inorganic fillers, coupling agents, and photo-initiating agent.

4.3.1 Dental Resin Matrix

Resin is a main matrix and is chosen in such a way that it has the ability to polymerize. There are different polymerizing reactions that can initiate polymerization in the presence of light, heat, cold conditions and/or in the presence of catalyst, etc. Polymerization depends upon the type of resin material. Polymerization is a chemical reaction that transforms small molecules into large polymer chains [12]. The polymerization of composite materials is never complete. The large percentage of reactive groups does not participate in polymerization. In addition, any surface layer exposed to air is incompletely polymerized. Polymerization shrinkage occurs when the dental composite is cured and changes the properties of the resin by chemical reaction, such as condensation, polymerization, or addition. Resin systems shrink during polymerization mainly because the monomer molecules are located at van der Waals distances (4 angstrom) from one another, while the corresponding polymers are within a covalent bond distance (1.5 angstrom) to each other. This accounts for the shrinkage during the polymerization process [13].

As the light-curing resin materials are very much convenient to use, they are in use on regular basis extensively. The resin monomer matrix should possess C=C group in their structures that can polymerize in the presence of photoinitiation to give a rigid polymer. Methacrylate polymeric networks with good resistance to aqueous milieu are sought for use in both prophylactic and restorative dentistry [14]. Dimethacrylate-based resins are widely used for manufacturing dental resin composite materials. They possess excellent mechanical properties, rapid polymerization, aesthetic quality, ease of handling, and the ability to bond to enamel surface [15]. Several commonly used dimethacrylate monomers include secondary functionality such as hydroxyl or urethane groups. Hydrogen with nitrogen and/or oxygen in resin structures remains engaged in intramolecular or intermolecular hydrogen bonding with different functional groups in resin materials. These types of interactions are responsible for viscosity differences on macroscopic scale of bulk monomer [16]. The mixture of different resins improves intermolecular hydrogen bonding and thus in turn improves compatibility in the final polymerized states.

The commonly used resin is Bis-GMA that possesses almost the desirable properties of dental materials like natural appearance, the absence of color, taste, or toxic agents. Many commercially available materials, at present, use Bis-GMA as major monomer matrix. Its distinct property is its high viscosity. Similarly, it shows high values of strength, surface hardness, low specific gravity, good thermal conductivity, good shelf life, good retention than other materials, and mainly useable for all types of prostheses. Its structure suggests that hydroxyl groups are the backbone and the π–π interactions of the aromatic rings increase the viscosity that helps in polymer conversion. This restricts the incorporation of fillers into it. It has also the ability to dilute easily in typical solvents or diluents.

To overcome the drawbacks of Bis-GMA (bisphenol a glycidyl methacrylate) resin, generally it is to be diluted to enhance the handling of composites. Due to the high viscosity of Bis-GMA (up to 1200 Pa.), it remains a challenge to add inorganic particles into the resin matrix. Without the inorganic fillers, the dental resin composites tend to generate high polymerization shrinkage during the curing stage and low mechanical properties, which can lead to secondary caries. It is therefore important to devise ways of introducing inorganic fillers into resin composites. Bis-GMA combines with low viscosity monomer like TEGDMA (triethylene glycol dye methacrylate). Diluting the composite with TEGDMA has been shown to have less desirable effects on the properties of the resin, because it increases water absorption and polymerization shrinkage [13, 17]. On the other hand, lower polymerization shrinkage stresses may result from the relative ease of flow of diluted resin composites during the early stages. TEGDMA is a long molecule which terminates with two functional methacrylate groups identical to those of Bis-GMA.

For having an attempt to replace TEGDMA or to combine with TEGDMA, short-chain polymers such as urethane dimethacrylate, i.e., UDMA (urethane dimethacrylate), came into picture. The study of UDMA as a base monomer and TEGDMA as a low viscosity monomer in different molar concentrations is also carried out with photopolymerization. Increasing the concentration of TEGDMA resulted in a decrease in the viscosity of the UDMA/TEGDMA mixture, a delay in the time to maximum polymerization rate, and an increase in the extent of polymerization (Ep) values of the resin mixtures. Furthermore, Ep values decreased with increasing filler content between 0 and 60 wt% [18]. Table 4.2 shows the chemical structures of Bis-GMA, TEGDMA, and UDMA.

Table 4.2 Dental resin materials with their structures

No	Resins	Structures
1	Structure of Bis-GMA bisphenol a glycidyl methacrylate ([2-hydroxy-3-[4-[2-[4-[2-hydroxy-3-(2-methyl prop-2-enoyloxy) propoxy] phenyl] propan-2-yl]phenoxy] propyl] 2-methylprop-2-enoate)	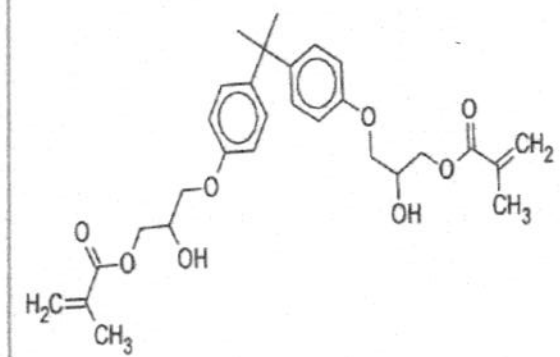 Bis-GMA
2	Structure of TEGDMA 2-[2-[2-(2-methylprop-2-enoyloxy)ethoxy]ethoxy]ethyl 2-methylprop-2-enoate commonly known as triethyleneglycol dimethacrylate	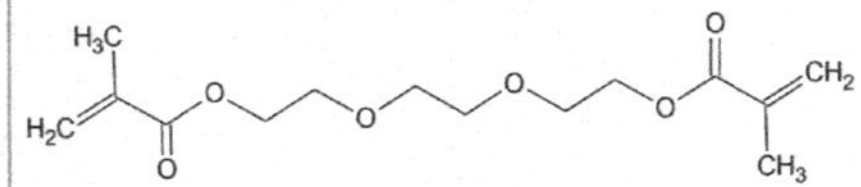 TEGDMA
3	Structure of UDMA urethane dimethacrylate, (2-[[3,5,5-trimethyl-6-[2-(2-methylprop-2 enoyloxy)ethoxycarbonylamino]hexyl]carbamoyloxy]ethyl2-methylprop-2-enoate)	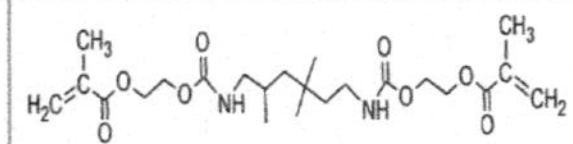 UDMA

4.3.2 Fillers

Generally, inorganic fillers, having constant volume, are added to the monomer system to attain the degree of strength which allows resin-based composites to be used in stress-bearing posterior areas. Thus, the residual polymerization shrinkage is reduced to a minimum. Final properties of the resin matrix depend upon the particle size, its distribution, etc. It is obvious that bigger and harder the size, greater is the strength. But at the same time, polishability decreases. The look is also the major concern of the patients. Only the correct combination of different filler particle fractions with respect to their size produces optimum mechanical and polishing properties.

To meet the physical properties of the amalgam with these resin composite materials, many attempts were made in the field of research worldwide. For improving the properties of these materials, there are plenty of opportunities to alter in these three fields, i.e., organic phase, inorganic phase, and coupling phase. Inorganic fillers are of two types, traditional and modern. Traditional includes macrofilled and microfilled materials, whereas modern ones are hybrid and nanofilled resin materials. Among all these types, nanofilled resin materials have gained much importance because of their improved applicability.

Nowadays, patients are equally cautious about the look of their teeth. Constant improvement in the composite materials is going on since then. Composite resins are classified according to their various characteristics (i.e., curing mechanism and particle type). The most commonly used classification considers mainly the distribution and average particle size of a given composite's filler phase [19]. These are thus divided as macrofilled, microfilled, hybrid, and finally nanofilled (Table 4.3) which correspond to the terminology of nanocomposites [20].

4.3.2.1 Macrofilled Composites

Macrofilled composites were the first resin composites marketed in the 1960s to fill front teeth [21]. Large crystalline particles were ground into a fine powder containing particles 5–50 μm in diameter. Particles larger than this are visible to the naked eye. Due to large particle size, macrofilled composites are not very polishable. They feel rough and can easily accumulate plaque and stain. Abrasion and wear are the major disadvantage of macrofilled composites. Loss of filler particles exposes more and more of the soft plastic matrix to abrasive force, and the restoration wears away leaving behind the potholes of that macrosized particles. Occlusal surfaces of back teeth receive many abrasive challenges. A filling that wears excessively causes the bite to change and may affect the other tooth by shifting a bite slightly [22]. Macrofilled resins are available in the market, but they are not of much interest to dentists because of their limited properties, poor clinical performance such as roughness, staining, wear. Figure 4.2 shows SEM image of macrofilled composite.

Table 4.3 Different types of resin composites with size

Types	Sizes	
(1) Macrofilled composites	5–50 μm	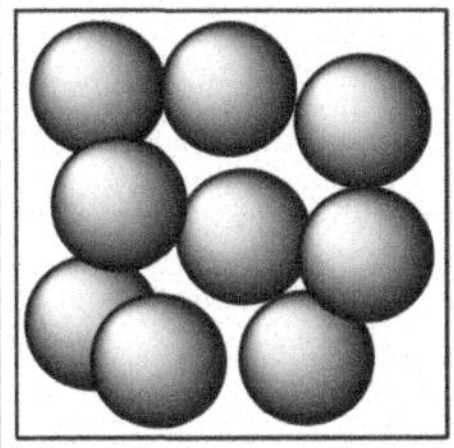
(2) Microfilled composites	0.1–10 μm	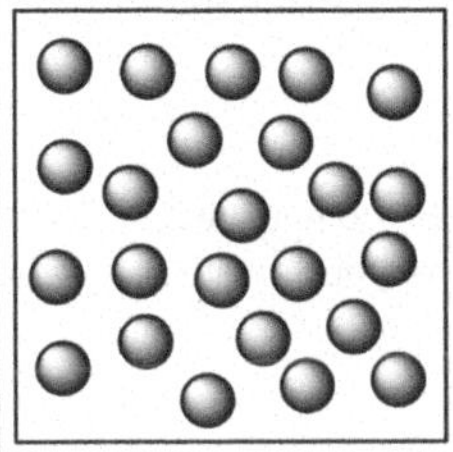
(3) Nanocomposites (i) Agglomerates	20–100 nm	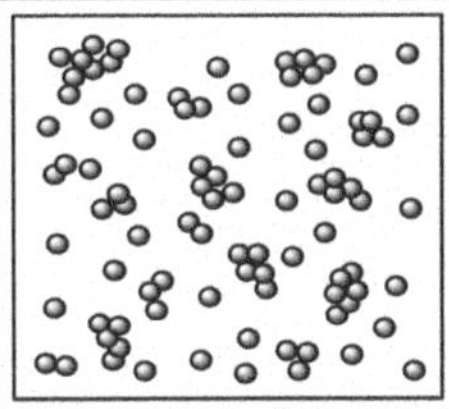

(continued)

Fig. 4.2 Ground quartz filler particles (with diameters of about 1–30 μm). Such relatively large fillers were used in early formulations of traditional composites. The smaller particles seen in the background contribute to a broad particle size distribution

Table 4.3 (continued)

Types	Sizes	
(ii) Isolated discrete particles	1–100 nm	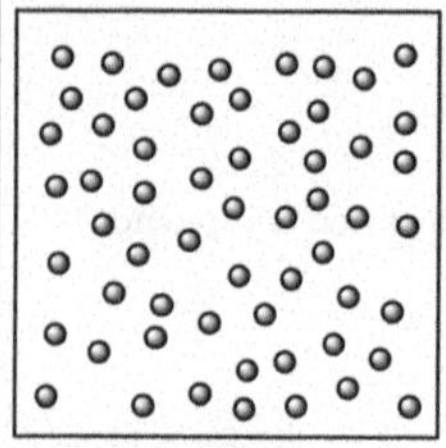
(4) Hybrid composites (i) Macrohybrid	1.0–5.0 μm and smaller filler particles of 0.05 μm	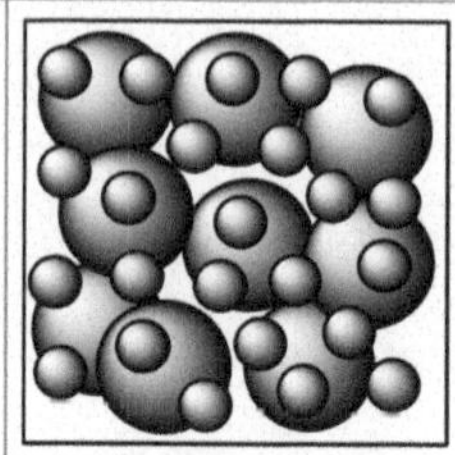
(ii) Microhybrid	Average 0.6 μm	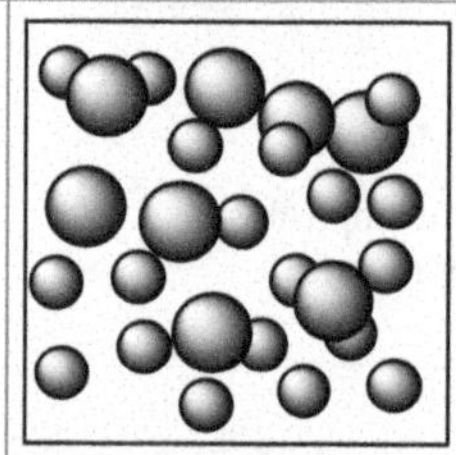
(iii) Nanohybrid	Less than 0.1 μm or 100 nm	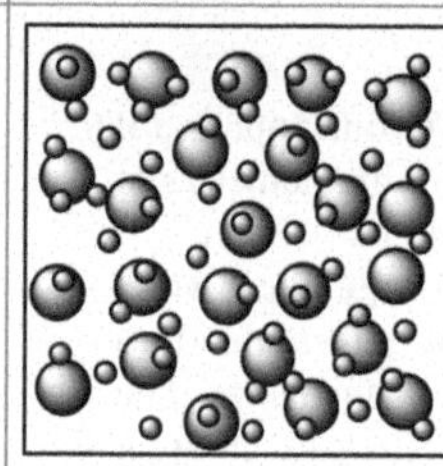

4.3.2.2 Microfilled Composites

Microfilled composites are in use in dentistry for nearly 30 years [23]. They maintained the aesthetics and polishability that macrofills lack. Thus, main characteristics of these composites are the high polish that can be achieved and maintained over time and excellent enamel-like translucency. Therefore, they are indicated for the restoration of anterior teeth and cervical abfraction lesions [15]; they should not be used in heavy stress-bearing areas because they frequently exhibit marginal chipping and bulk fracture [23]. As their name implies, the particle size in fillers over here ranges from 0.1 to 100 μm. Size varies due to the agglomeration of particles if present. These properties have limited the usefulness of microfills, particularly

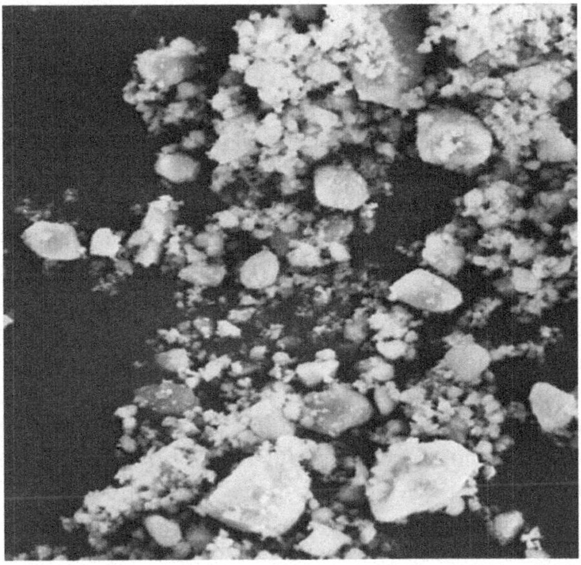

Fig. 4.3 Typical particles of a small particle-filled composite, with sizes in the range of 0.1–10 μm

in the posterior area while taking X-rays. Microfilled composites with smaller particle fillers allowed better polish retention while enhancing aesthetics, but offered less strength. Additionally, traditional microfills containing only silica filler are not radiopaque. The surface area of fillers is large; thus, the composites are least filled and highly viscous. Lower concentration of fillers results in decreasing the strength of final composites when compared with macrofills. They are not suitable for the places that are exposed to high-stress areas. Two examples of microfill in the market today are the various versions of Heliomolar from Ivoclar, Vivadent and Renamel from Cosmedent. Figure 4.3 shows SEM image of microfilled composite.

4.3.2.3 Nanocomposites

In the last few years, the nanotechnology has played an important role in improving the clinical performance of dental resin composites. Nanotechnology deals with chemical and physical methods to produce nanoscale operational materials ranging in size between 0.1 and 100 nm [24]. Striking enhancements in optical, chemical, physical, and mechanical properties can be attained using nanomaterials [25]. A large amount of examinations is being dedicated to the development of nanocomposites of distinct kinds for several programs, involving high-performance coatings, biomedical systems, structural materials, catalysts, photonics, and electronics. The nanocomposites are the newest addition to composite filling materials. They differentiate from hybrid composites by their particle size, i.e., nanosize, and it is the smallest size that can be used in dental resin materials as fillers. They are much more popular than the previous additions because they have superior wear characteristics, high polishability, and better handling characteristics. Nanoparticles also achieve enhanced transparency and aesthetics. They are now termed as universal composites [26]. Synthesis of filler nanoparticles is carried out using liquid precursors by bottom-up approach. Silicon nanoparticles show cage-like structure with 8 silicon and 12 oxygen atoms. Particles are so small (within the range of 1–100 nm) that they become a part of resin matrix. They can be synthesized in two ways where controlled synthesis gives discrete nanoparticles and the other shows clusters of nanoparticles along with discrete one. A nanocluster possesses reduced interstitial spacing of the filler particles giving higher filler loading. Increased filler loading in turn increases mechanical properties and ability to resist wear. Nanoclusters thus show better performance than discrete nanoparticles by having the capability of achieving smoother surface area and high translucency. Nanocomposites with nanoclusters allow smaller particles to come together and function as larger particles, resulting in increased strength, wear resistance, and polishability. The nanoclusters provide a distinct reinforcing mechanism compared with the microfill or nanohybrid systems resulting in significant improvements to the strength and reliability.

In dentistry, posterior restoration class I and II need composites that show high mechanical properties, whereas anterior restoration requires composites with superior aesthetics. The resin composite that offers all the requirements of both posterior and anterior restorations has not emerged yet. Nanotechnology has great impact on restorative dentistry by offering refinements to already resin-based composite system [25, 27] and the novel nanocomposite expected to be useful for all posterior and anterior restorative applications.

4.3.2.4 Hybrid Composites

This category of composite resins began to appear in the market about 10 years ago [28] to retain the advantages of macro- and microfilled resin, i.e., strength of macrofills and polishability of microfills, and to minimize or repair the problems faced by them. Hybrid composites blended in larger particle fillers to improve strength but were less polishable, resulting in more of a matte rather than gloss finish. They were chosen for their durability for posterior restorations in which strength was essential. It was developed to improve the hybrid composites and create a more universal material. The improvements in this category were focused on enhancing wear resistance and aesthetic properties, without sacrificing strength. Most composite resins in this category exhibit an average particle size of 0.5 μm. Figure 4.4 shows the SEM image of hybrid composites.

Hybrid composites contain various particle sizes, ranging from 0.02–0.04 to 1–3 μm. Hybrid composites do not retain a high polish for long, because large particles pop out from tooth surfaces. However, hybrid composites are easy to work with and are resistant to wear. They include smaller and submicron-sized particles which are more difficult to dislodge than large particles. Hybrid composites can be filled more densely with filler particles than composites containing only micro-sized particles. Large particles keep the paste consistency from becoming too stiff.

Submicron-sized particles take up space between large particles. The highest particle density attained with hybrids is 90% by weight. Because of high particle density, hybrids were the first composites promoted for posterior use, and they remain one of the most wear-resistant posterior composites in the market. On comparing it with other types, their physical strengths like tensile strength, compressive strength are almost 25% increased than macrofilled, and polymerization shrinkage decreases substantially [29].

Tetric EvoCeram and IPS Empress Direct from Ivoclar Vivadent along with Herculite Ultra from Kerr are the typical examples of the nanohybrids available in the market today. As an example, Herculite XRV from Kerr is one of the most popular microhybrids in the market.

4.3.3 Coupling Agents

Coupling agents are used to improve adherence of resin by modifying the surface of inorganic fillers. For good restoring properties, it is desirable to have strong bonding and good adherence between the interfaces like inorganic fillers and resin materials. Thus, coupling agents are a kind of hybrid inorganic–organic compounds that are specifically used to promote adhesion. This linker has been reported to slow the degradation process, to protect filler against fractures, and to improve distribution and stress transfer from flexible organic matrix to stiffer and stronger inorganic filler particles [28]. It also decreases water intake capacity of composites and minimizes wear. The purpose of coupling agents coupled with inorganic fillers is to help the inorganic particles to mix with organic matrix and form homogeneous mixture in same phases based on the principle 'Like dissolves Like' [28]. They have been used to coat fillers for over 50 years in industrial plastics and later in dental fillers. Today,

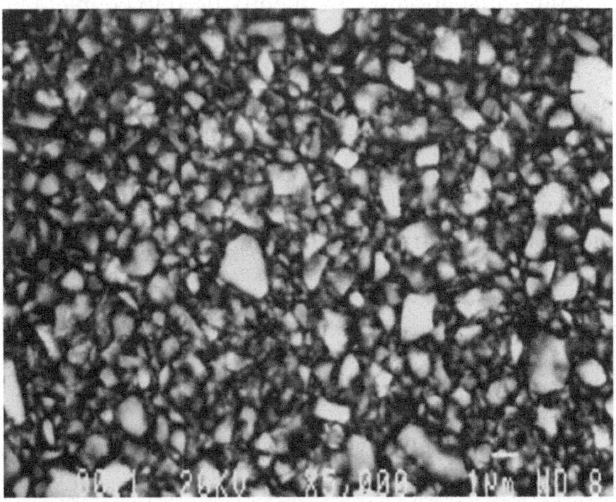

Fig. 4.4 Polished surface of a hybrid composite. Particle sizes range from about 0.1 to 3 μm (SEM magnification × 5000)

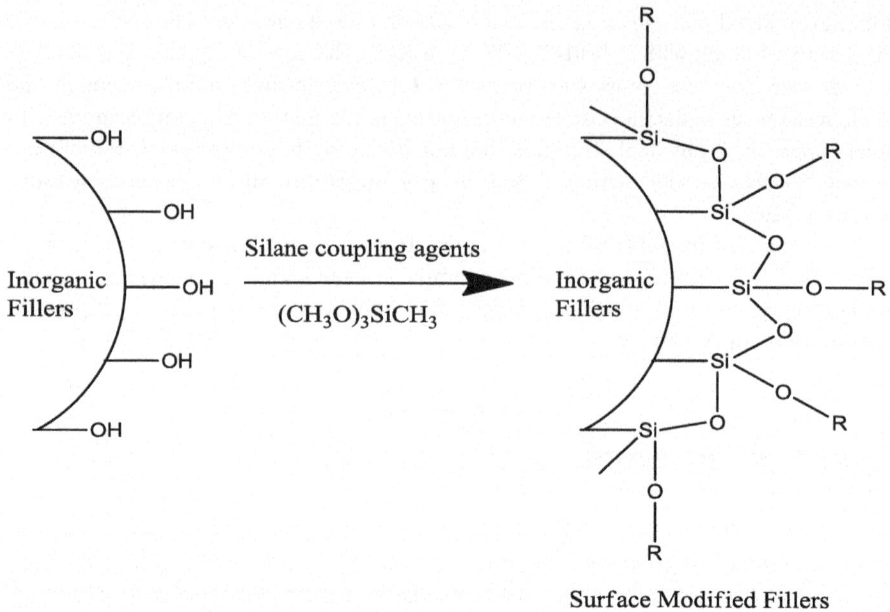

Fig. 4.5 Synthesis of surface-modified fillers with silane coupling agent

they are still state of the art. The history of coupling agents is not so old. In the mid-1940s, three patent applications were filed that laid the foundation for coupling agents. Steinman filed patent on the use of methallyl silicate, Te Grotenhuis on vinyl siloxanes, and Goebel and Iler on methacrylato chromic chloride. These patents were all concerned with improving the adhesion of matrix resins to glass fiber by the use of a coupling agent. After that, due to various applications, their use in chemistry and dentistry was increased. Here are some different types of coupling agents with their structures that are most frequently used to couple with inorganic fillers as shown in Table 4.4.

These are bifunctional molecules capable of reacting via their methoxysilane or ethoxysilane groups with the fillers and with the resin matrix. The covalent bond that forms with M–O–Si has a significant ionic character [30]. Better dispersion, smaller clusters improved microhardness and flexural strength. One side of the coupling agent tends to bond with hydroxyl groups of silica particles, and other is copolymerized with polymer matrix [31]. Chemical reaction of silane coupling agent with inorganic fillers is shown in Fig. 4.5.

Table 4.4 Different types of coupling agents with examples

Coupling agents	Examples
(1) Silane	(i) Trichloro vinyl silane 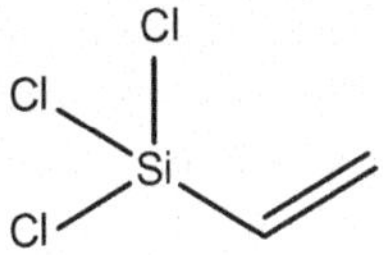
	(ii) Triethoxy vinyl silane 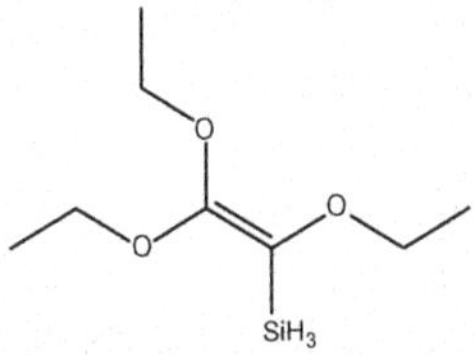
	(iii) 3-methacryloxypropyl trimethoxysilane (MPTS) 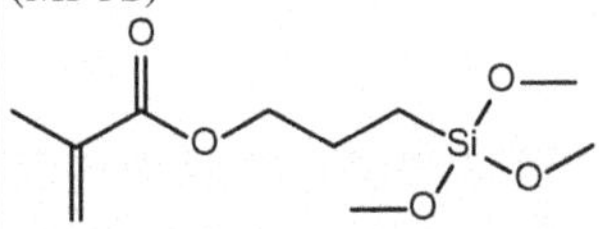
(2) Organic chromium	(i) Methacrylic acid chromium complex
(3) Titanate	(i) Triisostearyl titanate 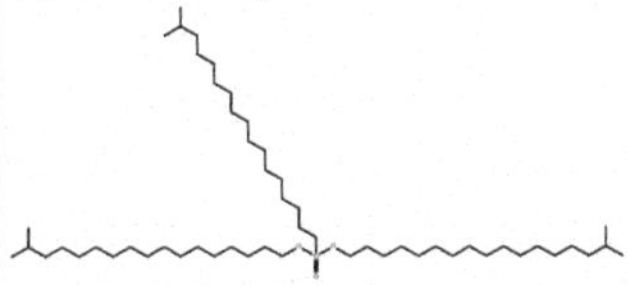
	(ii) Triisopropyl titanate 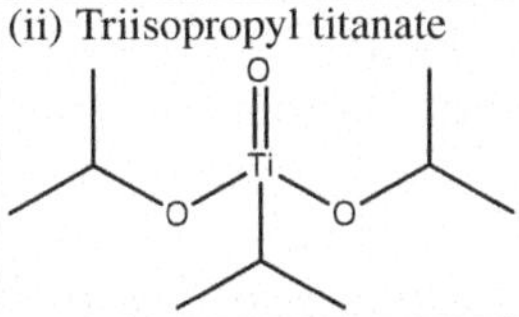
(4) Aluminate zirconium	(i) Tetrapropoxy zirconium 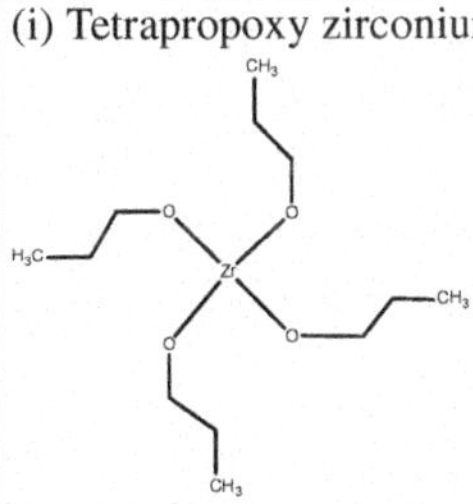

4.3.4 Photo-Initiating Agents

Photoinitiators are responsible for photoreaction on absorption of light, producing reactive species. These are capable of initiating or catalyzing chemical reactions that result in significant changes in the solubility and physical properties of suitable formulations like resins or polymer materials. Camphorquinone (CQ), a blue light photoinitiator, is commonly used in dental resin formulations. CQ is said to produce highly reactive polymerization that requires very small amount to be used in tooth-colored materials with enamel-like translucency [32]. Different resin composites require different light energy levels for proper curing. Manufacturers are now producing composite materials with more than one initiator, e.g., EDB, i.e., ethyl-4-dimethylamino benzoate, diphenyliodonium chloride, diphenyliodonium hexafluorophosphate, etc. The type, amount, etc. of photo initiators are important factors that affects the polymerization process during curing of composite materials. Typical structure of CQ is shown in Fig. 4.6.

4.4 Nanotechnology in Dentistry

4.4.1 Need of Nanotechnology

Nanotechnology is a science that works at the molecular level. It deals with structures ranging in the size of 100 nm or smaller in at least one dimension and developing materials or devices within that size [33]. Nanotechnology deals with the nanostructures, nanomaterials, nanoparticles, etc. Simply, this technology gives manipulation of matter on molecular and atomic levels to bring out enormous changes in applications of various fields. One such important field is dentistry [5]. As you go down with the size of particles, the physical and chemical properties change that can be useful to apply in different fields. Dentistry is one such field where nanotechnology plays various roles such as tissue engineering [6], nanodental implant [14], dental nanorobots [15], nanocomposites [16], dental nanostructures [13], nanomaterials for periodontal drug delivery [13], dental imaging [28], nanomedicines such as toothpastes etc. [30]. In recent years, the increasing interest in aesthetic restoration has led to further development of materials that have the same color as that of teeth. Nanotechnology allows the production of nanosized filler particles that are compatible with dental

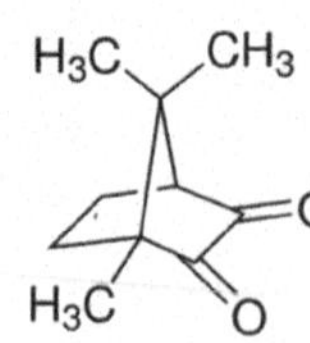

Fig. 4.6 Chemical structure of CQ (4, 7, 7-trimethylbicyclo [2.2.1] heptane-2, 3-dione)

composites; therefore, a greater amount of filler can be added into the composite resin matrix [16]. Composite resins containing such particles are easy to shape and have a high degree of strength and resistance to abrasion. Restorative dentistry comes up with the modern advancement where aesthetic filing comes into picture. These filling materials are nothing but the resin matrix with nanofillers mixed evenly into it. Nanofiller plays an important role in enhancing the properties of dental restorative filling materials in many ways, like its physical, mechanical, biological properties, its appearance. Nanofillers due to its smaller size, larger surface area, and high surface energy make it convenient to capture reactive monomer or polymer segment on its surface [34]. They also favored stressing transfer from soft resin matrix to hard organic nanofiller. Thus, resin-based composite materials have acquired nanotechnology for future progress and upcoming years will elaborate the topic further for more advancement.

4.4.2 Nano in Composite Resins

No synthetic nanocomposite material is better enough to be ideal for dental applications. Thus, in an attempt to improve the applicability of these nanocomposite materials, they have undergone progressive evolution from bulk-filled dental composites [35] to nanofilled dental composites [36]. The development in the use of nanocomposites patented in response to the persistent and discouraging issues of polymerization shrinkage, strength, microhardness, and wear resistance essential in posterior occlusal applications. Around in the same era where Bowen developed the resins specially for restorative dentistry [37], a new word 'nano' was introduced by the scientist Nobel Laureate Sir Richard Feynman in 1959. This discovery was a landmark for advances in dental composites and nanoparticles as fillers. Choice of nanoparticles is based on their physical properties, tackle issues like polymerization shrinkage, wear resistance, microhardness, and achieve patient satisfaction in terms of the aesthetic appearance [38]. Despite many improvements in this field, dental composites do not have enough toughness, strength, and durability in order to be used in stress-bearing areas. A variety of nanofillers was chosen to figure out their effects on final products such as nanoparticles of silica, glass, carbon nanotubes, alumina, zirconia, titania.

4.4.2.1 Nano-silica as Fillers

Crystalline quartz or silica is the first kind of fillers that were used in resin nanocomposite materials. Natural silica is nonreinforcing and has been used as a filler, only to reduce the cost. Silica and glass fillers are used in a wide range of dental applications because of its biocompatibility, wear resistance, and aesthetics. They are, however, subject to brittle failure [39]. The positive effect of nanosized silica and silicate-based fillers on flexural strength, surface hardness, fracture toughness, and optical properties has already been reported in the literature [40].

4.4.2.2 Carbon Nanotubes as Fillers

Carbon nanotubes are used in the preparation of composite materials that are biocompatible in nature and give high mechanical strength and resilience [41]. But these materials showed lack of hydrophobicity and chemical inertness that limits their applications [42].

4.4.2.3 Nano-zinc as Fillers

Zinc oxide is naturally white in color, and its incorporation into this new dental amalgam might fade the metallic color of the conventional amalgam out. Specialty of zinc oxides is their antibacterial properties [43]. ZnO probably has the richest family of nanostructures of all materials, both in structures and in properties. Its percentage loading is high which increases the mechanical properties of the resin [44]. This affects the polishability of the final composite and thus finally its look.

4.4.2.4 Nano-alumina as Fillers

Nano- and microsized alumina nanoparticles were used for improving the mechanical properties of the materials. Arora et al. [45] recently reviewed the effect of alumina addition and reported a positive impact on the properties of acrylic resin. It was chosen due to their unique properties like high elastic modulus, thermal and chemical stability, high strength, toughness. Comparing the two, nanosized filler was proved to be more effective [40]. Chen and coworkers found, that larger Al_2O_3 concentrations (9 wt%) caused a pronounced weight loss even at low temperatures [42].

4.4.2.5 Zirconia as Fillers

A lot of work has been already done with zirconia as fillers in composite materials. In addition to that, zirconia–silica nanofibres were used [39] in the past few years. Zirconia is biocompatible, chemically inert, corrosion resistant, opaque in nature, low thermal conductivity, etc. They need to be present in higher amount as fillers ~60–70% in content which is responsible for the toughness, whereas research is now in focus on minimum amount of filler content, i.e., less than 10%.

4.4.2.6 Nanohydroxyapatite as Fillers

Hydroxyapatite (HA) is a natural mineral containing calcium and phosphate. It is very much suitable for dental applications as it is biologically compatible and substituent for dentin [46]. It gives excellent bioactivity, enhances physical strength, and is the most stable form of calcium phosphate. More work has been done on nanohydroxyapatite and glass ionomer cement for restoration [47]. It resulted in an increased resistance to demineralization and acceptable bonding strength. Nanophase crystals of HA can bind to the bone and stimulate bone healing by the stimulation of osteoblastic activity [48]. However, the setting time of nano-HA cement exceeded the clinically suitable maximum setting time.

4.4.2.7 Nano-TiO_2 as Fillers

Titanium dioxide (TiO_2) nanoparticles, for example, are very fine, agglomerate easily in practical applications, and are especially difficult to disperse in organic solvents [48]. But it shows excellent properties if synthesized in a controlled manner. Adding surface-modified TiO_2 nanoparticles improves the microhardness and flexural strength of dental resin-based composites [49]. Though the researchers have not yet achieved the level of increasing mechanical strength with reduction in polymerization shrinkage, it is mainly due to the size and type of nanoparticles they have used.

4.4.3 Current Progress

Increasing demand for aesthetic dentistry has led to the development of resin composite materials for direct restorations with improved physical and mechanical properties, aesthetics, and durability. It seems that no single material is better enough to satisfy all its desired properties of dental composites. It is required to have a kind of material which gives high mechanical strength, good aesthetic properties, etc. To come up with this conclusion, fillers should have good chemical and physical properties, optimum size distribution of smaller and larger particles, etc. Nanohybrid is the mixture of smaller and larger particles where smaller particles reside in between the gaps of larger particles and larger particles take care of the consistency of the paste and provide wear resistance. Smaller particle size also helps in reducing the polymerization shrinkage. Other factors are color (shade), translucency, fluorescence, and opalescence (optical properties) that give natural tooth its vital-looking appearance [50]. The rest of the impressive features of TiO_2 nanoparticles are that they are nontoxic and have high refractive index and antibacterial properties, white in color, chemically inactive, corrosion resistance, high microhardness, etc. Furthermore, the literature has also shown that nanoscale TiO_2 reinforcement agents bring new optical, electrical, physiochemical properties attained at very low TiO_2 content,

Table 4.5 Comparison of physical and mechanical properties of nanocomposite resins of various filler sizes [53]

Characteristics	Macrofilled	Microfilled	Microhybrid	Nanohybrid
Size (μm)	8–12	0.04–0.4	0.4–1.0	0.5–3
Inorganic filler (wt%)	60–70	35–67	75–80	80–90
Compressive strength (MPa)	250–300	250–350	300–350	350–400
Tensile strength (MPa)	50–65	30–50	40–50	75–90
Elastic modulus (GPa)	8–15	3–6	11–15	15–20
Curing shrinkage (vol.%)	–	2–3	2–3	2–3
Water sorption (mg/cm^2)	0.5–0.7	1.4–1.7	0.5–0.7	0.5–0.6

which makes polymer–TiO_2 nanocomposites a promising new class of materials. It can be anticipated that it will be commercially beneficial for widespread fields [51]. The major concern is synthesis of nanohybrid TiO_2 particles and its high cost. Green synthesis together with microwave-assisted nanoparticles synthesis can be a protocol to overcome this problem. Recently, green synthesis using fruit peel extract, an agricultural waste, and rapid microwave synthesis with respect to size and shape of nanostructures is done by Raorane et al. [52]. This may help in reducing the cost for TiO_2 nanoparticles by keeping the quality unaltered by using microwave technique. Fruit peel contains phytochemicals that act synergistically as capping agent in biosynthesis of nanoparticles in hybrid form. Agricultural waste reduces the cost of process, hazardous chemical waste, etc., and microwave synthesis reduces the time from several hours to few minutes. Dentistry and green chemistry both are the most important and promising fields. This protocol will be the best example of green chemistry and dentistry that will move forward with hand in hand. Major importance goes to the final physical and mechanical properties of the material. Comparison of many such physical and mechanical properties with respect to size is given in Table 4.5. It is seen that the properties of nanohybrid materials are superior to macrofilled, microfilled, macrohybrid, microhybrid, etc. Thus, TiO_2 nanohybrid synthesized by green and microwave method would be an appropriate solution.

4.5 Challenges and Future Outlook

The performances of a dental composite depend on filler type, resin composition, filler matrix bonding, and cure conditions. Primary challenge in this field is to achieve manufacturing ability to produce discrete nonagglomerated nanoparticles that are evenly distributed in resin matrix. Conventional particles that are in macrofilled or microfilled resin-based composites are prepared by top-down procedures where larger particles are ball milled or ground into smaller ones. The nanohybrids are produced by manipulating the structure of materials to provide dramatic improvements in the electrical, chemical, mechanical, and optical properties. Nanofillers have higher contact surface with the organic phase as compared to macro-/mini-filled composites to improve the material strength. More invitro studies should be carried out followed by invivo trails to examine the feasibility of the nano filler composite materials to use on regular basis.

References

1. WHO I What is the burden of oral disease? In: WHO. http://www.who.int/oral_health/disease_burden/global/en/
2. Health C for D and R Dental Amalgam—About Dental Amalgam Fillings. https://www.fda.gov/MedicalDevices/ProductsandMedicalProcedures/DentalProducts/DentalAmalgam/ucm171094.htm
3. Sidhu S (2011) Glass-ionomer cement restorative materials: a sticky subject? Glass-ionomer cement restorative materials. Aust Dent J 56:23–30
4. Shenoy A (2008) Is it the end of the road for dental amalgam? A critical review. J Conserv Dent JCD 11:99–107
5. Soares AC, Cavalheiro A (2010) A review of amalgam and composite longevity of posterior restorations. Rev Port Estomatol Med Dentária E Cir Maxilofac 51:155–164
6. Willems G, Lambrechts P, Braem M, Celis JP, Vanherle G (1992) A classification of dental composites according to their morphological and mechanical characteristics. Dent Mater Off Publ Acad Dent Mater 8:310–319
7. Densified Radiata Pine For Structural Composites. http://www.scielo.cl/scielo.php?script=sci_arttext&pid=S0718-221X2006000200002. Accessed 6 Dec 2017
8. Lin J-C, Chang LC, Nien MH, Ho HL (2006) Mechanical behavior of various nanoparticle filled composites at low-velocity impact. Compos Struct 74:30–36
9. Heterogeneous Mixture and Homogeneous Mixture I Difference I Byjus. https://byjus.com/chemistry/heterogeneous-mixture-homogeneous-mixture/. Accessed 6 Dec 2017
10. Grumezescu A (2016) Nanobiomaterials in dentistry: applications of nanobiomaterials. William Andrew
11. Prashanth S, Subbaya KM, Nithin K, Sachhidan S (2017) Fiber reinforced composites—a review. J Mater Sci Eng 6:1–6
12. Anusavice KJ, Shen C, Rawls HR (2013) Phillips' science of dental materials. Elsevier Health Sciences
13. Peutzfeldt A (1997) Resin composites in dentistry: the monomer systems. Eur J Oral Sci 105:97–116
14. Morgan DR, Kalachandra S, Shobha HK, Gunduz N, Stejskal EO (2000) Analysis of a dimethacrylate copolymer (bis-GMA and TEGDMA) network by DSC and 13C solution and solid-state NMR spectroscopy. Biomaterials 21:1897–1903
15. Sarode GS, Sarode SC (2013) Abfraction: A review. J Oral Maxillofac Pathol JOMFP 17:222–227
16. Jung M, Sehr K, Klimek J (2007) Surface texture of four nanofilled and one hybrid composite after finishing. Oper Dent 32:45–52

17. Kalachandra S, Taylor DF, DePorter CD, Grubbs HJ, McGrath JE (1993) Polymeric materials for composite matrices in biological environments. Polymer 34:778–782
18. Tanimoto Y, Hayakawa T, Nemoto K (2005) Analysis of photopolymerization behavior of UDMA/TEGDMA resin mixture and its composite by differential scanning calorimetry. J Biomed Mater Res B Appl Biomater 72:310–315
19. Ritter AV (2005) Direct resin-based composites: current recommendations for optimal clinical results. Compend Contin Educ Dent Jamesburg NJ 1995 26:481–482, 484–490; quiz 492, 527
20. MS AC By Gary M Radz, DDS, Karl F Leinfelder, DDS Clinical Materials Review: The Current State of Composite Resins | ID | dentalaegis.com. https://www.dentalaegis.com/id/2008/12/the-current-state-of-composite-resins
21. Marshall GW (1993) Dentin: microstructure and characterization. Quintessence Int Berl Ger 1985 24:606–617
22. Bruscino T (2012) Basic ethics in dentistry. The Academy of Dental Learning OSHA Train. CE Support Dent. Org October 2010 Available Online Www Dent. OrgcourseEthicsEthics Pdf
23. Sirin Karaarslan E, Bulucu B, Ertas E (2014) Clinical evaluation of direct composite restorations and inlays: results at 12 months. J Restor Dent 2:70
24. Editor DJGS (1999) A review of: "Kirk-Othmer encyclopedia of chemical technology". Pet Sci Technol 17:673
25. Terry DA (2004) Direct applications of a nanocomposite resin system: Part 1-the evolution of contemporary composite materials. Pract Proced Aesthetic Dent 16:417–432
26. Dental Composite Biomaterials. https://www.cda-adc.ca/jcda/vol-64/issue-10/732.html. Accessed 24 Aug 2017
27. Beun S, Glorieux T, Devaux J, Vreven J, Leloup G (2007) Characterization of nanofilled compared to universal and microfilled composites. Dent Mater Off Publ Acad Dent Mater 23:51–59
28. Selective solubility: "Like dissolves like"—Journal of Chemical Education (ACS Publications). http://pubs.acs.org/doi/abs/10.1021/ed054p228.2. Accessed 21 Sep 2017
29. Goldman M (1983) Polymerization shrinkage of resin-based restorative materials. Aust Dent J 28:156–161
30. Antonucci JM, Dickens SH, Fowler BO, Xu HHK, McDonough WG (2005) Chemistry of silanes: interfaces in dental polymers and composites. J Res Natl Inst Stand Technol 110:541–558
31. Saunders S (2009) Current practicality of nanotechnology in dentistry. Part 1: focus on nanocomposite restoratives and biomimetics. Clin Cosmet Investig Dent 1:47–61
32. Santini A (2013) Photoinitiators in dentistry: a review. Primary Dent J **2**(4)
33. What is Nanotechnology? | National Nanotechnology Infrastructure Network. http://www.nnin.org/news-events/spotlights/what-nanotechnology. Accessed 31 Aug 2017
34. Wu M, Zhang F, Yu J, Zhou H, Zhang D, Hu C, Huang J (2014) Fabrication and evaluation of light-curing nanocomposite resins filled with surface-modified TiO_2 nanoparticles for dental application. Iran Polym J 23:513–524
35. Dionysopoulos D (2016) Bulk fill composite resins. A novelty in resin-based restorative materials. ARC J Dent Sci 2:1–3
36. Monfared M, Mirdamadi S, Khavandi A (2013) Synthesis of new dental nanocomposite with glass nanoparticles. Nanomedicine J 1:107–111
37. Söderholm K-J, Mariotti A (1999) BIS-GMA–based resins in dentistry: are they safe? J Am Dent Assoc 130:201–209
38. Khurshid Z, Zafar M, Qasim S, Shahab S, Naseem M, AbuReqaiba A (2015) Advances in nanotechnology for restorative dentistry. Mater Basel Switz 8:717–731
39. Guo G, Fan Y, Zhang J-F, Hagan JL, Xu X (2012) Novel dental composites reinforced with zirconia-silica ceramic nanofibers. Dent Mater Off Publ Acad Dent Mater 28:360–368
40. Foroutan F, Javadpou J, Atai M, Rezaie HR et al. (2011) Mechanical properties of dental composite materials reinforced with micro and nano-size Al_2O_3 filler particles. Iran J Mater Sci Eng 8:25–33

41. Kealley C, Ben-Nissan B, van Riessen A, Elcombe M (2006) Development of Carbon Nanotube Reinforced Hydroxyapatite Bioceramics. Key Eng Mater 309–311:597–600
42. Permal A, Devarajan M, Hung HL, Zahner T, Lacey D, Ibrahim K (2016) Thermal and mechanical properties of epoxy composite filled with binary particle system of polygonal aluminum oxide and boron nitride platelets. J Mater Sci 51:7415–7426
43. Sevinç BA, Hanley L (2010) Antibacterial activity of dental composites containing zinc oxide nanoparticles. J Biomed Mater Res B Appl Biomater 94:22–31
44. Yahya N, Puspitasari P, Latiff NRA (2013) Hardness improvement of dental amalgam using zinc oxide and aluminum oxide nanoparticles. In: Öchsner A, da Silva LFM, Altenbach H (eds) Characterization and Development of Biosystems and Biomaterials. Springer, Berlin, pp 9–32
45. Arora P, Singh SP, Arora V (2015) Effect of alumina addition on properties of polymethylmethacrylate–a comprehensive review. Int J Biotech Trends Technol 9:1–7
46. Besinis A, De Peralta T, Tredwin CJ, Handy RD (2015) Review of nanomaterials in dentistry: interactions with the oral microenvironment, clinical applications, hazards, and benefits. ACS Nano 9:2255–2289
47. Chen L, Xu C, Wang Y, Shi J, Yu Q, Li H (2012) BisGMA/TEGDMA dental nanocomposites containing glyoxylic acid modified high-aspect ratio hydroxyapatite nanofibers with enhanced dispersion. Biomed Mater Bristol Engl 7:045014
48. Ogle O, Byles N (2014) Nanotechnology in dentistry today. West Indian Med J 63:344–348
49. Xia Y, Zhang F, Xie H, Gu N (2008) Nanoparticle-reinforced resin-based dental composites. J Dent 36:450–455
50. Yu B, Ahn J-S, Lim JI, Lee Y-K (2009) Influence of TiO_2 nanoparticles on the optical properties of resin composites. Dent Mater 25:1142–1147
51. Ahmed MA, El-Shennawy M, Althomali YM, Omar AA (2016) Effect of titanium dioxide nano particles incorporation on mechanical and physical properties on two different types of acrylic resin denture base. World J Nano Sci Eng 6:111–119
52. Raorane D, Chavan P, Pednekar S, Chaughule R (2017) Green and rapid synthesis of copper-doped TiO_2 nanoparticles with increased photocatalytic activity. Adv Chem Sci 6:13
53. Themes UFO (2015) Resin-based composites. Chapter 13. Pocket Dent

Chapter 5
Applications of Nanoparticles in Orthodontics

Panchali Batra

5.1 Background

5.1.1 Introduction

Orthodontic treatment of malaligned teeth and jaws is being carried out since ages. Advances in material sciences have lead to development of materials that enable treatment to be carried out in all walks of life and in mutilated dentitions. Despite their broad success, the orthodontic treatment suffers from a range of limiting factors like friction, long duration of treatment, and complications like inflammation of the biological tissues, decalcification of teeth due to bacterial growth, and root resorptions. To reduce these limitations and complications researchers and manufacturers of devices are focusing on modification of surface characteristics of materials so as to conquer these problems. The research is being focused on development of coatings that are stable in the oral environment and noncytotoxic. The coatings developed are either lubricants to reduce friction or antibacterial to reduce the microbial count and problems associated with it like caries, periodontal diseases, and decalcification around the brackets. Developments are also taking place to improve the bond strength of the brackets to the enamel. Nanoparticles are also being used in photodynamic therapy to control oral infections. The purpose of this article is to examine the various applications of nanoparticles in the field of orthodontics. This chapter will guide a young researcher to invent novel coatings and materials and to experiment in the area of material sciences.

P. Batra (✉)
Faculty of Dentistry, Jamia Millia Islamia, New Delhi, India
e-mail: panchali.batra@gmail.com

R. S. Chaughule (ed.), *Dental Applications of Nanotechnology*,
https://doi.org/10.1007/978-3-319-97634-1_5

5.1.2 Definitions

British Standards Institution defines nanoparticles as those particles in which all the fields or diameters are in the nanoscale range. Whereas, nanomaterials are those material for which at least one side or internal structure is in the nanoscale [1]. An engineered nanoparticle may be defined as any intentionally produced particle that has a characteristic dimension from 1 to 100 nm and has properties that are not shared by non-nanoscale particles with the same chemical composition [2]. Nanotechnology [3] is enabling technology that deals with nanometer-sized objects. Bionanotechnology and nanobiotechnology are terms that refer to the intersection of nanotechnology and biology [4]. Given that the subject is one that has only emerged very recently, bionanotechnology and nanobiotechnology serve as blanket terms for various related technologies. These two terms are often used interchangeably. When a distinction is intended, though, it is based on whether the focus is on applying biological ideas or on studying biology with nanotechnology. Bionanotechnology generally refers to the study of how the goals of nanotechnology can be guided by studying how biological "machines" work and adapting these biological motifs into improving existing nanotechnologies or creating new ones [5, 6]. Nanobiotechnology, on the other hand, refers to the ways that nanotechnology is used to create devices to study biological systems [7].

5.1.3 Historical Background

In 1959, the late Nobel Prize-winning physicist Richard P Feynman presented a talk entitled "There's plenty of room at the bottom" at the annual meeting of the American Physical Society at the California Institute of Technology, Pasadena, CA [3, 8]. This was the first time when the vision of nanotechnology was introduced. Prof. Kerie E. Drexeler introduced the term "nanotechnology," which was defined by Norio Taniguchi [9] as follows: "Nanotechnology mainly consists of the processing, separation, integration, and deformation of materials by one atom or one molecule."

The optical characteristics of nanoparticles have been used in sculptures paintings even before the fourth century AD. The most famous example is the Lycurgus cup, known as dichroic glass, that changes color when held up to the light. The opaque green cup turns to a glowing translucent red when light is shone through it internally (i.e., light is incident on the cup at 90° to the viewing direction). Analysis of the glass revealed that it contains a very small quantity of tiny (70 nm) metal crystals of Ag and Au in an approximate molar ratio of 14: 1, which give it these unusual optical properties [1, 10].

The soluble gold was mostly used for its fabulous curative powers of various diseases. The first book on colloidal gold was published in 1618 by the philosopher and medical doctor Francisci Antonii. A complete treatise on colloidal gold was published in 1718 by Helcher [11]. Nanotechnology is easily evident in various old churches where ruby red color had been used in making of stained glass window. Science had not advanced much at that time to answer this phenomenon. It was later when developments in the field of chemistry took place that an answer to this phenomenon was revealed. These vivid colors were controlled by the size and the form (or shape) of the nanoparticles of gold and silver.

Industrial production of nanomaterials saw its origins in the twentieth century. For example, nanoparticles of carbon black (tire soot) have been used in the fabrication of rubber tires of automobiles from the beginning of the twentieth century [8]. Later developments in the chemical and physical properties led to new possibilities in various fields.

Nanotechnology has come a long way to find its application in supramolecular chemistry self-assembling drug carriers and gene delivery systems [12], nanoparticles and nanocapsules, antibody technologies, polymer-drug conjugates, polymer-protein and antibody conjugates [13], nanoprecipitation, nanocrystals, emulsification technologies, liposome technology [14], in situ polymerization, tissue engineering and repair [15], dendrimer technologies [16], molecular imprinting including recent innovations in dental diagnostics, material and therapeutics [17].

The confederation of nanotechnology with the field of dentistry has given rise to new stream "nanodentistry." The important historical events in the development of Nanotechnology are summarized in Table 5.1.

5.2 Shapes and Types of Nanoparticles

Different shapes of nanoparticles such as rod, rectangle, hexagon, cube, triangle, and star-shaped nanoparticles can be produced by variation of experimental parameters such as concentration of the metal precursor, reducing agents, and stabilizers and reaction conditions such as temperature, time [19]. Bulk solution synthetic methods often produce nanoparticles of multiple sizes and shapes, and low yield of the desired size and shape. Colloidal solution can generally produce particle of desired shape and size [20]. Controlling size, shape, and structural architecture of the nanocrystals requires manipulation of the kinetic and thermodynamic parameters [21]. Following are certain types of nanoparticles;

5.2.1 Carbon-Based Materials

Carbon nanomaterials are one of the broadly discussed, researched, and applied of synthetic nanomaterials. Carbon nanotubes (CNTs), nanofibers (CNFs), and graphene are promising components for next-generation high-performance structural and multifunctional composite materials. The carbon nanomaterials have exceptional tensile strength, elastic modulus, electrical and thermal conductivity, and unique electronic, magnetic and optical properties. Apart from their above properties, their robust

Table 5.1 Important historical breakthroughs in nanotechnology: [18]

Year	Event/remarks	People/country
1931	Transmission electron microscope (TEM)	M. Knoll and E. Ruska (Technical University of Berlin, Germany)
1937	Scanning electron microscope (SEM)	M. vonArdenne (Forschungslaboratorium für Elektronenphysik, Germany)
1954	Targeted drug delivery concept by miniature particles –magic bullets	Paul Ehrlich
1959	Feynman's Lecture on "There' s Plenty of Room at the Bottom"	R.P. Feynman (California Institute of Technology, Pasadena, CA, USA)
1960's	Investigation of polyacrylic beads for oral administration	Peter Paul Speiser, Zurich
Late 1960's	First nanoparticles for drug delivery purposes and for vaccines	Paul Speiser's research group
1969	"Micelle polymerization"—First work employed toward processing sustained release nanoparticles for vaccination purpose	Gerd Birrenbach, Germany
1974	Concept of nanotechnology proposed	N. Taniguchi (Tokyo University of Science, Japan)
1977	Lysosomotropic effects of the nanoparticles discovered	Patrick Couvreur, Zurich
1978	First review article about nanoparticles	Kreuter, Germany
1979	First rapidly biodegradable acrylic nanoparticles	Patrick Couvreur, Zurich
1981	First poly(lactic acid) nanoparticles for drug delivery developed	Robert Gurny, Switzerland
1986	Three-dimensional space manipulation of atoms demonstrated	S. Chu (Bell Lab., USA)
1984,1986	Nanoparticles can improve the oral bioavailability of drugs	Maincent et al., France
1986	Nanocapsules	Al Khouri Fallouh et al.
1993	Concept of nanomedicine	Robert A. Freitas Jr.
1995	Nanoimprinting	S.Y. Chou, USA

(continued)

Table 5.1 (continued)

Year	Event/remarks	People/country
1996	Nanosheets	T. Sasaki (National Institute for Research in Inorganic Materials, Japan)
2000	Term "nanodentistry" introduced	Robert Freitas, USA
2000	National Nanotechnology Initiative (NNI)	USA
2003	Twenty-first Century Nanotechnology Research and Development Act	USA
2005	Nanosciences and Nanotechnologies: An action plan	Europe

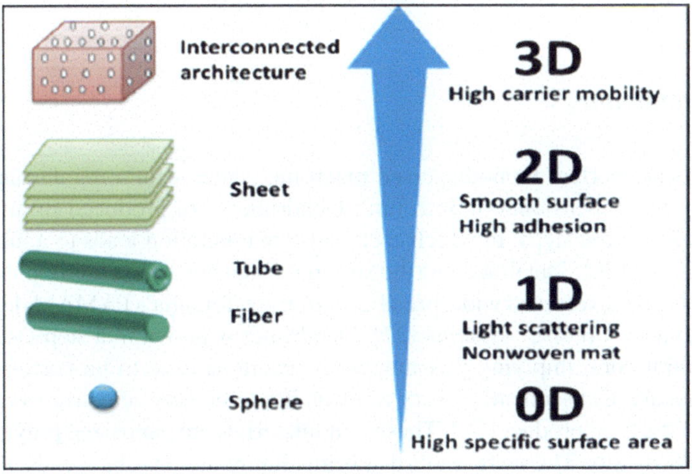

Fig. 5.1 Various kinds of nanomaterials and their specific properties

chemistry and ease of manipulation make them attractive candidates for diagnostic applications [22]. Carbon nanomaterials exist in allotropic forms such as diamond, graphene, amorphous carbon, and single-walled carbon nanotubes (SWNTs). These materials can be classified according to the number of dimensions, which are not confined to the nanoscale range (0–100 nm), i.e., zero-dimensional (0-D) nanoparticles, one-dimensional (1-D) nanotubes, and two-dimensional (2-D) such as graphene [23] (Fig. 5.1).

5.2.2 *Metal-Based Materials*

Metal nanoparticles are the building blocks of the next generation of electronic, optoelectronic, biomedical, and chemical sensing devices. Among several metal nanomaterials, silver and gold nanoparticles are one of the most commercialized NPs because of their unique optical, electrical, and photothermal properties. They have wide applications in bio-sensing, diagnostic imaging, wastewater treatment, chemo-catalyst, cancer diagnosis, and therapy [24].

Metallic nanoparticles can be used as effective growth inhibitors of various microorganisms and thereby are applicable to diverse medical devices. Nanotechnology discloses the use of elemental nanoparticles as active antibacterial ingredient for dental materials. In dentistry, both restorative materials and oral bacteria are believed to be responsible for restoration failure. Metal nanoparticles can also control the formation of biofilms within the oral cavity, as a result of which it is often used as topically applied agents within dental materials [25].

5.2.3 *Dendrimers*

Dendrimers are perfect monodispersed macromolecules with a regular and highly branched three-dimensional architecture. Dendrimers are produced in an iterative sequence of reaction steps, in which each additional iteration leads to a higher generation material [26]. The first dendritic structures that have been thoroughly investigated and that have received widespread attention are Tomalia's PAMAM dendrimers and Newkome's "arborol" systems [27]. Dendrimer is grown in a stepwise manner from a central core, implying that numerous reactions have to be performed on a single molecule. Consequently, every reaction has to be very selective to ensure the integrity of the final product [28]. These nanomaterials are nanosized polymers built from branched units. The surface of a dendrimer has numerous chain ends, which can be tailored to perform specific chemical functions. This property could also be useful for catalysis. Also, because three-dimensional dendrimers contain interior cavities into which other molecules could be placed, they may be useful for drug delivery [29]. Dendritic molecules have been tested in supramolecular polymer chemistry, in medicinal chemistry, and in catalysis.

5.2.4 *Nanocomposites*

A nanocomposite is a multiphase solid material where one of the phases has one, two, or three dimensions in the nanometer scale, which is of less than 100 nm. Nanocomposites show unique properties, because of the nanometric size effect, compared to conventional composite even at low filler content. The organic–inorganic nanocomposites are often developed by grafting synthetic polymers on inorganic particles or by adding modified nanoparticles (NPs) into polymer matrices [30]. This leads to composite materials with improved properties. Nanoparticles can be incorporated into polymeric nanocomposites. Polymeric nanocomposites consisting of inorganic nanoparticles and organic polymers represent a new class of materials that exhibit improved performance compared to their microparticle counterparts. It is therefore expected that they will advance the field of engineering applications. The type of nanoparticles incorporated determines the properties of polymer composites. It depends on the size, shape, and concentration of the nanoparticles and their interactions with the polymer matrix [31].

5.3 Properties of Nanoparticles

The characteristic of nanoparticles like the size, shape, and surface characteristics determine the properties of the nanoparticles. Nanoparticles have properties different from microparticles due to their small size and relatively large surface area. When the size of a particle is close to or smaller than the de Broglie wavelength of the charge carrier (electrons and holes) or the wavelength of light, the periodic boundary conditions of the crystalline particle are destroyed, or the atomic density on the amorphous particle surface is changed [32]. Due to these, a lot of the physical properties of nanoparticles are quite different from bulk materials, yielding a wide variety of new applications. Optical properties of nanoparticles are due to the excitation of surface plasmons in metallic nanoparticles, this property can be used in biomedicine, energy, and environment protection technologies [33]. Magnetic properties of nanoparticles are by virtue of its external magnetic field and hence can be used for biomedical imaging and information storage technology.

The adhesion and the friction of nanoparticles play important roles in nanofabrication, lubrication, the design of micro/nano devices, colloidal stabilization, and drug delivery. Controlling the size, shape, and surrounding media of metal nanoparticles are important as many of their intrinsic properties are determined by these parameters. Particular emphasis has recently been placed on the control of shape, because, in many cases it allows properties to be fine-tuned with a greater versatility that gives the particles a unique nature. It is only within the past decade that it has become possible to control the shape of particles synthesized in solution, and numerous methods have been developed for this. Stabilizing agent also plays a role in the size of nanoparticle. The key effect of the stabilizer on the nanoparticle size lies in the initial particle nucleation stage [34]. A thermodynamically stable and mature nanoparticle can only be formed when a nucleus grows into a cluster that is larger than a certain critical size. Therefore, a fast initial nucleation is critical to the production of stable nuclei and subsequently smaller nanoparticles. The higher the temperature, larger and more polydisperse nanoparticles are obtained. Similarly, change in pH or H^+ activity can impact the reduction of ions. As the reduction of H^+ proceeds, the solution pH goes up, which favors the reduction of metal ions.

5.4 Synthesis of Nanoparticles

Nanoparticles may be synthesized by physical, chemical, or biological methods. Physical methods include evaporation, condensation, high gravity reactive precipitation, and laser ablation procedures. Chemical synthesis includes solvothermal methods, sol–gel conversions, chemical reduction, electrochemical techniques, photochemical reduction, and pyrolysis.

Biosynthesis of nanoparticles may be done via microorganisms, enzymes, fungi, plants, and plant extracts. Depending upon the location of nanoparticles, their synthesis via biological mode may be intracellular or extracellular. Intracellular method involves transport of ions into microbial cells to form nanoparticles in the presence of enzymes while extracellular synthesis is not within the cellular components of the organism [35] (Fig. 5.2).

There are two alternative approaches for synthesis of metallic nanoparticles: the "bottom-up" approach and the "top-down" approach. In bottom-up approach, atoms or molecules are assembled to molecular structures in nanometer range. Bottom-up approach is commonly used for chemical and biological synthesis of nanoparticles. Advantage of the bottom-up approach is the enhanced possibility of obtaining metallic nanoparticles with comparatively lesser defects and more homogeneous chemical composition(s). In top-down approach, the bulk materials are gradually broken down to nanosized materials using physical (e.g., mechanical) or chemical means. A major drawback of the top-down approach is the imperfection of the surface structure. Such

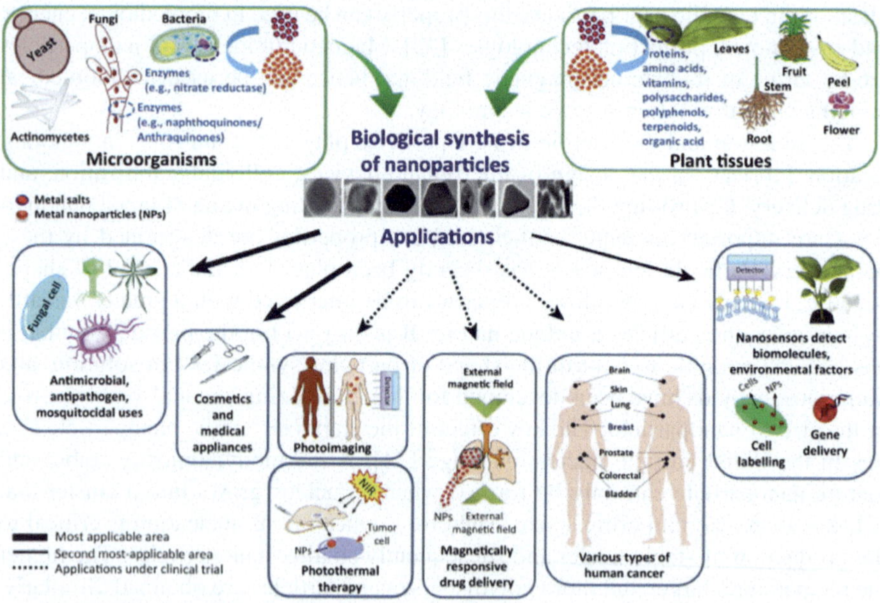

Fig. 5.2 Biosynthesis of nanoparticles

Table 5.2 Various approaches for synthesis of nanoparticles

Method	Process involved	
Physical method	Ball Milling	Top down approach
	Mechanochemical synthesis	
	Plasma vapor deposition	Bottom – up approach
	Laser pyrolysis	
	Flash spray pyrolysis	
	Inert gas condensation	
	Flame hydrolysis	
Chemical	Sol-gel or gel-sol conversions	
	Plasma/laser/flame enhanced chemical vapor deposition	
	Thermal decomposition	
	Solvo-thermal synthesis	
Biological	Fungi	Intracellular & Extracellular synthesis
	Bacteria	
	Yeast	
	Plant extracts	

defects in the surface structure can have a significant impact on physical properties and surface chemistry of the metallic nanoparticles due to the high aspect ratio [36].

Synthesis by various methods by top-down and bottom-up approach has been summarized as (Table 5.2):

Nanoparticles can be made from different materials composition with different physical and chemical properties. They can be attached with a various ligands for biological targeting for different functions like contrasting agents, drug delivery vehicles, and therapeutics (Fig. 5.3).

5.5 Applications of Nanoparticles in Dentistry

Nanoparticles are used in various forms in dentistry for diagnostic and therapeutic purposes. They have been used for administering local anesthesia, for cure of dentinal hypersensitivity, to diagnosis and cure oral cancers. Nanoneedles and nanofibers have been employed for wound dressings [37]. Nanoparticles due to their property of biocidal, anti-adhesive, and delivery capabilities are being explored to prevent the formation of biofilms within the oral cavity [38]. As nanoparticles possess a greater surface-to-volume ratio, they can interact more efficiently with microbial membranes and provide considerably larger surface area for antimicrobial activity. They have been used as device coatings [39] as topically applied agents, within materials like in dental resin composites [40], cavity liners, pit and fissure sealants, cores and buildups, indirect restorations, cements, acrylic resin denture bases [41] mouth rinses, and toothpastes [39].

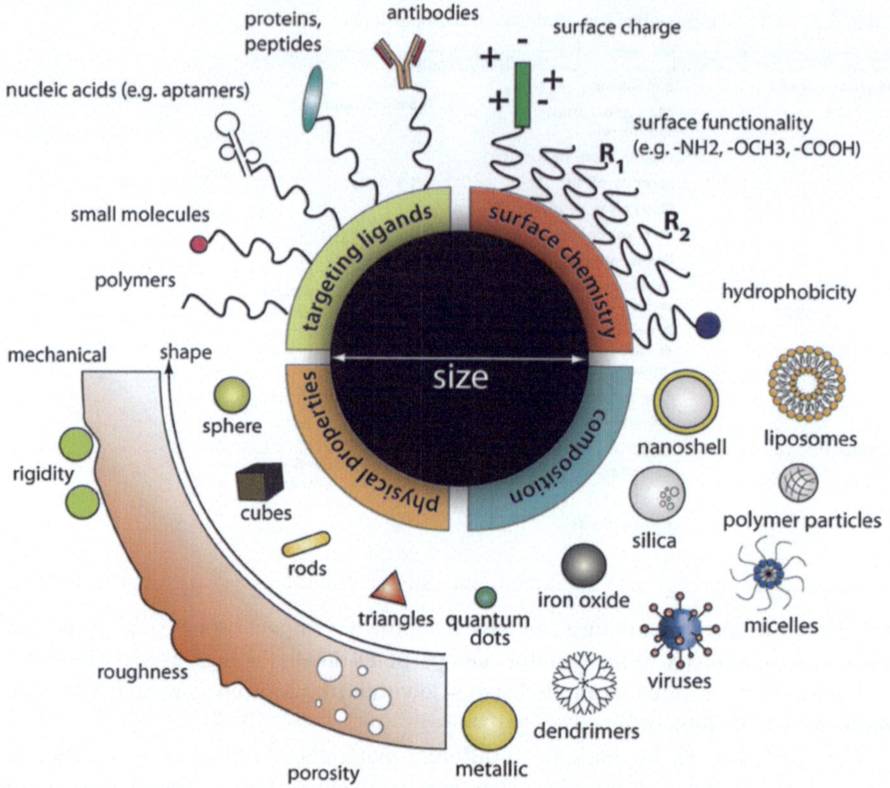

Fig. 5.3 Designing nanoparticles for intracellular applications

5.6 Applications of Nanoparticles in Orthodontics

Nanoparticles have been used for various below-mentioned purposes in orthodontics (Fig. 5.4).

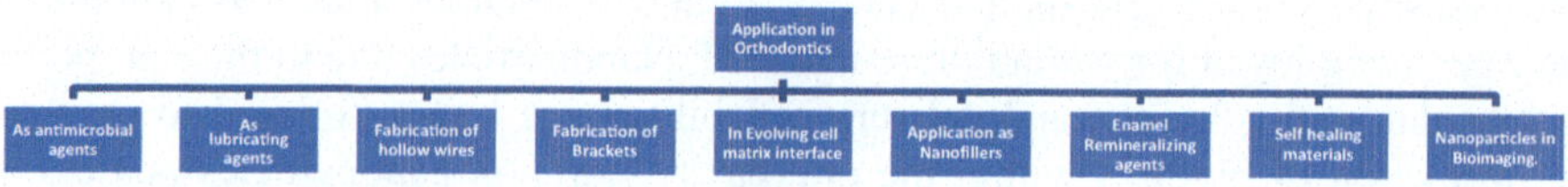

Fig. 5.4 Applications of nanoparticles in orthodontics

5.6.1 *Application as Antimicrobial Agent*

Since ages efforts are being made to reduce the microbial activity in the oral cavity. Inventions are being done in the methods of sterilization or the materials are being coated or incorporated into particles thus making them self-disinfecting [42]. Plaque accumulation on fixed and removable appliances is a common problem encountered during orthodontic treatment. This bacterial growth can lead to many complications like gingivitis, periodontitis, white spot lesions, increased risk of caries, halitosis, superimposed infections, failure of TAD's, and delayed tooth movement. To minimize these problems, nanoparticles are being incorporated in various materials.

5.6.1.1 Materials Modified by Nanoparticles

Nanoparticles as antimicrobial agents can be incorporated or applied on:

Bracket Surface

Orthodontic brackets have been coated with nitrogen-doped titanium dioxide. The activation of nitrogen-doped titanium dioxide leads to the formation of OH free radicals, superoxide ions (O_2), peroxyl radicals (HO_2) and hydrogen peroxide (H_2O_2). These chemicals, through a series of oxidation reactions, react with biological molecules such as lipids, proteins, enzymes and nucleic acids, damage biological cell structures, and also exert antimicrobial activity [43].

Resin Composite during Bonding

TiO_2 nanoparticles of size 21 ± 5 nm have been blended to light cure orthodontic composite paste (Transbond XT) in 1, 2, and 3%. All the three concentrations had similar antibacterial effects [44]. ZnO-NPs when mixed with Chitosan NP in the ratio 10% (w/w) have also shown antibacterial activity when added to resin [45]. Silver and HA nanoparticles have also been added to the primer of Transbond XT in 1 and 5% concentrations and have shown good antibacterial properties [46]. 0.0100 wt% of copper NPs have also been used in resins [47].

Resin-Modified GIC

Resin-modified GIC has been improved by incorporating nanosized fluoroapatite (NFA) or fluorohydroxyapatite (NFHA) particles at 25% concentration. However, there was a significant reduction in shear bond strength. The fluoride release nearly tripled after 70 days [48]. Nanohydroxyapatite (Nano-HA) has also been added to orthodontic banding cement to prevent microleakage under orthodontic bands [49]. Zinc oxide has been added to light-cured resin-modified glass ionomer to create mixtures of 13% ZnO and 23.1% zinc oxide [50, 51].

Microimplants

Since bacterial infection has been identified as one of the major causes of titanium microimplant failures, a novel antibiotic vehicle composite, TiO_2NT–PSPMA, has been synthesized via atom transfer radical polymerization; this method improved the local antibiotic concentration and prolonged its sustainable release by loading larger amounts of antibiotic into Titanium nanotubes (TiO_2 NTs) arrayed on Ti implants. Ag nanoparticles loaded into TiO_2 NTs with the assistance of the ionic polymer 3-sulfopropyl methacrylate potassium salt (PSPMA) have been used. This composite perhaps could be used in future to prevent implant infection [52].

Acrylic Resins
TiO_2, SiO_2, and silver NPs have been added to cold-cure acrylic resins that are mainly made of polymethyl methacrylate (PMMA). However, the cytotoxicity of this resin over long duration still needs to be verified [53–55].

5.6.1.2 Tests Carried Out

The following laboratory tests are carried out to check for the properties of material:

Antimicrobial activity: Disk agar diffusion (DAD) test
Cytotoxicity on cell lines: Live and dead staining, lactate production, spectroscopy
Distribution of nanoparticles: SEM
Shear bond strength: Ionstron machine and adhesive remnant index (ARI) scores
Microleakage: Microleakage under the bands can be assessed by the methylene blue dye
Insolubility of NP: Atomic absorption test
Bacteria test: Streptococcus mutans since it causes caries, Aggregatibacter actinomycetemcomitans since it causes periodontal diseases.

5.6.2 Application as Lubricating Agent to Reduce Friction

Friction is one of the major deterrents present in alignment or retraction of teeth during orthodontic treatment. High friction leads to anchorage loss, increased duration of treatment, and increased waiting list in hospitals [56, 57]. Numerous attempts were made to overcome this problem, by using wires of a different metal, shape, and size [58, 59]. Not only the wires even the bracket surfaces are coated and the ligation methods are also varied to reduce the problem of friction [60].

5.6.2.1 Nanoparticles Used

Various nanoparticles are being used as a coating in an effort to reduce friction [61–63].

The nanoparticles used are as follows:

1. Nickel-phosphorus and tungsten disulfide (WS_2)
2. Co + fullerene-like WS_2

3. Carbone nitride (CNx)
4. ZnO
5. Molybdenum disulfide
6. Diamond-like carbon coating and nitrocarburizing
7. Polysulfone embedded with hard alumina nanoparticles for brackets

5.6.2.2 The Method of Coating

The orthodontic wires are inserted into the electroless solution of the nanoparticles which needs to be coated usually for about 30 min. A short (1 min) sonication can be used to disperse the agglomerates and ensure the stability of the suspension. Some scientists have even etched the wire with HF (20%) acid to improve the adherence of the coatings on wires. The researchers interested in this method can do further reading on electroless plating [64].

5.6.2.3 Tests Used to Determine the Efficiency of Coatings

Whenever one is coating the wires, it needs to check the strength/quality of coating in terms of adherence and the effectiveness of coating in terms of the purpose for which the coating is applied. One also needs to check how much of ions are being released once they are coated because the release of free ions can lead to cytotoxicity [65].

(a) **Tests used to determine reduction in friction**

Tribiological assays are done using a ball on flat tribometer and the friction coefficient is determined after various cycles. Dry and wet friction tests with paraffin oil lubricant can be carried out during 50–200 cycles. Apparatus used is universal mechanical testing apparatus.

(b) **Tests to determine the quality of film**

- SEM/TEM/micrographs can be taken of the coated wires.
- X-ray photoelectron spectroscopy (XPS) analyses can be done
- Energy dispersive x-ray spectroscopy, x-ray powder diffractometry

(c) **Tests to determine the adequate adherence of the film**

Scotch-tape test or bending test of the wire with a 2 cm radius of curvature

(d) **Test to determine corrosion behavior**

Potentiodynamic polarization test and electrochemical impedance spectroscopy

(e) **Test to check cytotoxicity**

The cytotoxicity of the coating needs to be checked on various cell lines like the human periodontal ligament fibroblast cells (HPLF), human gingival fibroblast cells (HGF), human pulp cells (HPC).

5.6.2.4 Mechanism of Action of Coating

The mechanism by which the reduction in friction is achieved can be explained by the theories suggested by Rapoport et al. [66] and Cizaire et al. [67]. At the first stage, when there is no angle between the slot and wire, the nanoparticles act as spacers and reduce the number of asperities that come in contact, resulting in a lower coefficient of friction. The friction at the wire increases as the angle increases. At this point on the coated wire, the nanoparticles from the coating are released into the tribological interface and their exfoliation occurs, resulting in the formation of a solid lubricant film on the sliding wire. The higher load at this point brings the asperities of the mating surfaces in straight contact causing the fluid (saliva in the mouth) to be squeezed out of the gap between the wire and slot, relying on the excellent tribological behavior of the solid lubricant film to allow the sliding of the archwire. When the two materials are SS, as is the case with the uncoated wire, the friction coefficient is high. The presence of nanosheets at the interface under high loads leads to a very facile sliding between these sheets, thereby reducing the coefficient of friction.

5.6.3 Fabrication of Hollow Wires

NiTi/Ni-TiO_2 composite nanoparticles are being used for the fabrication of hollow wires. These wires are hollow from inside but retain the properties of NiTi wires. They are synthesized via the synthesis method called the ultrasonic spray pyrolysis (USP). The orthodontic wire is used to obtain the precursor solution for the synthesis of spherical NiTi particles. These spherical NiTi particles are then coated over a textile or a polymer fiber via electrospinning. Then, the fiber is removed from inside thus producing a hollow wire. Bending properties of these hollow NiTi wires were performed by the three-point bending test and compared with conventional NiTi wires [68, 69].

Advantages of hollow wires:

1. This wire could potentially have the shape-memory and superelasticity properties, while possibly reducing the material needed for the wire production.
2. They may deliver lighter and more continuous force.
3. The bending properties can be customized by inserting another wire into the hollow core.

Limitation:

It is still difficult to obtain pure NiTi particles so research is being done on different precursor solutions, gases and collection media so as to obtain pure NiTi particles.

5.6.4 *Fabrication of Orthodontic Brackets*

Nanoparticles are being used for fabrication of brackets like the hard alumina nanoparticles embedded in polysulfone. These brackets have strength, reduced friction and biocompatibility while maintaining the transparency of the bracket.

Brackets are also being coated with nanoparticles to improve their properties like:

- Titanium dioxide because of its photocatalytic properties
- Nickel-phosphorous and tungsten disulfide (WS_2) nanoparticles to reduce friction.
- ZnO and CuO nanoparticles for antibacterial properties [70, 71].

Once they are coated, the following tests are carried out to check for their properties:

- Surface roughness is seen on atomic force microscope.
- Compressive strength, maximum strain, and elastic modulus are checked on Instron testing machine.
- Photocatalytic ability is checked by decolorization method with toluidine blue followed by measurement on absorption spectrometer.

5.6.5 *Use of Nanoparticles in Evolving the Cell-Matrix Interface*

5.6.5.1 To Improve the Primary Stability of TADs

The surface of the microimplants is being modified with the help of nanoparticles. The following nanoparticles are being used:

(a) Nanostenciled rgd-gold patterns have been used to control and influence the differentiation of mesenchymal cells with implant surface. Manipulating the maturation of cell-matrix adhesions by nanopatterned surfaces allows influencing morphology, actin dynamics, migration, and ECM assembly of adhering fibroblasts [72].

(b) Primary stability and partial osseointegration of TADs could be achieved by the synergistic effects from nanoclay reinforced tricalcium phosphate (TCP) nanocoating on titanium miniscrews [73].

(c) Ultrafine grain-sized titanium (UFG Ti) obtained by severe plastic deformation presents a bright potential for biomedical applications because it provides the strength of titanium alloys without toxic alloying elements, such as Al and V that, by dissolving away from the implant, may be harmful to human health.

The osseointegration of these UFG Ti microimplants is also found to be superior [74].

5.6.5.2 To Improve the Stability of the Newly Forming Bone

Patients with clefts many a times require a bone graft to fill the defects. Bioactive and biodegradable poly (lactide-co-glycolide)/bioactive glass/hydroxyapatite (PBGHA) and poly (lactide-co-glycolide)/bioactive glass (PBG) nanocomposite coatings have been tested. They have an ability to serve as a scaffold or template to guide the newly forming bone along its surfaces thus promoting its stability. BGs also serve as synthetic biocompatible osteoconductive bone substitutes, with bone bonding capacity and documented antibacterial and angiogenesis-promoting properties. The following laboratory tests are carried out to check for the quality of coating:

- The nanocomposite coatings are characterized by scanning electron microscopy, X-ray diffraction, and atomic force microscopy.
- Mechanical stability of the prepared nanocomposite coatings can be studied by intramedullary implantation of coated Kirschner wires (K-wires) into rabbit tibia [75].

5.6.5.3 To Fill Defects in Damaged Bone

Nanosized hydroxyapatite particles can be converted into injectable paste with the help of neutral phosphate buffer which is self-setting at 37 °C in 20 min. Stability of the injectable hydroxyapatite has been confirmed in aqueous medium as well as in human blood [76].

5.6.5.4 To Enhance the Formation of Bone

Titanium nanotubes with crystallized Ag_2O nanoparticles with diameters ranging from 5 nm to 20 nm embedded in them have been found to significantly enhance the functions of many cell types including osteoblasts thus having promising applications orthodontics. This leads to controlled release of Ag and hence long-lasting antibacterial activity without showing cytotoxicity. It has even shown some favorable effects on promoting cell spreading and can be used as a biomedical coating on devices [77].

5.6.6 Nanomaterials as Nanofillers in Orthodontics

Nanoparticles of reduced size are being used as fillers to reduce polymerization shrinkage and to improve the mechanical properties of strength. There are two types of fillers: nanoclusters and nanoparticles [78]. They are synthesized by techniques such as flame pyrolysis, flame spray pyrolysis, and sol–gel processes. Nanosized filler particles have been incorporated into the composite matrix and glass ionomer cements to form nanocomposites and nanoionomers [78–81]. The following nanoparticles are being used as fillers:

- Silica nanosized filler particles (10 wt%, particle diameter <7 nm) are being added to orthodontic adhesives [40].
- Titanium dioxide and zirconia due to their very high refractive indices and less weight of material are very useful [82].
- Nanozirconia is also being used in ionomer cements to improve properties like esthetics (e.g., low visual opacity), polish retention, and radiopacity as compared to previously known glass ionomer compositions. The nanozirconia is surface modified with silanes to aid in the incorporation of the nanozirconia into ionomer compositions [83].

5.6.7 Nanoparticles as Enamel Remineralizing Agents

Demineralization and white spot lesions are common problems encountered during and post-orthodontic treatment. Nanoparticles are being used for remineralization of decalcified enamel.

Nanohydroxyapatite has been developed as a paste. Calcium nanophosphate crystals which are smaller than 100 nm, lead to improved bioactivity of the product, resulting from the increase in surface area and wettability of HA (hydroxyapatite) nanoparticles and thus form a protective layer on the enamel surface and provides protection against erosion [84].

5.6.8 Self-healing Materials

Research is being directed toward the fabrication of self-healing materials or materials that could mimic the biologic system and fill the cracks or damages on their own. When a crack appears near the network, the healing fluid or precursor can flow to the damaged region and fill the fissure. This fluid can be stored in bubbles which can be incorporated in the material.

This reservoir, upon exposure to air, polymerizes as a result of crack formation and closes the crack, thus maintaining the structural integrity of the material. This concept can be applied to polymer brackets and archwires. The autopolymerized monomer can be incorporated in nanosized bubbles and can be integrated with the material. Fracture of the bracket or wire would induce bursting of the nanobubbles and exposure of the monomer to air, thereby resulting in polymerization and filling of the crack-induced gap [85, 86].

5.6.9 Nanoparticles in Bioimaging

One of the potential applications of nanoparticles is also in the field of bioimaging as contrast agents. The advantages of nanoparticles over existing contrast agents include tunable physical (e.g., optical and magnetic) properties, high stability (e.g., against photobleaching), possibility of targeted delivery, and specific binding via chemical functionalization, multimodality (ability to combine several functions in one particle), high sensitivity, and selectivity. They can be used in bioimaging techniques such as optical and confocal microscopy, NIR imaging, magnetic resonance imaging (MRI), computed tomography (CT), positron emission tomography (PET), single-photon emission CT (SPECT), and ultrasound imaging. The current challenges include potential nanotoxicity and biocompatibility issues [87].

5.7 Mechanism of Action of Various Nanoparticles

Mechanism of antibacterial activity of metallic nanoparticles is not exactly known. However, the presence of positive charge over the metal ion is crucial for this antimicrobial action which aids in the electrostatic attraction between the negatively charged bacterial wall and the positively charged metallic nanoparticles [88]. Nanoparticles can enter the cell after binding to a specific receptor target. There are several factors which can regulate the behavior of nanomaterials at the nano–biointerface. Such as the shape of nanoparticles directly influences uptake into cells: Rods show the highest uptake, followed by spheres, cylinders, and cubes when the synthesized nanoparticles are larger than 100 nm [89]. In studies with sub-100-nm nanoparticles, spheres show an appreciable advantage over rods. In fact, at this size range, increasing the aspect ratio of nanorods seems to decrease total cell uptake. The size-dependent uptake of nanoparticles is likely related to the membrane-wrapping process. Small nanoparticles have less ligand-to-receptor interaction than do larger nanoparticles (Table 5.3).

5.8 Toxicity

The rapid developments in the field of nanotechnology also bring with them the concerns related to toxicity through new sources of exposure like inhalation, ingestion,

Table 5.3 Different nanoparticles with their mechanism of action

Name	Mechanism of action
Silver	Increase cell permeability → pitting of cell surface → agglomeration → cause cell death [90]. Free radical formation. [91, 92]
Copper	Combines with—SH groups of key microbial enzymes
Zinc oxide	Induction of reactive oxygen species(ROS)
Quaternary ammonium compounds	Transfusion across and damage to the bacterial cell membrane
Gold	1. Collapse membrane potential inhibit ATPase activities to decrease the ATP level; 2. Inhibit the subunit of ribosome from binding t RNA; 3. Enhance chemotaxis in the early-phase reaction
Titanium dioxide	Generation of ROS along with superoxide, peroxide, and hydrogen peroxide radicles

and injection. Not only that there are serious environmental implications but also associated with the manufactured nanomaterials.

Nanomaterials have unique physicochemical properties, such as ultra-small size, large surface area to mass ratio, and high reactivity and these properties also influence and determine the biological responses and can often lead to toxicity. Nanoparticles have been found to cross the transplacental membrane or cross the peritoneal cavity into uterus and may affect the cranial development of embryo [93]. Endocytosis of nanoparticles can also lead to oxidative stress, which can be a concern for a number of autoimmune diseases [94]. Cytotoxicity of silver nanoparticles has also been reported a number of times, but the exact mechanism of action of silver nanoparticle is yet to be elucidated. It has been reported that silver nanoparticle is associated with the depletion of glutathione (GSH) level, increased level of ROS [95]. Ag NPs can enter via the blood-brain barrier and accumulate in different regions of the brain and this may be beneficial for drug delivery, but at the same time can increase a risk to the patient. Copper nanoparticles are often metabolized in liver and are reported to have toxic effects to hepatic and renal tissues [96]. Gold nanoparticles are also associated with male sterility, it is reported that they can affect the motility of spermatozoa by penetrating penetrate sperm cells, which could result in fragmentation [97]. Surface charge of Si NPs is also associated with the cytotoxicity toward human cells and positively charged Si NPs are 250-fold more cytotoxic compared to their negatively charged counterparts [98] (Fig. 5.5).

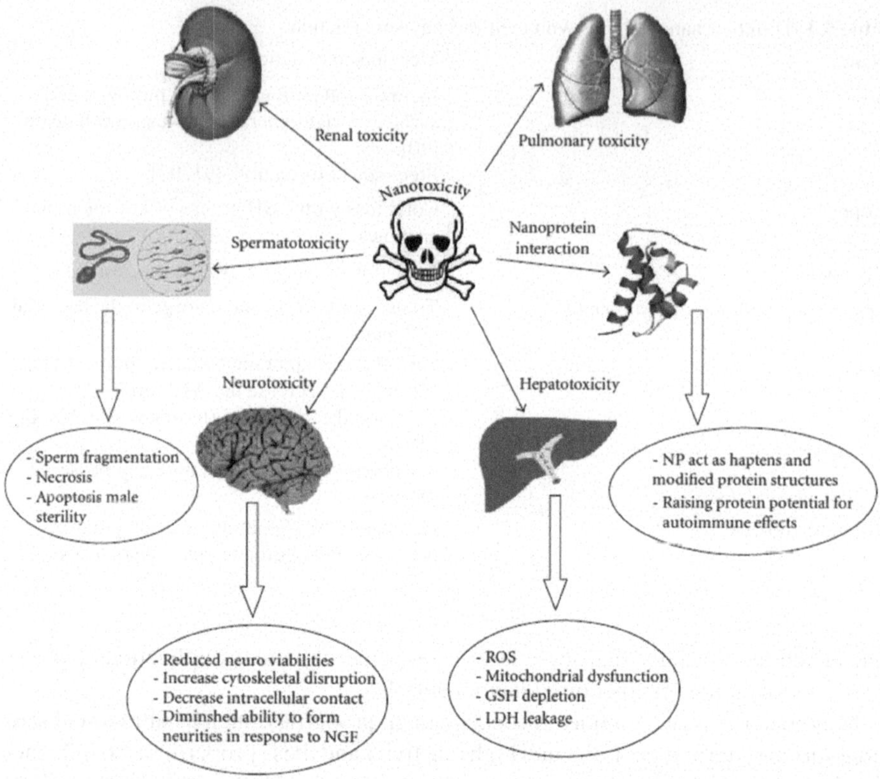

Fig. 5.5 Most important recorded toxic effects of therapeutically used nanoparticles

5.9 Conclusion

Nanoparticles have lot of potential in the field of dentistry and orthodontics per se. There is lot of research being focused on development of newer materials by the application of nanoparticles. Majority of the research is at the level of publications. The translation of this research to reality in the form of commercial products is a long journey. It is high time that researchers focus and encash on the potential of these vibrant particles as their applicability needs to be explored further in the field of dentistry.

References

1. Horikoshi S, Serpone N (2013) Introduction to nanoparticles. In: Microwaves in Nanoparticle Synthesis: Fundamentals and Applications, pp 1–24
2. Nanoscale Science Engineering and Technology Subcommittee (2004) National nanotechnology initiative: strategic plan. In: US National Science and Technology Council, 2004, pp 1–48
3. Feynman R (1991) There's plenty of room at the bottom. Science 254:1300–1301
4. Gazit Ehud (2007) Plenty of room for biology at the bottom: an introduction to bionanotechnology. Imperial College Press, London
5. Imperial College Press (2007) ISBN 9781860946776 http://www.wordiq.com/definition/Bionanotechnology
6. Nolting B (2005) Biophysical nanotechnology. In: Methods in modern biophysics. Springer, Berlin. ISBN 3-540-27703-X
7. Fakruddin M, Hossain Z, Afroz H (2012) Prospects and applications of nanobiotechnology: a medical perspective. J Nanobiotechnol 10(1):31
8. Maheshwari S, Verma SK, Tariq M, Gaur A (2014) Nano-orthodontics revolutionizing oral health care. Indian J Oral Sci 5(3):109
9. Taniguchi N (1974) On the basic concept of nanotechnology. In: International Conference on Production Engineering, Tokyo, Part II, Japan Society of Precision Engineering 1974 Feb, pp 18–23
10. British museum collection online. Google arts and culture: The Lycurgus cup. Available from https://www.google.com/culturalinstitute/beta/asset/the-lycurgus-cup/aQG7BXUoa7xwug. Accessed 17 Jun 2016
11. Helcher HH (1718) Aurum Potabile oder Gold Tinstur. Johann Herbord Klossen, Breslau und Leipzig
12. Tian H, Chen J, Chen X (2013) Nanoparticles for gene delivery. Small 9(12):2034–2044
13. Jain KK (2005) Nanotechnology in clinical laboratory diagnostics. Clin Chim Acta 358(1):37–54
14. Malam Y, Loizidou M, Seifalian AM (2009) Liposomes and nanoparticles: nanosized vehicles for drug delivery in cancer. Trends Pharmacol Sci 30(11):592–599
15. Zhang L, Webster TJ (2009) Nanotechnology and nanomaterials: promises for improved tissue regeneration. Nano Today. 4(1):66–80
16. Scott RW, Wilson OM, Crooks RM (2005) Synthesis, characterization, and applications of dendrimer-encapsulated nanoparticles. J Phys Chem B 109(2):692–704
17. Zhang L, Gu FX, Chan JM, Wang AZ, Langer RS, Farokhzad OC (2008) Nanoparticles in medicine: therapeutic applications and developments. Clin Pharmacol Ther 83(5):761–769
18. Kreuter J (2007) Nanoparticles—a historical perspective. Int J Pharm 331(1):1
19. Sajanlal PR, Sreeprasad TS, Samal AK, Pradeep T (2011) Anisotropic nanomaterials: structure, growth, assembly, and functions. Nano Rev 2(1):5883
20. Sau TK, Murphy CJ (2004) Room temperature, high-yield synthesis of multiple shapes of gold nanoparticles in aqueous solution. J Am Chem Soc 126(28):8648–8649
21. De Santis CJ, Peverly AA, Peters DG, Skrabalak SE (2011) Octopods versus concave nanocrystals: control of morphology by manipulating the kinetics of seeded growth via co-reduction. Nano Lett 11(5):2164–2168
22. Baptista FR, Belhout SA, Giordani S, Quinn SJ (2015) Recent developments in carbon nanomaterial sensors. Chem Soc Rev 44(13):4433–4453
23. Ionescu MI (2011) Synthesis of One-Dimensional and Two-Dimensional Carbon Based Nanomaterials. Ir.lib.uwo.ca
24. Mazumder JA, Ahmad R, Sardar M (2016) Reusable magnetic nanobiocatalyst for synthesis of silver and gold nanoparticles. Int J Biol Macromol 31(93):66–74
25. Hamouda IM (2012) Current perspectives of nanoparticles in medical and dental biomaterials. J Biomed Res 26(3):143–151
26. John PR (2006) New nanotechnology research. 2006. Chapter 6: Preparation of nanomaterials in presence of branched molecules. Nova science publisher, New York, p 188

27. Bosman AW, Janssen HM, Meijer EW (1999) About dendrimers: structure, physical properties, and applications. Chem Rev 99(7):1665–1688
28. Carlmark A, Hawker C, Hult A, Malkoch M (2009) New methodologies in the construction of dendritic materials. Chem Soc Rev 38(2):352–362
29. Nanjwade BK, Bechra HM, Derkar GK, Manvi FV, Nanjwade VK (2009) Dendrimers: emerging polymers for drug-delivery systems. Eur J Pharm Sci 38(3):185–196
30. Kango S, Kalia S, Celli A, Njuguna J, Habibi Y, Kumar R (2013) Surface modification of inorganic nanoparticles for development of organic–inorganic nanocomposites—a review. Prog Polym Sci 38(8):1232–1261
31. Camargo PH, Satyanarayana KG, Wypych F (2009) Nanocomposites: synthesis, structure, properties and new application opportunities. Mater Res 12(1):1–39
32. Guo D, Xie G, Luo J (2013) Mechanical properties of nanoparticles: basics and applications. J Phys D Appl Phys 47(1):013001
33. García MA (2011) Surface plasmons in metallic nanoparticles: fundamentals and applications. J Phys D Appl Phys 44(28):283001
34. He F, Zhao D (2007) Manipulating the size and dispersibility of zerovalent iron nanoparticles by use of carboxymethyl cellulose stabilizers. Environ Sci Technol 41(17):6216–6221
35. Malam Y, Loizidou M, Seifalian AM (2009) Liposomes and nanoparticles: nanosized vehicles for drug delivery in cancer. Trends Pharmacol Sci 30(11):592–599
36. Thakkar KN, Mhatre SS, Parikh RY (2010) Biological synthesis of metallic nanoparticles. Nanomed Nanotechnol Biol Med 6.2:257–262
37. Shalumon KT, Anulekha KH, Nair SV, Nair SV, Chennazhi KP et al (2011) Sodium alginate/poly (vinyl alcohol)/nano ZnO composite nanofibers for antibacterial wound dressings. Int J Biol Macromol 49(3):247–254
38. Allaker RP (2010) The use of nanoparticles to control oral biofilm formation. J Dent Res 89(11):1175–1186
39. Boldyryeva H, Umeda N, Plaksin OA, Takeda Y (2005) Kishimoto N High-fluence implantation of negative metal ions into polymers for surface modification and nanoparticle formation. Surf Coat Technol 196(1–3):373–377
40. Ahn SJ, Lee SJ, Kook JK (2009) Lim BS Experimental antimicrobial orthodontic adhesives using nanofillers and silver nanoparticles. Dent Mater 25(2):206–213
41. Lee CJ, Lee MS, Nam KY (2008) Inhibitory effect of PMMA denture acrylic impregnated by silver nitrate and silver nano-particles for Candida Albicans. J Korean Chem Soc 52(4):380–386
42. Batra P, Jyothikiran H (2013) Tips for maintaining sterilization in your orthodontic work station. Int J Orthod (Milwaukee, Wis.) 25(2):21–30
43. Cao B, Wang Y, Li N, Liu B, Zhang Y (2013) Preparation of an orthodontic bracket coated with an nitrogen-doped $TiO_{(2-x)}N(y)$ thin film and examination of its antimicrobial performance. Dent Mater J 32(2):311– 316
44. Poosti M, Ramazanzadeh B, Zebarjad M, Javadzadeh P, Naderinasab M et al (2013) Shear bond strength and antibacterial effects of orthodontic composite containing TiO2 nanoparticles. Eur J Orthod 35(5):676–679
45. Mirhashemi A, Bahador A, Kassaee M, Daryakenari G, Ahmad Akhoundi M et al (2013) Antimicrobial effect of nano-zinc oxide and nano- chitosan particles in dental composite used in orthodontics. J Med Bacteriol 2(3–4):1–10
46. Akhavan A, Sodagar A, Mojtahedzadeh F, Sodagar K (2013) Investigating the effect of incorporating nanosilver/nanohydroxyapatite particles on the shear bond strength of orthodontic adhesives. Acta Odontol Scand 71(5):1038–1042
47. Argueta Figueroa L, Scougall Vilchis RJ, Morales Luckie RA, Olea Mejia OF (2015) An evaluation of the antibacterial properties and shear bond strength of copper nanoparticles as a nanofiller in orthodontic adhesive. Aust Orthod J 31(1):42–48
48. Lin J, Zhu J, Gu X, Wen W, Li Q et al (2011) Effects of incorporation of nano-fluorapatite or nano- fluorohydroxyapatite on a resin-modified glass ionomer cement. Acta Biomater 7(3):1346–1353

49. Enan ET, Hammad SM (2013) Microleakage under orthodontic bands cemented with nano-hydroxyapatite-modified glass ionomer. Angle Orthod 83(6):981–986
50. Spencer CG, Campbell PM, Buschang PH, Cai J, Honeyman AL (2009) Antimicrobial effects of zinc oxide in an orthodontic bonding agent. Angle Orthod 79(2):317–322
51. Ramazanzadeh B, Jahanbin A, Yaghoubi M, Shahtahmassbi N, Ghazvini K, et al (2015) Comparison of antibacterial effects of ZnO and CuO nanoparticles coated brackets against streptococcus mutans. J Dent 16(3):200–205
52. Zhang M, Wei M, Wang D, Duan Y (2015) Preparation and characterization of a drug vehicle: polymer brush immobilized Ag nanoparticles onto titanium nanotubes. Mater Lett 135:51–54
53. Farhadian N, Mashoof RU, Khanizadeh S, Ghaderi E, Farhadian M et al (2016) Streptococcus mutans counts in patients wearing removable retainers with silver nanoparticles vs those wearing conventional retainers: a randomized clinical trial. Am J Orthod Dentofacial Orthop 149(2):155–160
54. Batra P, Miglani R (2016) Antibacterial properties of retainers with silver nanoparticles. Am J Orthod Dentofac Orthop 150(2):208–209
55. Sodagar A, Bahador A, Khalil S, Shahroudi AS, Kassaee MZ (2013) The effect of TiO2 and SiO2 nanoparticles on flexural strength of poly (methyl methacrylate) acrylic resins. J Prosthodont Res 57(1):15–19
56. Wichelhaus A, Geserick M, Hibst R, Sander FG (2005) The effect of surface treatment and clinical use on friction in NiTi orthodontic wires. Dent Mater 21(10):938–945
57. Batra P (2017) Eisenhower box for prioritising waiting list of orthodontic patients. OHDM 16(1):1–3
58. Kusy RP (2000) Ongoing innovations in biomechanics and materials for the new millennium. Angle Orthod 70(5):366–376
59. Jyothikiran H, Shantharaj R, Batra P, Subbiah P, Lakshmi B, Kudagi V (2014) Total recall: an update on orthodontic wires. Int J Orthod (Milwaukee, Wis.) 25(3):47–56
60. Hain M, Dhopatkar A, Rock P (2003) The effect of ligation method on friction in sliding mechanics. Am J Orthod Dentofac Orthop 123(4):416–422
61. Katz A, Redlich M, Rapoport L, Wagner HD, Tenne R (2006) Self-lubricating coatings containing fullerene-like WS2 nanoparticles for orthodontic wires and other possible medical applications. Tribol Lett 21(2):135–139
62. Redlich M, Katz A, Rapoport L, Wagner HD, Feldman Y, Tenne R (2008) Improved orthodontic stainless steel wires coated with inorganic fullerene-like nanoparticles of WS 2 impregnated in electroless nickel–phosphorous film. Dent Mater 24(12):1640–1646
63. Samorodnitzky-Naveh GR, Redlich M, Rapoport L, Feldman Y, Tenne R (2009) Inorganic fullerene-like tungsten disulfide nanocoating for friction reduction of nickel–titanium alloys. Nanomedicine 4(8):943–950
64. Mallory GO, Hajdu JB (1990) Electroless plating: fundamentals and applications. William Andrew
65. Panchali B, Anam M, Jahirul m, Meryam SR, Ragini M (2016) Nanoparticles and their applications in orthodontics. Adv Dent Oral Health 2(2):1–10
66. Rapoport L, Leshchinsky V, Lapsker I, Volovik Y, Nepomnyashchy O, Lvovsky M, Popovitz-Biro R, Feldman Y, Tenne R (2003) Wear 255:785
67. Cizaire L, Vacher B, Le-Mogne T, Martin JM, Rapoport L, Margolin A, Tenne R (2002) Mechanisms of ultra-low friction by hollow inorganic fullerene-like MoS2 nanoparticles. Surf Coat Technol 160:282–287
68. Zadno-Azizi GR, Muni KP, Bagaoisan CJ, inventors; Medtronic PercuSurge, Inc., assignee. Hollow medical wires and methods of constructing same. United States patent US 6,375,628. 2002 Apr 23
69. Shima Y, Otsubo K, Yoneyama T, Soma K (2002) Bending properties of hollow super-elastic Ti–Ni alloy wires and compound wires with other wires inserted. J Mater Sci - Mater Med 13(2):169–173
70. Ramazanzadeh B, Jahanbin A, Yaghoubi M, Shahtahmassbi N, Ghazvini K, Shakeri M, Shafaee H (2015) Comparison of antibacterial effects of ZnO and CuO nanoparticles coated brackets against Streptococcus mutans. J Dent 16(3):200

71. Sato Y, Miyazawa K, Sato N, Nakano K, Takei Y, Kawai T, Goto S (2009) Study on fabrication of orthodontic brackets with the photocatalytic function of titanium dioxide. Dent Mater J 28(4):388–395
72. Lutz R, Pataky K, Gadhari N, Marelli M, Brugger J et al (2011) Nano-stenciled RGD-gold patterns that inhibit focal contact maturation induce lamellipodia formation in fibroblasts. PLoS ONE 6(9):e25459
73. Zhou WY, Rabie AB, Wong RW, Tang B (2010) Nanocoating of montmorillonite/Mg-β-tricalcium phosphate on orthodontic titanium miniscrews. In: 2010 3rd International Nanoelectronics Conference (INEC), IEEE, pp 817–818
74. Elias CN, Meyers MA, Valiev RZ, Monteiro SN (2013) Ultrafine grained titanium for biomedical applications: an overview of performance. J Mater Res Tech 2(4):340–350
75. Mehdikhani-Nahrkhalaji M, Fathi MH, Mortazavi V, Mousavi SB, Akhavan A, Haghighat A, Hashemi-Beni B, Razavi SM, Mashhadiabbas F (2015) Biodegradable nanocomposite coatings accelerate bone healing: in vivo evaluation. Dent Res J 12(1):89
76. Varma NP, Garai S (2012) Sinha a synthesis of injectable and cohesive nano hydroxyapatite scaffolds. J Mater Sci Mater Med 23(4):913–919
77. Gao A, Hang R, Huang X, Zhao L, Zhang X et al (2014) The effects of titania nanotubes with embedded silver oxide nanoparticles on bacteria and osteoblasts. Biomaterials 35(13):4223–4235
78. Uysal T, Yagci A, Uysal B, Akdogan G (2010) Are nano-composites and nano-ionomers suitable for orthodontic bracket bonding? Eur J Orthod 32(1):78–82
79. Hosseinzadeh Nik T, Karimzadeh A, Ayatollahi MR (2013) Bond strength of a nano-composite used for bonding ceramic orthodontic brackets. Mater Des 51:902–906
80. Chalipa J, Akhondi MS, Arab S, Kharrazifard MJ, Ahmadyar M (2013) Evaluation of shear bond strength of orthodontic brackets bonded with nano-filled composites. J Dent (Tehran) 10(5):461–465
81. Coutinho E, Cardoso MV, De Munck J, Neves AA, Van Landuyt KL et al (2009) Bonding effectiveness and interfacial characterization of a nano-filled resin-modified glass-ionomer. Dent Mater 25(11):1347–1357
82. Craig BD, Kolb BU, Oxman JD, Peez RF, Frank SA (2006) Inventors; 3M Innovative Properties Company, assignee. Use of nanoparticles to adjust refractive index of dental compositions. United States patent US 7090721
83. Kolb BU, Bui HT, Thalacker JP, Kangas LS, Oxman JD, et al (2010) Inventors; 3M Innovative Properties Company, assignee. Dental compositions containing nanozirconia fillers. United States patent US 7649029
84. Carvalho FG, Brasil VL, Silva Filho TJ, Carlo HL, Santos RL et al (2013) Protective effect of calcium nanophosphate and CPP-ACP agents on enamel erosion. Braz Oral Res 27(6):463–470
85. Cordier P, Tournilhac F, Souli_e-Ziakovic C, Leibler L (2008) Self-healing and thermoreversible rubber from supramolecular assembly. Nature 451:977–980
86. Eliades T (2015) Orthodontic material applications over the past century: evolution of research methods to address clinical queries. Am J Orthod Dentofac Orthop 147(5):S224–S231
87. Gun'ko YK (2016) Nanoparticles in bioimaging. Nanomaterials (Basel). 6(6):105
88. Fan TX, Chow SK, Zhang D (2009) Biomorphic mineralization: from biology to materials. Prog Mater Sci 54(5):542–659
89. Gratton SE, Ropp PA, Pohlhaus PD, Luft JC, Madden VJ et al (2008) The effect of particle design on cellular internalization pathways. Proc Natl Acad Sci USA 105:11613–11618
90. Gu H, Fan D, Gao J, Zou W, Peng Z, Zhao Z, Ling J, LeGeros RZ (2012) Effect of ZnCl 2 on plaque growth and biofilm vitality. Arch Oral Biol 57(4):369–375
91. Sodagar A, Bahador A, Khalil S, Shahroudi AS, Kassaee MZ (2013) The effect of TiO_2 and SiO_2 nanoparticles on flexural strength of poly (methyl methacrylate) acrylic resins. J Prosthodont Res 57(1):15–19
92. Cushing BL, Kolesnichenko VL, O'Connor CJ (2004) Recent advances in the liquid-phase syntheses of inorganic nanoparticles. Chem Rev 104(9):3893–3946

93. Vega-Villa KR, Takemoto JK, Y´a˜nez JA, Remsberg CM, Forrest ML, Davies NM (2008) Clinical toxicities of nanocarrier systems. Adv Drug Deliv Rev 60(8):929–938
94. Klimuk SK, Semple SC, Nahirney PN et al (2000) Enhancedanti-inflammatory activity of a liposomal intercellular adhesion molecule-1 antisense oligodeoxynucleotide in an acutemodel of contact hypersensitivity. J Pharmacol Exp Ther 292(2):480–488
95. Chung Y-C, Chen I-H, Chen C-J (2008) The surface modification of silver nanoparticles by phosphoryl disulfides for improved biocompatibility and intracellular uptake. Biomaterials 29(12):1807–1816
96. Meng H, Chen Z, Xing G et al (2007) Ultrahigh reactivity provokes nanotoxicity: explanation of oral toxicity of nanocopper particles. Toxicol Lett 175(1–3):102–110
97. Manin OI, Nikolaev VA, Kolomi˘ıtsev AA, Lebedenko II (2007) Comparative toxicological evaluation of domestic golden alloys for soldering. Stomatologiia 86(1):64–67
98. Bhattacharjee S, de Haan LH, Evers NM, Jiang X, Marcelis AT, Zuilhof H et al (2010) Role of surface charge and oxidative stressin cytotoxicity of organic monolayer-coated silicon nanoparticles towards macrophage NR8383 cells. Part Fibre Toxicol 7:25–36

Chapter 6
Nanomaterials: A Boon to Prosthodontics

Rajashree Dhananjay Jadhav

6.1 Introduction

Prosthodontics is an important branch of dentistry. The main purposes of oral prosthesis are to restore function, facial appearance and maintain the patient's health. In recent years, nanomaterials have captured more and more attention because of their unique structure and properties. Researchers started studying and using various nanomaterials not only in medical field but also in dentistry particularly in prosthodontics. Nanomaterials are mainly used in ceramic, resin, and metals providing a huge space for the improvement and innovation of dental material. The term nanotechnology was first introduced by Richard Feynman in 1959. This was practically made possible by Eric Drexler in the mid-1980s when he emphasized the potential of molecular nanotechnology [1, 2]. The term "nano" is derived from the Greek word "dwarf." One nanometer is one-billionth or 10^{-9} of a meter. Single hair strand has a thickness of 100,000 nm [3]. Nanotechnology is the direct manipulation of materials at the nanoscale [4]. Nanotechnology is translated as "the science of the small" [5]. Nanotechnology modulates metals into their nanosize. This drastically changes the physical, chemical, and optical properties of metals. Nowadays, nanoparticles have been known as materials with good potential to be extensively used in biological and medical applications. Inorganic nanoparticles and their nanocomposites are used as antibacterial agents [6]. An important aspect of nanoscience is the formation of toxicity-free synthesis of metal nanoparticles, which is a great challenge. The interaction of metal nanoparticles with microorganisms is a vast field of research. In recent years, a rapid increase in microbes that are resistant to antibiotics has been observed [7]. Due to the emergence and increase of microbial infections which is resistant to

R. D. Jadhav (✉)
Department of Prosthodontics and Crown and Bridge,
Bharati Vidyapeeth (Deemed to be) University
Dental College and Hospital, Pune, India
e-mail: rajashreedjadhav@gmail.com

R. S. Chaughule (ed.), *Dental Applications of Nanotechnology*,
https://doi.org/10.1007/978-3-319-97634-1_6

antibiotics, many researchers have tried to develop new and effective antimicrobial reagents free of resistance. This led to the resurgence in the use of nanosized metals used in medical and dental fields.

In recent years, new methods have been proposed to synthesize nanoparticles so that size and shape of the various nanoparticles can be tailored as per requirement. These nanomaterials are present with different interesting morphologies such as spheres, tubes, rods, and prisms. Inorganic nanoparticles based on metal oxides, such as zinc oxide, iron oxide, titanium dioxide, and cerium oxide, and metals, such as gold, silver and iron, copper, and magnesium, and quantum dots such as cadmium sulfide and cadmium selenide and also silicon dioxide and aluminum oxide are available and used freely [8]. Alginate nanomaterials can also be used as antimicrobial agents [9].

Metals have been used for centuries as antimicrobial agents. Gold, copper, titanium, and zinc have attracted the attention of many researchers and used extensively having different properties and spectra of antimicrobial activities [10, 11]. But silver nanoparticles have been proved to be most effective as they have good antimicrobial efficacy against bacteria, viruses, and other eukaryotic microorganisms [12]. For thousands of years, silver has been used in medicines. Silver has been known for its significant broad-spectrum antimicrobial activity against gram-positive and gram-negative bacteria, fungi, and viruses [13], including antibiotic-resistant strains [14]. The use of silver as an antibacterial agent decreased with the discovery of antibiotics, but because of the antibiotic-resistant pathogens there is revival in silver-based applications. Particle size, size distribution, shape, and surface chemistry of silver nanoparticles determine their performance that means in vivo distribution, biological fate, toxicity, and the targeting ability of nanoparticles. Their small size and large surface area can lead to particle–particle aggregation and that makes physical handling of nanoparticles difficult in liquid and dry forms. Aggregation of the silver nanoparticles may lead to the loss of properties associated with the nanoscale nature of the particles. Agglomeration and particle sizes of AgNPs are responsible for cytotoxicity. Smaller AgNPs (3 nm) are more cytotoxic than larger particles (25 nm) at a concentration of 10 μg/mL [15]. For better efficacy, size, shape, and morphology are very important. Pal et al. found that triangular silver nanoparticles showed greater biocidal action than rod or spherical nanoparticles. Recent advances in nanotechnology help in modulation of size and shape of nanoparticles [16].

Thus, various nanoparticles are incorporated in the acrylic resin, tissue conditioner, dental composites, dental adhesives, dental cements, dental porcelain, maxillofacial prosthesis, and implants to improve their antimicrobial and mechanical properties.

6.2 Nanoparticles in Denture Base Resins (Polymethyl Methacrylate)—PMMA

Most of the oral prosthesis are made up of polymethyl methacrylate (PMMA) was introduced as a denture base material by Wright in 1937. Its overall performance is satisfactory; it is widely used for fabrication of oral prosthesis like removable partial denture, complete dentures, and maxillofacial prosthesis [17]. It has been shown by researchers that PMMA denture base material may act as reservoir for many microorganisms and has the potential to support biofilm formation [18, 19]. Placement of a prosthesis in the oral cavity results in changes of the oral environmental conditions as the prosthesis becomes colonized with oral microorganisms. It isolates the underlying mucosa from the cleansing effect of the tongue and free flow of saliva. Also, the porous surface of denture base material (PMMA) and irregularities on anatomical surface of the prosthesis favor the accumulation of various microorganisms. This leads to candidiasis or denture stomatitis [20, 21].

6.2.1 Denture Stomatitis or Candidiasis

Candida species are ubiquitous, human fungal pathogens. They are responsible of initiating a variety of recurring diseases in the oral and vaginal mucosae. Many species of Candida have been involved in pathogenesis, but among them *Candida albicans* has been shown to be most opportunistic pathogen causing infection in the oral cavity and to be able to colonize acrylic materials [22–25]. Candida species form biofilm on acrylic denture surfaces. This biofilm is a network of yeasts, pseudohyphae, and hyphae surrounded by an extracellular matrix and logged into irregularities of anatomical surface of acrylic prosthesis [26]. Tissue invasion by these species causes infection of the oral mucosa. *C. albicans* has the ability to degrade the proteins in both yeast and hyphal forms. They can induce a chronic inflammatory response in the oral mucosa, described as denture stomatitis [27, 28]. Denture-induced stomatitis is an inflammatory reaction of the denture-bearing mucosa that affects approximately 60–70% of complete denture wearers [29]. Denture stomatitis is a challenge for the dental field. Methods for prevention of this such as the incorporation of antimicrobial agents into the denture base resin and into tissue conditioners have been developed.

Incorporation of nanoparticles into the acrylic resin denture base material is mainly in the form of silver and platinum nanoparticles as an effective antimicrobial agent. Some researchers showed that the addition of metal nanoparticles such as TiO_2, Fe_2O_3, and silver to PMMA materials could increase the surface hydrophobicity [30].

Wady [21] in 2012 evaluated the activity of a silver nanoparticle solution against *C. albicans* and then the effect of incorporation of AgNPs on the material's hydrophobicity, Candida adhesion, and biofilm formation. But they concluded that although the AgNPs had antifungal activity, there was no effect on *C. albicans* adherence and biofilm formation.

But Li et al. evaluated the effect of denture base resin containing silver nanoparticles on *C. albicans* adhesion and biofilm formation and concluded that AgNPs had antifungal activity and inhibited *C. albicans* biofilm formation [31].

Acosta-Torres et al. [32] developed a PMMA containing 1 μg/mL of AgNPs, and they compared this new PMMA–Ag compound to PMMA. It has been observed that PMMA–AgNPs specimens showed significantly less *C. albicans* adherence compared to PMMA. Besides that, they evaluated the activity of mouse fibroblasts and human lymphocytes, and it has been shown that PMMA–AgNPs compound does not present cytotoxicity or genotoxicity. These results suggest that the acrylic resin incorporated with AgNPs could be developed as an antifungal denture base resin.

In a study performed by Monteiro et al. [33], AgNPs were incorporated in a commercial denture base resin, in 0.05, 0.5, and 5% of AgNPs (by mass) concentrations. The authors evaluated the mechanical properties of this modified denture base resin, as well of the unmodified one (0% of AgNPs). The flexural strength test was performed, and it was observed that all the groups presented very similar flexural resistance values, suggesting that AgNPs incorporation does not affect the mechanical properties of acrylic resin.

PMMA/AgNPs nanocomposite shows antifungal efficacy against *C. albicans* and can be evaluated by zone of growth inhibition method. Figure 6.1 shows silver nanoparticle's inhibited bacterial growth by the clear fungal inhibition zone (concentration of silver nanoparticles taken 1 mg/disk), which is correlated with the antifungal activity of AgNPs denture base resin nanocomposite.

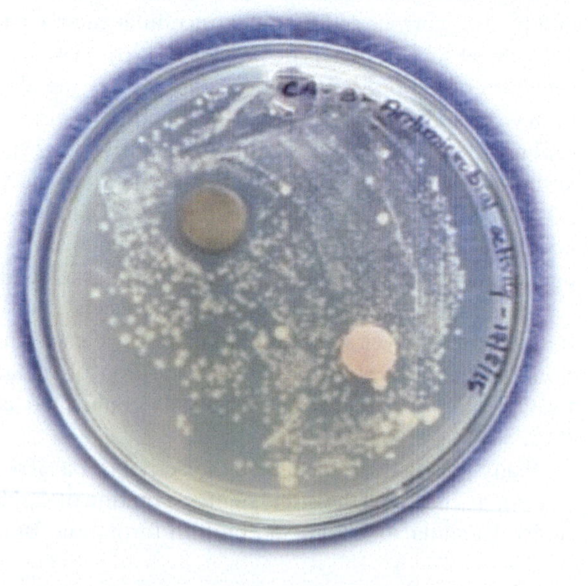

Fig. 6.1 PMMA–Ag nanocomposite acts as an antifungal material. Pink colored denture base resin shows adherence of *Candida Albicans*

Humada et al. investigated the effect of silver nanoparticles (AgNPs) incorporation on viscoelastic properties of acrylic resin denture base material. They concluded that incorporation of AgNPs within the acrylic denture base material can improve its viscoelastic properties [34].

Castro et al. assessed the antimicrobial activity and the mechanical properties of an acrylic resin embedded with nanostructure silver vanadate (β-$AgVO_3$). They concluded that incorporation of β-$AgVO_3$ has the potential to promote antimicrobial activity in the acrylic resin. At reduced rates, it improves the mechanical properties, and at higher rates, it does not promote changes in the control [35].

Other nanoparticles such as ZrO_2, TiO_2, and carbon nanotubes have been used to improve the performance of PMMA, and the results showed that desired mechanical property enhancement can be achieved in those composites with small amounts of nanoparticles added [36].

6.3 Silver Nanoparticles in Tissue Conditioners

Relining the ill-fitting denture with a tissue conditioner allows tissues to return to normal health. Tissue conditioners are used in the treatment of chronic soreness from complete dentures [37]. Tissue conditioners are degradable with time and susceptible to microbial colonization mainly *C. albicans*. Incorporation of AgNPs in the tissue conditioners could help in the reduction of microbial colonization. Zeolites have been incorporated by many researchers into tissue conditioner to make it antimicrobial [38–41]. Zeolites are aluminum silicate crystalline structures having void spaces measuring 3–10 angstroms. Antimicrobial cations, such as silver and zinc, may be lodged within the void spaces of the zeolites. As this ion availability occurs, the free cations come into contact with the microorganisms, suppressing their development by inactivating microbial enzymes, interrupting RNA replication, and blocking their respiration by an oxidative process [40, 41].

Abe et al. [39] studied the effects of incorporating antimicrobial silver zeolite 2 and 5 wt% on in vitro cytotoxicity of five tissue conditioners against the living dermal model, which consisted of human dermal fibroblast in a collagen lattice. Samples made up of each tissue conditioner containing silver zeolite were tested for cytotoxicity. Cell viabilities for Visco-Gel, FITT, GC Soft-Liner, and SR-Ivoseal decreased with increasing content of silver zeolite in the tissue conditioner. Cell viabilities of Shofu tissue conditioner showed no significant difference with degree of silver zeolite incorporation and had better results than other tissue conditioners. The result suggests that Shofu tissue conditioner has the highest cell viability.

Matsuura et al. [40] in 1997 studied the antimicrobial effect of tissue conditioners containing silver zeolite (SZ). Tissue conditioners containing SZ have been proved that they have antibacterial effects. They showed that silver zeolite continuously releases a small amount of silver ions into water resulting in antimicrobial activity which is not harmful to cells. SZ appears to be an appropriate material to use in tissue conditioners. Cell viability of tissue conditioners containing silver zeolite was tested.

The results suggested that Shofu tissue conditioner has the highest cell viability as compared to other selected tissue conditioners.

Nikawa et al. [41] studied the antifungal effect of zeolite-incorporated tissue conditioner against *C. albicans* growth, and they said that Zeomic combined tissue conditioner would be a potential aid to improve affected oral tissues of denture-induced stomatitis or candidiasis patients.

Nam KY in 2011 has incorporated AgNPs into a tissue conditioner, in 0.1, 0.5, 1.0, 2.0, and 3.0% concentrations. Their inhibitory effect was studied against *Staphylococcus aureus*, *Streptococcus mutans*, and *C. albicans* after 24 and 72 h. The authors have reported that the modified tissue conditioner showed antimicrobial properties even at lower concentrations, that is, 0.1% for *S. mutans* and *S. aureus* and 0.5% for *C. albicans* [42].

6.4 Nanoparticles in Impression Materials

Physical properties of polyvinylsiloxane impression materials were improved by the addition of nanosized silica. The study was carried out in which six types of fillers including nanosized fumed silica were added to polyvinylsiloxane impression materials. They had reported that the fumed silica was effective in increasing the viscosity, tensile strength, and maximum % strain.

Irreversible hydrocolloid, i.e., alginate impression powders can be mixed with water that contains silver hydrosol nanoparticles to make an alginate impression. This will reduce microbial cross-contamination by bacteria, yeasts, fungi, and viruses to the stone model from the infected impression [43].

Nanofillers are added to vinyl poly siloxanes, thus producing novel addition silicone impression materials. This material is claimed to have better properties such as flow and improved hydrophilic properties that helps in avoiding voids in the impression and model as well [44].

Ginjupalli et al. evaluated the antimicrobial activity and properties of two commercially available irreversible hydrocolloid impression materials (Zelgan and Tropicalgin) after incorporating varying concentrations of AgNPs. They concluded that adding silver nanoparticles to irreversible hydrocolloid impression materials alginate resulted in superior antimicrobial activity without having any adverse effect on their properties. Addition of silver nanoparticles to Zelgan alginate impression material significantly increased the gel strength compared with the control group. An increase in the permanent deformation was found with the incorporation of AgNPs of both Zelgan and Tropicalgin. The flow of Zelgan increased with the incorporation of AgNPs, whereas a decrease in the flow of Tropicalgin was observed at 1 and 2 wt% of AgNPs. An increase in the gelation time of both Zelgan and Tropicalgin was observed with the incorporation of AgNPs. They concluded that AgNPs can be incorporated into irreversible hydrocolloid impression materials as antimicrobial agents without adversely affecting their properties [45].

Table 6.1 Properties of prosthodontics material

Dental materials	PMMA	Ceramics ZrO_2 Al_2O_3 glass ceramic	Metal $Ti_6A_{14}V$ CoCrMo CoCr
Advantage	Good biocompatibility, esthetics processability, and reparability	High strength, suitable color, and low thermal and electrical conductivity	Titanium alloy has high strength, low density, light weight, low shrinkage, nonmagnetic, good mechanical properties and corrosion resistance, and nonallergic, teratogenic, and carcinogenic. CoCr alloy has high strength, wear resistance, and less tooth tissue cutting, with good biological safety. CoCrMo has good corrosion resistance, wear resistance, ductility, gloss, anti-plaque adhesion, and biosafety
Disadvantage	Poor strength, low fracture resistance, radiopacity behavior, and microbial adhesion	Low ductility and brittleness	Further improvement is desired to improve the corrosion resistance and biocompatibility of the Ti and CoCrMo alloy. CoCr alloy easily leads to sensitive symptoms
Nanoresearch	TiO_2 nanoparticle reinforced the mechanical behavior of PMMA. Well-dispersion nano-ZrO_2 particles can improve the modulus and strength and maintain or even improve ductility. Ag TiO_2 and Fe_2O_3 particles significantly reduce adherence of *C. albicans* of PMMA and do not affect metabolism or proliferation	The hardness and fracture toughness increased nanozirconia ceramics. Glass ceramics with nanosized grains showed excellent corrosion resistance, high fracture toughness, and translucency	Nanophase metals (specifically, Ti, $Ti_6A_{14}V$, and CoCrMo alloys) promote osteoblast adhesion, proliferation, differentiation, and mineralization

Various nanoparticles are added to dental materials used in prosthodontics. The properties of those prosthodontic materials are summarized in Table 6.1 [46].

6.5 Nanoceramics

At present, ceramic dental crown mainly includes alumina and zirconia ceramic. Nanoceramic refers to the ceramic material with nanoscale dimensions in the microstructures phase. Compared with the conventional ceramics, nanoceramics have a unique property, which makes it become the topic in the study of material science. Ceramic is brittle material; however, nanoceramic shows toughness and ductility. Compared to the conventional ceramics, nanoceramic has the superior mechanical properties, such as strength and hardness. The hardness, toughness, and strength of nanoceramics are four to five times higher than those of the traditional materials. Mitsunori et al. conducted a study on addition of silver nanoparticles in dental porcelain, and they showed that the fracture toughness and Vickers hardness of the porcelain were increased [47].

Tokushi et al. incorporated silver and platinum nanoparticles in dental porcelain and concluded that the addition of silver nanoparticles and platinum nanoparticles enhanced the mechanical properties of porcelain. The incorporation of silver and platinum nanoparticles increased Young's modulus and the fracture toughness of ceramics. Silver nanoparticles increased the fracture toughness more than platinum nanoparticles [48].

6.6 Luting Cements and Nanoparticles

Antibacterial activity of luting cement is advisable when luting crowns, bridges, inlay, onlay, and veneers. Silver nanoparticles are used in luting cements because of their advantage that they show antibacterial activity due to their higher surface-area-to-volume ratio. Yoshida et al. showed that resin composite cement incorporated with silver-containing materials had a long-term inhibitory effect against *S. mutans* and favorable mechanical properties [49].

6.7 Dental Adhesives

Dental adhesives are the materials used to increase adhesion or cohesion between two substances or between a material and tooth structure. Silane is added to dental adhesives in order to increase the cohesive strength. Since the adhesive liquid is not very viscous, the filler particles tend to settle out during storage which leads to inconsistency in their performance. To overcome this disadvantage, discrete silane-treated nanoparticles of silica or zirconia in the size range of 5–7 nm [50] are added to dental adhesives. According to a study by Silikas et al., no decrease in bond strength of dental adhesives after the incorporation of silica or zirconia nanoparticles was obtained [51].

6.8 Silver Nanoparticles in Maxillofacial Prosthesis

Maxillofacial prostheses are used to restore lost facial parts. *Candida albicans* infection remains a problem for maxillofacial prostheses made of silicone, as the microorganism causes degradation of the material and infection of the tissue. Fungal infection as a result of *Candida* adherence on the surface of the material is always a problem for maxillofacial patients [52].

The intaglio or anatomical surface of maxillofacial prostheses is exposed salivary and nasal secretions. That is why these prostheses are susceptible to biofilm formation and colonization by microorganisms, with subsequent degradation of the prosthesis [53]. The coating of silicone materials with AgNPs could be of great use to prevent fungal infection for patients with maxillofacial prostheses.

Zhala Meran et al. studied biocompatibility and antifungal properties of silicone facial prostheses coated with silver nanoparticles (AgNPs) . Medical-grade silicone disks were coated with 5 and 50 mg L21 dispersions of either AgNPs or $AgNO_3$. They showed that silicone prosthetic materials coated with AgNPs are biocompatible with fibroblast cells and showed antifungal properties. This study demonstrated that AgNPs-coated silicone elastomer had antifungal activity without appreciable adverse effects on human dermal fibroblast cells [54].

6.9 Silver Nanoparticles in Dental Implants

Titanium has been used for dental implants due to their good biocompatibility, better mechanical strength, and great corrosion resistance [55].

Failures in dental implants are due to lack of stability or misfit at the implant-abutment interface [56]. Micro-gap on the interface of two-piece implants with variable fluid flow at this interface leads to the bacterial infiltration and inflammatory cells that leads to the bone loss around this area. Some failures are due to the infection and inflammation [57–59].

Surface modification of titanium-based implant is done to avoid implant failure. Due to the interaction of the implant surface with bone and gingival and also, partial exposition of implant to the oral cavity that includes bacteria, fabricating a coating material with both antimicrobial activity and biocompatibility is important and leads to implant success [60]. Nanoparticles have also been used as coatings to the implant surface to improve osseointegration. Coating of bioinert nanomaterials like zirconium and aluminum over titanium implant surfaces may change the physical properties and enhance the osteogenic potential of implants [61]. It has been also reported that materials such as titanium nanotubes are able to increase the density of osteoblast cells on the implant that may lead to better implant stability [62].

Al_2O_3 nanoparticles help in osseointegration and in the healing process around dental implant. It helps in new bone formation with haversian canals, osteoblast, and osteocyte and also improves mechanical properties of dental implants [63, 64].

Silver nanoparticles coating helps to prevent the growth of bacteria and thus helps in ossiointegration. Titanium is a biocompatible material widely used in medical and dental implants. After implantation, titanium surfaces are prone to bacterial colonization that could lead to bacterial infection, inflammation, and finally lead to implant failure. Silver nanoparticles (AgNPs) are known as antibacterial agents, and their integration to titanium surfaces decreases the risk of implant failure.

Flores et al. studied the antibacterial potential of silver nanoparticles against *Pseudomonas aeruginosa*. Their data proved that the incorporation of AgNPs on Ti implants is an effective method to protect implant surface against pathogen colonization [65].

Zhao et al. studied titania nanotubes (TiO_2 NTs) incorporated with silver nanoparticles fabricated on Ti implants. The silver nanoparticles adhere to the wall of the TiO_2 NTs prepared by immersion in a silver nitrate solution followed by ultraviolet light radiation. Their results showed that TiO_2 NTs loaded with Ag nanoparticles can kill bacteria in the suspension [66].

6.10 Conclusion

Future development of materials used in prosthodontics has been recognized to be dependent mainly on the progress of nanoscience. Nanomaterials have been playing a significant role in basic scientific innovation and clinical technological changes of prosthodontics. Dental materials used in prosthodontics can be significantly improved after their scales were reduced from micron-size to nanosize. Nanoparticles have come up as one of the most effective antibacterial agents due to their large surface-area-to-volume ratios. They can be used as effective growth inhibitors of various microorganisms. Nanoparticles not only act as antibacterial agents but also improve mechanical and physical properties of the dental materials.

Although numbers of different nanoparticles are in use in prosthodontics, still AgNPs are extensively used and studied more. The uses of silver nanoparticles are many. They are already used for many commercial applications, certain medical applications, and in dentistry. Number of mechanical and physical properties of dental materials could be amended by adding AgNPs. In particular, suitable applications in acrylic resin, tissue conditioner, dental adhesives, dental porcelain, dental composites, dental cements, implants, and maxillofacial prosthesis are important.

References

1. Drexler KE (1980) Engines of creation, the coming era of nanotechnology. Anchor press, New York
2. Drexler KE (1981) Molecular engineering: an approach to the development of general capabilities for molecular manipulation. Proc Natl Acad Sci 78:5275–5278

3. Gumusderelioglu M, Mavis B, Karakecli A, Kahraman AS, Cakmak S, Tigli S, Demirtas TT, Aday S (2007) Doku mühendisliğinde nanoteknoloji. Bilim ve Teknik. 479 Ek
4. Kong LX, Peng Z, Li SD, Bartold M (2006) Nanotechnology and its role in the management of periodontal diseases. Periodontol 2000, 40:184–196. [PubMed]
5. Duke ES (2003) Has dentistry moved into the nanotechnology era? Compend Contin Educ Dent. 24:380–382. [PubMed]
6. Hamouda IM (2012) Current perspectives of nanoparticles in medical and dental biomaterials. J Biomed Res. 26(3):143–151
7. Goffeau A (2008) Drug resistance: the fight against fungi. Nature 452:541–542. [PubMed]
8. Ju-Nam Y, Lead JR (2008) Manufactured nanoparticles: an overview of their chemistry, interactions and potential environmental implications. Sci Total Environ 400:396–414. [PubMed]
9. Ahmad Z, Pandey R, Sharma S, Khuller GK (2005) Alginate nanoparticles as antituberculosis drug carriers: formulation development, pharmacokinetics and therapeutic potential. Ind J Chest Dis Allied Sci. 48:171–176. [PubMed]
10. Allaker RP (2010) The use of nanoparticles to control oral biofilm formation. J Dent Res 89:1175–1186. [PubMed]
11. Giertsen E (2004) Effects of mouth rinses with triclosan, zinc ions, copolymer, and sodium lauryl sulphate combined with fluoride on acid formation by dental plaque in vivo. Caries Res 38:430–435. [PubMed]
12. Rai M, Yadav A, Gade A (2009) Silver nanoparticles as a new generation of antimicrobials. Biotechnol Adv 27(1):76–83
13. Balazs DJ, Triandafillu K, Wood P, Chevolot Y, van Delden C, Harms H, Hollenstein C, Mathieu HJ (2004) Inhibition of bacterial adhesion on PVC endotracheal tubes by RF-oxygen glow discharge, sodium hydroxide and silver nitrate treatments. Biomaterials 25:2139–2151
14. Melaiye A, Youngs WJ (2005) Silver and its application as an antimicrobial agent. J Expert Opin Ther Pat 15:125–130
15. Humberto HH, Garza-Trevino EN, Ixtepan-Turrent L, Singh DK (2011) Silver nanoparticles are broad-spectrum bactericidal and virucidal compounds. J Nanobiotechnol 9:30. https://doi.org/10.1186/1477-3155-9-30
16. Pal S, Tak YK, Song JM (2007) Does the antibacterial activity of silver nanoparticles depend on the shape of the nanoparticle? A study of the Gram-negative bacterium *Escherichia coli*. Appl Environ Microbiol 73:1712–1720
17. Craig RG (1996) Restorative dental materials, 10th edn. Mosby, St Louis, p 500
18. Pereira-Cenci T, Del Bel Cury AA, Crielaard W, Ten Cate JM (2008) Development of Candida-associated denture stomatitis: new insights. J Appl Oral Sci 16:86–94
19. Radford DR, Sweet SP, Challacombe SJ, Walter JD (1998) Adherence of *Candida albicans* to denture-base materials with different surface finishes. J Dent 26:577–583
20. Vallentini F, Luz MS, Boscato N, Pereira Sensi T (2013) Biofilm formation on denture liners in a randomised controlled in situ trial. J Dent 41:420–427
21. Wady AF, Machado AL, Zucolotto V, Zamperini CA, Berni E (2012) Evaluation of *Candida albicans* adhesion and biofilm formation on a denture base acrylic resin containing silver nanoparticles. J Appl Microbiol 112:1163–1172
22. Oliveira HF, Silva-Lovato C, De Souza RF, Cruz PC, Freitas-Pontes De, Watanabe E (2009) Effect of three methods for cleaning dentures on biofilms formed in vitro on acrylic resin. J Prosthodont. 18:427–431
23. Jorgensen E (1990) Candida-associated denture stomatitis and angular cheilitis. In: Samaranayake LP, MacFarlane TW (eds) Oral Candidosis. Butterworth, London, UK
24. Ramage G, Tomsett K, Wickers BL, López-Ribot J, Redding SW (2004) Denture stomatitis: a role for Candida biofilms. Oral Surg Oral Med Oral Pathol Oral Radiol Endod 98:53–59
25. Pusateri C, Monaco E, Edgerton M (2009) Sensitivity of Candida albicans biofilm cells grown on denture acrylic to antifungal proteins and chlorhexidine. Arch Oral Biol 54:588–594
26. Redding S, Bhatt B, Rawls HR (2009) Inhibition of *Candida albicans* biofilm formation on denture material. Oral Surg Oral Med Oral Pathol Oral Radiol Endod 107:669–672
27. Arendorf TM, Walker DM (1987) Denture stomatitis: a review. J Oral Rehabil 14:217–227

28. Baena-Monroy T, Moreno-Maldonado V, Franco-Martínez F, Aldape-Barrios B, Quindós G, Sánchez-Vargas LO (2005) *Candida albicans, Staphylococcus aureus* and *Streptococcus mutans* colonization in patients wearing dental prosthesis. Med Oral Patol Oral Cir Bucal. 10(Suppl 1):e27–e39
29. Webb BC, Thomas CJ, Willcox MD, Harty DW, Knox KW (1998) Candida associated denture stomatitis. Aetiology and management: a review. Part 2. Oral diseases caused by Candida species. Aust Dent J 43:160–166
30. Freitas RA Jr (2005) Nanotechnology, nanomedicine and nanosurgery. Int J Surg. 3(4):243–246
31. Li Z, Sun J, Lan J, Qi Q (2014) Effect of a denture base acrylic resin containing silver nanoparticles on *Candida albicans* adhesion and biofilm formation. Gerodontology 25(2):204–207
32. Acosta-Torres LS, Mendieta I, Nuñez-Anita RE, Cajero-Juárez M, Castaño VM (2012) Cytocompatible antifungal acrylic resin containing silver nanoparticles for dentures. Int J Nanomed 7:4777–4786
33. Monteiro DR, Gorup LF, Takamiya AS, de Camargo ER, Filho ACR, Barbosa DB (2012) Silver distribution and release from an antimicrobial denture base resin containing silver colloidal nanoparticles. J Prosthodon 21(1):7–15
34. Hamada ZM, Kusai B (2015) Effect of a denture base acrylic resin containing silver nanoparticles on Candida albicans adhesion and biofilm formation. Eur J Dent 9(2):207–212
35. Castro DT, Holtz RD, Alves OL, Watanabe E, Valente ML, Silva Andréa Cândido dos Reis (2014) Development of a novel resin with antimicrobial properties for dental application. J Appl Oral Sci 22(5):442–449
36. Hua Y, Gu L, Watanabe H (2013) Micromechanical analysis of nanoparticle-reinforced dental composites. Int J Eng Sci 69–76
37. Gonzalez JB (1977) Use of tissue conditioners and resilient liners. Dent Clin North Am 21(2):249–59
38. Essentials of complete denture Prosthodontics, 2nd ed. Sheldon Winkler page 81
39. Abe Y, Ishii M, Takeuchi M, Ueshige M, Tanaka Akagawa Y (2004) Effect of saliva on an antimicrobial tissue conditioner containing silver-zeolite. J Oral Rehabil 31:568–573
40. Matsuura T, Abe Y, Sato Y, Okamoto M, Ueshige M, Akagawa Y (1997) Prolonged antimicrobial effect of tissue conditioners containing silver-zeolite. J Dent 25:373–377
41. Nikawa H, Yamamoto T, Hamada T, Rahardjo MB, Murata H, Nakanoda S (1997) Antifungal effect of zeolite incorporated tissue conditioner against Candida albicans growth and/or acid production. J Oral Rehabil 24:350–357
42. Nam KY (2011) In vitro antimicrobial effect of the tissue conditioner containing silver nanoparticles. J Adv Prosthodont 3(1):20–24
43. Rao KVP, Kumar JS (2013) Nanotechnology in dentistry 36:56–59
44. Hamissi H, Hamissi Z, Hamissi ZH (2016) Nanotechnology in dental practice: current achievement and prospects. Acta Medica Mediterranea 32:1441–1448
45. Ginjupalli K, Alla RK, Tellapragada C, Gupta L, Upadhya Perampalli N (2016) Antimicrobial activity and properties of irreversible hydrocolloid impression materials incorporated with silver nanoparticles. J Prosthet Dent 115(6):722–728
46. Wang W, Liao S, Zhu Y, Liu M, Zhao Q, Fu Y (2015) Recent applications of nanomaterials in prosthodontics. J Nanomaterials 1–15
47. Mitsunori U, Masakazu K, Nobukazu W, Yutaka D (2013) Effects of adding silver nanoparticles on the toughening of dental porcelain. J Prosthet Dent 109(4):241–247
48. Tokushi F, Mitsunori U, Hajime I, Masakazu K, Nobukazu W, Yutaka D (2012) Addition of platinum and silver nanoparticles to toughen dental porcelain. Dent Mater J 31(5):711–716
49. Yoshida K, Tanagawa M, Matsumoto S, Yamada T, Atsuta M (1999) Antibacterial activity of resin composites with silver containing materials. Euro J Oral Sci 107(4):290–296
50. Davidson RS, Kolb BU, Anderson DB, Higgins JA, Hendrickson MJ, Brady JT (2007) Zirconia particles. US patent 7; 241:437
51. Silikas N, Masouras K, Satterthwaite J, Watts DC (2007) Effects of nanofills in adhesives and aesthetic properties of dental resin composites. Int J Nano Biomater 1(2):116–127

52. Mattos BSC, de Sousa AA, de Magalhães MHCG, André M, Brito Dias R (2009) *Candida albicans* in patients with oronasal communication and obturator prostheses. Braz Dent J 20(4) (Ribeirão Preto)
53. Kurtulmus H, Kumbuloglub O, Özcana M, Ozdemir G, Vural C (2010) Candida albicans adherence on silicone elastomers: Effect of polymerisation duration and exposure to simulated saliva and nasal secretion. Dent Mater 26:76–82
54. Meran Zhala, Besinis Alexandros, De Peralta Tracy, Handy Richard D (1993) Antifungal properties and biocompatibility of silver nanoparticle coatings on silicone maxillofacial prostheses in vitro. J Biomed Mat Res B Applied Biomat. https://doi.org/10.1002/jbm.b.33917
55. Zhang P, Zhang Z, Li W (2013) Antibacterial TiO_2 coating incorporating silver nanoparticles by microarc oxidation and ion implantation. J Nanomaterials. https://doi.org/10.1155/2013/542878
56. Jansen VK, Conrads G, Richter E-J (1997) Microbial leakage and marginal fit of the implant-abutment interface. Int J Oral Maxillofac Implants 12(4):527–540
57. Paulo da Silva-Neto J, de Arruda Antônio, Nóbilo M, Paulo Amante Penatti M, Simamoto C Jr, Domingues das Neves F (2012) Influence of methodologic aspects on the results of implant-abutment interface microleakage tests: a critical review of in vitro studies. Int J Oral Maxiillofac Implants 27:793–800
58. Broggini N, McManus L, Hermann J, Medina R, Schenk R, Buser D, Cochran D (2006) Peri-implant inflammation defined by the implant-abutment interface. J Dent Res 85:473–478
59. Oshida Y, Tuna EB, Aktören O, Gençay K (2010) Dental implant systems. Int J Mol Sci 11(4):1580–1678
60. Chidambaranathan AS, Mohandoss K, Balasubramaniam MK (2016) Comparative evaluation of antifungal effect of titanium, zirconium and aluminium nanoparticles coated titanium plates against *C. albicans*. J Clin Diagn Res 10(1):56–59
61. Shokuhfar T, Hamlekhan A, Chang J-Y, Choi CK, Sukotjo C, Friedrich C (2014) Biophysical evaluation of cells on nanotubular surfaces: the effects of atomic ordering and chemistry. Int J Nanomed 9:3737–3748
62. Wood NJ, Jenkinson HF, Davis SA, Mann S, O'Sullivan DJ, Barbour ME (2015) Chlorhexidine hexametaphosphate nanoparticles as a novel antimicrobial coating for dental implants. J Mater Sci - Mater Med 26:1–10
63. Al-Haik M, Hanson C, Luhrs C, Tehrani M, Phillips J, Miltenberger S (2009) Mechanical performance of dental fillers based on alumina nanoparticles. In: Proceedings of the SEM Annual Conference, pp 1–4
64. Zhanhu G, Tony P, Oyoung C, Ying W, Hahn HT (2006) Surface functionalized alumina nanoparticle filled polymeric nanocomposites with enhanced mechanical properties. J Mater Chem 16:2800–2808
65. Flores CY, Diaz C, Rubert A, Benítez GA, Moreno MS, Fernández Lorenzo de Mele MA, Salvarezza RC (2010) Spontaneous adsorption of silver nanoparticles on Ti/TiO_2 surfaces. Antibacterial effect on *Pseudomonas aeruginosa*. J Colloid Interface Sci 350(2):402–8
66. Zhao L, Wang H, Huo K, Cui L, Zhang W, Ni H, Zhang Y, Wu Z, Chu PK (2011) Antibacterial nano-structured titania coating incorporated with silver nanoparticles. Biomaterials 32(24):5706–5716

Chapter 7
Nanomaterials: The Changing Phase of Prosthodontics

Himanshu Aeran and Jyotsna Seth

7.1 Introduction

Prosthodontics is a dental specialty pertaining to the diagnosis, treatment plans, rehabilitation and maintenance of oral function, comfort, appearance and health of patients with clinical conditions associated with missing or deficient teeth or oral and maxillofacial tissues using biocompatible substitutes. Due to the improvement in living standards, awareness, and promotion of oral health care, a prosthodontics with newer concepts and technologies received widespread attention.

Nano science primary deals with synthesis, characterization, exploration, and exploitation of nanostructured materials. These materials are characterized by at least one dimension in the nanometer range. A nanometer (nm) is one billionth of a meter, or 10^{-9} m.

Nano sized materials currently are being used in numerous industries; e.g., carbon black particles make rubber tires wear-resistant, nanofibers are used for insulation and reinforcement of composite, iron oxides create the magnetic material used in disk drives and audio–video tapes, nano-zinc oxides and titanium are used as sun blocks for ultraviolet rays. Nanomaterials are cornerstones of nanoscience and nanotechnology. Nanostructure science and technology is a broad and interdisciplinary area of research and development activity that has been growing explosively worldwide in the past few years. It has the potential for revolutionizing the ways in which materials and products are created and the range and nature of functionalities that can be accessed. New materials with outstanding mechanical properties are rapidly being developed for dental applications [1].

H. Aeran (✉) · J. Seth
Department of Prosthodontics and Oral Implantology,
Seema Dental College and Hospital,
Rishikesh 249201, Uttarakhand, India
e-mail: drhimanu@yahoo.com

R. S. Chaughule (ed.), *Dental Applications of Nanotechnology*,
https://doi.org/10.1007/978-3-319-97634-1_7

Over the past few decades, inorganic nanoparticles, whose structures exhibit significantly novel and improved physical, chemical, and biological properties, due to their nanoscale size, have elicited much interest [2]. Nanophasic and nanostructured materials are attracting a great deal of attention because of their potential for achieving specific processes and selectivity, especially in biological and pharmaceutical application. The use of nanoparticles has become very popular in design and development of many dental materials so as to improve their clinical, physical, and mechanical properties [3].

It has been shown that the performances of many biomaterials used in prosthodontics enhanced after their scales were reduced by nanotechnology from micron into nanosize. After the addition of nanomaterials various nanocomposites composed of nanomaterials and traditional metals, ceramic resin or other matrix which have been widely used in prosthodontics shows significant improvement in properties like modulus of elasticity, surface hardness, strength and filler loading [4].

7.1.1 Approaches in Nanotechnology

Nanotechnology works mainly by four approaches [5] as follows:

- Bottom-up approach: This approach arranges smaller components into more complex assemblies. This begins by designing and synthesizing custom-made molecules that have the ability to self-assemble or self-organize into higher-order structures. This is a chemical approach.
- Top-down approach: This approach creates smaller devices by using larger ones to direct their assembly. This approach requires larger amounts of materials and can lead to waste if excess material is discarded. Here, larger materials are patterned and carved down to make nanoscale structures in precise patterns. Materials reduced to the nanoscale can suddenly show very different properties, enabling unique applications. This is a physical method. Figure 7.1 shows the methods of chemical as well as physical approach.
- Functional approach: This approach develops components of the desired functionality without much importance to their assembly or structure.
- Speculative approach: This approach often takes a big picture view of nanotechnology, with more emphasis on its societal implications than the details of how such inventions could actually be created.

7.1.2 Synthesis of Nanomaterials in Prosthodontics

Nanomaterials, being a billionth of a meter, indeed allow us to think in both the "bottom-up" and the "top-down" approaches to synthesize nanomaterials, i.e., either to assemble atoms together or to disassemble bulk solids into finer pieces until they

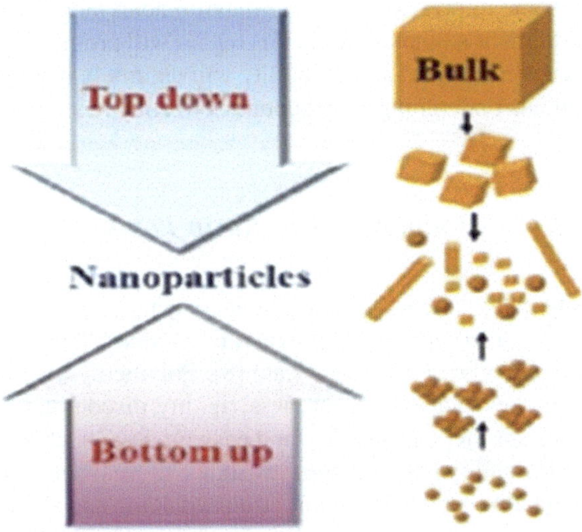

Fig. 7.1 Schematic illustration of the preparative methods of nanoparticles

are constituted of only a few atoms. This domain is a pure example of interdisciplinary work encompassing physics, chemistry, and engineering up to medicine.

The nanomaterials in prosthodontics are mainly used in top-down approach. In top-down approach, various different techniques are used. Mechanical grinding is one of these techniques where the material is prepared not by cluster assembly but by the structural decomposition of coarser-grained structures as the result of sever plastic deformation. This has become a popular method to make nanocrystalline materials because of its simplicity, the relatively inexpensive equipment needed, and the applicability to essentially the synthesis of all classes of materials.

The other approach used in top-down approach is wet chemical synthesis. In top-down method, single crystal is etched in an aqueous solution for producing nanomaterials, e.g., the synthesis of porous silicon by electrochemical etching [6].

Nanomaterials have been developed promptly, and some researches of nanomaterials have been carried out in prosthodontics. Many of the current dental materials are available through nanocrystallization to improve their original performance and play continuously a key role in oral applications. Research of nanotechnology in dental materials is mainly focussed on two ways [7]: one is the preparation of new inorganic nanoparticles, and the other is to modify the surface with inorganic nanofillers and thereby to develop ultralow shrinkage rate of repair resin.

7.2 Application of Nanodentistry in Prosthodontics

In prosthodontics, various types of nanomaterials are added to improve the properties of commonly used materials like resin denture base material, ceramics, polyvinyl siloxane impression material, maxillofacial materials, luting cements. Research on addition of nanoparticles in this regard will promote the usage of such materials with greater efficiency, and durability will definitely be of great advantage to dentists and patients undergoing prosthodontics treatment.

7.2.1 Nanoparticles in Resin Polymethyl Methacrylate

Polymethyl methacrylate is obtained by polymerization of acrylic acid and its esters. The main component is polymethyl methacrylate, also containing small amount of ethylene glycol dimethacrylate [8]. Polymethyl methacrylate has good mechanical properties such as high hardness, rigidity, discontinuity deformation, biological properties, aesthetic properties, and easy processing characteristics. Main disadvantages are instability of color, poor resistance to wear and tear volume shrinkage after polymerization, oral mucosa initiation and aging and staining or discoloration that easily occurs. Nanoparticles are added to polymethyl methacrylate as antimicrobial agents to increase the viscoelastic properties of resin [4].

Epidemiological studies report that approximately 1% of removable denture wears suffer from denture stomatitis and oral pathological conditions like denture stomatitis mainly caused by adherence of biofilm onto the denture base [9]. Incorporation of nanoparticles into the denture base materials is mainly in the form of silver and platinum nanoparticles as an effective antimicrobial agent [10]. Some researchers showed that addition of metal nanoparticles such as titanium oxide, ferric oxide, and silver to polymethyl methacrylate materials could increase the surface hydrophobically to reduce biomolecular adherence [11]. Polymethyl methacrylate containing nanoparticles shared a lowered *Candida albicans* (*C. albicans*) cell adhesion and a lower porosity, compared to standard polymethyl methacrylate [12]. Li et al. [13] evaluated the effect of denture base resin containing silver nanoparticles on *C. albicans* adhesion and biofilm formation and an inhibitory effect on adhesion and biofilm formation by denture base resin containing nanosilver discovered especially at a higher concentration.

Monterio et al. [14] evaluated a denture base resin containing silver nanoparticles through morphological analysis to check the distribution and dispersion of the particles in the polymer and by testing the silver release in deionized water at different time periods. The silver nanoparticles were incorporated in the polymethyl methacrylate denture resin to attain an effective antimicrobial material to help control common infections involving oral mucosal tissues in complete denture wearer, because nanocomposites have good efficacy against *C. albicans*.

Nam [15] did a study to evaluate the antimicrobial property of a denture base as characteristics by the synthesis of a modified polymethyl methacrylate denture acrylic loading platinum nanoparticles. The study showed that platinum nanoparticles were successfully loaded and uniformly immobilized into polymethyl methacrylate denture acrylic with a proper thermal stability and similar surface morphology and platinum nanoparticles expressed significant bacterial anti-adherent effect rather than bactericidal effect above 50 mg/L.

Many products used for dental restoration have been produced from acrylic resins based on heat-cured polymethyl methacrylate, due to its optical properties, biocompatibility, and aesthetics. However, it has a long-standing drawback that is lack of strength particularly under fatigue failure inside the mouth and also shows low abrasion resistance and microbial adhesion onto polymethyl methacrylate to long-term polymethyl methacrylate wearers [16].

Carbon nanotubes and carbon nanofibrils have been used as reinforcement or additives in various materials to improve the properties of the matrix materials. Cooper et al. [17] prepared the polymethyl methacrylate matrix with different quantities of carbon nanotubes or carbon nanofibrils using a dry powder mixing method. It showed that impact strength of the composites is significantly improved by even small amounts of single-wall nanotubes.

Acosta-Torres et al. [11] reinforced polymethyl methacrylate with titanium dioxide and ferric oxide nanoparticles for improving coloring and antimicrobial property of polymethyl methacrylate. These results indicated that nanostructured metal coloring additives produce nontoxic hybrid materials with antimicrobial property for prosthetic applications.

Clinical Implication: With addition of nanoparticles in resin polymethyl methacrylate, the resultant material achieved is with better color stability, good resistance to wear and tear, lesser volume shrinkage after polymerization, and better antimicrobial efficacy.

7.2.2 *Nanoceramic Materials*

Dental ceramics is the most aesthetic material used in dentistry. These materials are a part of systems designed with the purpose of producing dental prostheses that in turn are used to replace missing or damaged dental structures [18]. They are used because of their high aesthetic value, high strength, suitable color, and low thermal and electrical conductivity. Various materials such as hydroxyapatite ceramic, glass ceramic, alumina, and zirconia are used for ceramic materials. Alumina ceramics have good aesthetics, high gloss, chemical stability, wear resistance, high hardness, good biocompatibility, but the biggest drawback is porcelain crack [19]. Due to its low tensile strength, brittleness of ceramic directly influence and limit the use of traditional ceramic materials and scientists believe that nanoceramics may offer better specific improvements. Nanoceramics also may meet the need for translucency of dental restoration. Nanoceramics refer to ceramic material with nanoscale dimensions in the microstructured phase. As compared to conventional ceramic, nanoceramics have superplasticity. Ceramic is brittle material; however, nanoceramics shows good toughness and ductility. In nanoceramics, the atoms are very easy to migrate under the conditions of force deformation. Secondly, compared

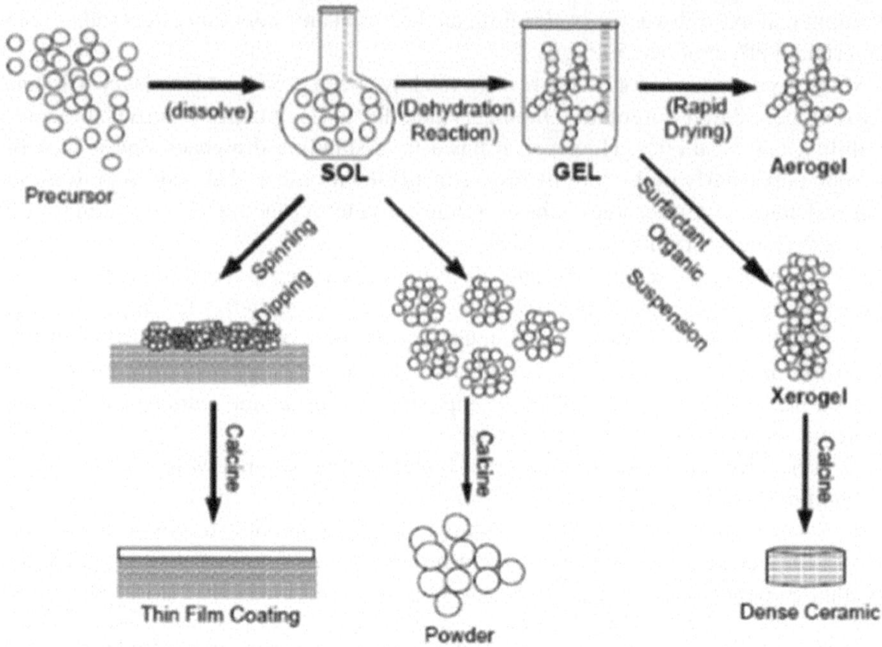

Fig. 7.2 Schematic representation of solgel process of synthesis of nanomaterials

to conventional ceramics, nanoceramics have the superior mechanical properties, such as strength and hardness increasing significantly [20]. The studies show that microhardness of nano-titanium dioxide ceramics is 13,000 kN/mm^2, while that of ordinary titanium dioxide ceramics is lower than 2000 kN/mm^2.

Li et al. [21] reported the different physical properties of nano-zirconium dioxide ceramic material from the conventional ones. Conventional zirconium dioxide hardness is around 1500, and its fracture toughness is quite low, and so break easily occurs in processing. However, nano-zirconia ceramic hardness reaches more than 1750, increased by about 20%. Not only its hardness is improved, but also the fracture toughness is increased. Wang et al. [22] reported the influence of nano-ZrO_2 control on the mechanical properties and found that composite has better toughness.

Another form of glass ceramic which is based on lithium disilicate with lack of mechanical properties used commonly in dental veneers and crowns, but due to insufficient mechanical properties, failures have been reported. Person et al. [23] used a solgel method to produce glass ceramic in zirconia–silica system with nanosized grains, which was found to be translucent, with a transmittance of over 70% and possessed excellent corrosion resistance as shown in Fig. 7.2. The hardness of nanomaterial increased more than the conventional lithium disilicate. Carbon nanotubes nowadays are widely used as nanofiller in nanomaterial to improve mechanical and electronic properties [24].

An et al. [25] produced alumina–carbon nanotube composites by hot pressing and studied mechanical and tribological properties of alumina–carbon nanotube composites. They concluded that wear and mechanical properties were enhanced by 30% by adding 0–4% carbon nanotube content.

Clinical Implication: Resultant nanoceramics show excellent translucency and superior mechanical properties such as strength, hardness, corrosion resistance, toughness, and ductility.

7.2.3 Nanoparticles in Impression Material

Improved physical properties of polyvinyl siloxane impression material are done by addition of nanosized fumed silica. Choi et al. [26] have used dental impression materials formulated with loading combination of six types of fillers including nanosized fumed silica. They concluded that the test group in which quarter of quartz was replaced with fumed silica showed the most ideal working and setting time for clinical use.

Combining the nanosized fumed silica, it was effective in increasing the viscosity, tensile strength, and maximum strain producing better detail reproduction of the material. The best modulation of physical properties in polyvinyl siloxane was seen after combining fillers during the formulation.

Clinical Implication: Addition of nanoparticles improved impression material with most ideal working and setting time along with the effective viscosity, tensile strength, and maximum strain for best clinical use to record better tissue details.

7.2.4 Nanoparticles in Maxillofacial Materials

Pesqueira et al. [27] evaluated the effect of disinfection and accelerated aging on the dimensional stability and detail reproduction of a facial silicone with different types of nanoparticles. Samples were divided into three groups: colorless nanoparticles, pigmented make-up powder nanoparticles, and pigmented ceramic powder nanoparticles. Half specimens were disinfected, and half were kept neutral; all specimens were subjected to accelerated aging. Tests for both dimensional stability and detail reproduction were performed. Chemical disinfection and also accelerated aging affected adversely the dimensional stability of facial silicone with significant results. The silicone's detail reproduction was not affected by these factors regardless of nanoparticles type, disinfection, and accelerated aging.

Clinical Implication: Maxillofacial material with nanoparticles produces better detail reproduction of tissue.

7.2.5 *Nanoparticles in Hybrid Resin Luting Cement*

Habekost et al. [28] investigated the influence of nanoparticle loading level on properties of experimental hybrid resin luting agent. The properties evaluated were flexural strength (σ), modulus (E_+), Knoop hardness number (KHN), and film thickness (FT). It was concluded that modest incorporation of nanoparticles may improve the properties of resin luting materials. Nanofiller mass fractions above 25% should however be avoided because they may be detrimental to the properties of the resin luting cement (Fig. 7.3).

Clinical Implication: Nanoparticles in hybrid resin luting cement improved the properties of the cement like flexural strength, modulus, Knoop hardness, and reducing film thickness.

7.2.6 *Nanometal Material and Dental Implants*

Most metal stents of partial denture are made by using cobalt–chromium alloy or cobalt–chromium–molybdenum alloy and titanium alloy [29]. The initial cobalt-based alloy is cobalt–chromium binary alloy and is then modified to cobalt–chromium–tungsten and later developed into cobalt–chromium–molybdenum alloy. Its mechanical properties and corrosion resistance are better than stainless steel or gold alloy [30].

Another metal material that is nowadays used in dentistry [31] is titanium (Fig. 7.4). With excellent mechanical properties and good biological security, biological integration of titanium is usually satisfactory but some patients are prone to allergies, causing skin, mucous membrane inflammation [32]. Biological integration

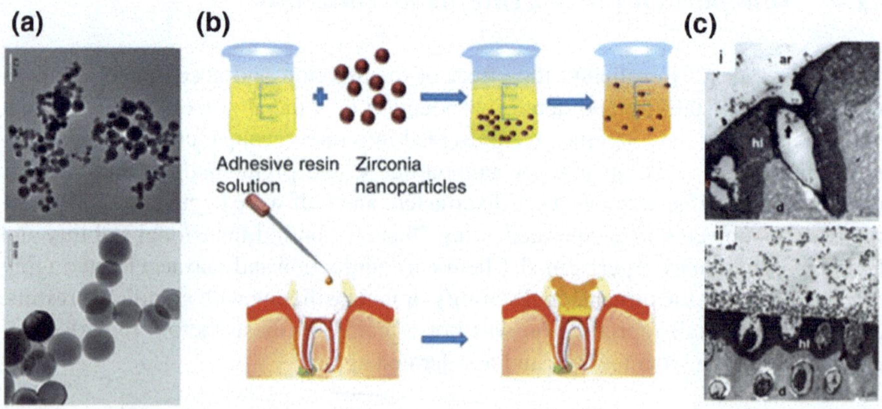

Fig. 7.3 Addition of nanoparticles in resin luting cement

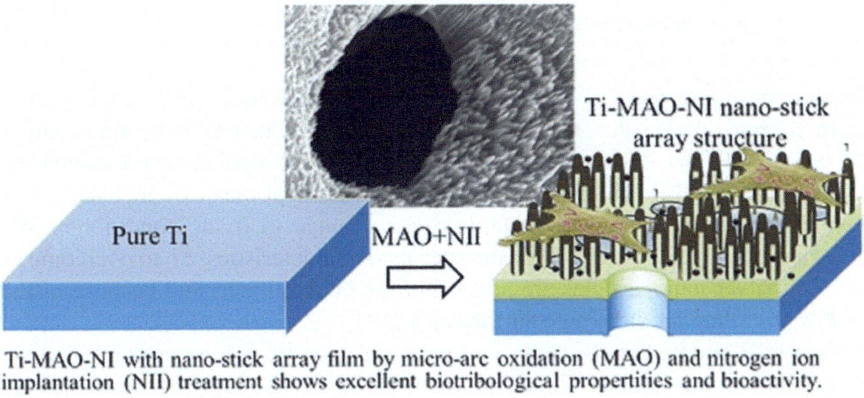

Fig. 7.4 Addition of nanoparticles in titanium

with surrounding host tissue is one of the most important factors for long-term success of dental implants, and modification of titanium implant surface improves this integration. Dorkhan et al. [33, 34] modified surface of titanium surface by anodic oxidation into nanoscales with pores in the 50 nm concluded that these implants with nanostructured surface improve surrounding host tissue cell adherence.

Other property to improve dental implants was wear resistance. To overcome wear resistance, Sathish et al. [35] coated a novel nanostructured bilayered zirconium dioxide/aluminum trioxide–titanium dioxide on biomedical titanium–niobium–zirconia alloy. The bilayered coating exhibits 200–500 fold better in the wear resistance with lower porosity.

Lan et al. [36] prepared a nanotextured titanium surface using a chemical etching technique and concluded that the effects of nanotextured titanium made the surface rough of pure titanium which further increased pre-osteoblastic cells adherence, proliferation, differentiation, and mineralization leading to increase in osseointegration of dental implant to underlying bone.

Yao et al. [37] created nanometer surface features on titanium and titanium–aluminum–vanadium implants by anodization and concluded that osteoblast adhesion was enhanced on the anodized metal substrates compared to unanodized substrates.

Webster and Ejiofor [38] further provided the evidence of increased osteoblast adhesion on titanium, titanium–aluminum–vanadium and cobalt–chromium–molybdenum compacts with nanometer compared to conventionally sized metals.

Clinical Implication: Nanometals in implants show excellent physical properties like corrosion resistance, wear resistance, adhesion strength, lower porosity along with increased osteoblast adhesion which ultimately increases the longevity of dental implants.

7.2.7 Nanoparticles in Tissue Conditioner

Tissue conditioner used to enhance the recovery of denture bearing tissues from trauma, damage, or residual ridge resorption are usually caused by ill-fitting dentures. These materials are degenerated with time and are susceptible to colonization by microorganisms [39]. Tissue conditioners could be kept clean by mechanical and chemical methods, but these methods cause damage to tissue conditioners [40]. Silver has been well known for its antimicrobial characteristics. So to overcome this problem, silver nanoparticles are added in tissue conditioner, and because of their smaller size, they provide large surface area [41].

Nam [42] conducted the study and modified tissue conditioner combined with silver nanoparticles that displayed antimicrobial properties against *Staphylococcus aureus*, *Staphylococcus mutans* at 0.1% and *C. albicans* at 0.5% after a 24 h and 72 h incubation periods.

Clinical Implication: As disinfecting tissue conditioner damages the material, upcoming nanoparticles added antimicrobial activity in this material for improving its clinical use.

7.3 Summary

Latest research progress on applications of nanometals, nanoceramics, nanoresins, and other nanomaterials in prosthodontics shows that materials in prosthodontics can be significantly improved after their scales were reduced from micron to nanosize by nanotechnology. The performances of composites can also be enhanced by adding appropriate nanomaterials. Nanomaterials have been playing a significant role in basic scientific innovation and clinical technological change of prosthodontics.

References

1. Petrie CS, Walker MP, Williams K (2005) A survey of U.S Prosthodontics and dental schools on the current materials and methods for final impressions for complete denture prosthodontics. J Prosthodont 14(4):253–262
2. Mehra M, Farhad V, Robert W (2014) A complete denture impression technique survey of post doctoral prosthodontics programs in the United States. J Prosthodont 23:320–327
3. Cuy JL, Mann AB, Livi KJ, Teaford MF, Wehs TP (2002) Nano indentation mapping of the mechanical properties of human molar tooth enamel. Arch Oral Biol 47(4):281–291
4. Saavedra G, Valandro LF, Leite FP et al (2007) Bond Strength of acrylic teeth to denture base resin after various surface conditioning methods before and after thermocyclic. Int J Prosthodont 20(2):199–201
5. Jhaveri HM, Balaji PR (2005) Nanotechnology: the future of dentistry. J Ind Prosthodont Soc 5:15–17
6. Whitesides GM, Love JC (2001) The art of building small. Sci Am 285:33–41

7. Trapalis CC, Keivanidis P, Kovdas G et al (2003) TiO_2 (Fe^{3+}) nanostructured thin films with antibacterial properties. Thin Solid Films 433(1–2):186–190
8. Xia Y, Zhang F, Xie H, Gu N (2008) Nanoparticle-reinforced resin-based dental composite. J Dent 36(6):450–455
9. Gendreau L, Lowey ZG (2011) Epidemiology and etiology of denture stomatitis. J Prosthodont 20(4):251–260
10. Yamauchi M, Yamamoto K, Wakabayashi M, Kawano J (1990) In vitro adherence of microorganisms to denture base resin with different surface texture. Dent Mater J 9(1):19–24
11. Acosta-Torres LS, Lpez-Marin LM, Nunez-Anita RE, Hernandez-Padron G, Castano VM (2011) Biomaterials metal-oxide nanoparticles: nanotechnology improvement of conventional prosthetic acrylic resins. J Nanomater 2011:Article ID 941561)
12. Acosta-Torres LS, Lpez-Marin LM, Nunez-Anita RE, Cajero-Juarez M, Castano VM (2012) Cytocompatible antifungal acrylic resin containing silver nanoparticles for dentures. Int J Nanomed 7:4777–4786
13. Li Z, Sun J, Lan J, Qi Q (2014) Effect of a denture base acrylic resin containg silver nanoparticles on *Candida albicans* adhesion and biofilm formation. Gerodontology
14. Monterio DR, Group LF, Takamiya AS, de Camargo ER, Filho AC, Barbosa DB (2012) Silver distribution and release from an antimicrobial denture base resin containing silver colloidal nanoparticles. J Prosthodont 21(1):7–15
15. Nam KY (2014) Characterization and bacterial anti adherent effect on modified PMMA denture acrylic resin containing platinum nanoparticles. J Adv Prosthodont 6(3):207–214
16. Jagger DC, Harrison A, Jagger RG, Milward P (2003) The effect of the addition of poly(methyl methacrylate) fibers on some properties of high strength heat cured acrylic resin denture base material. J Oral Rehabil 30(3):231–235
17. Cooper CA, Ravich D, Lips D, Mayer JJ, Wagner HD (2002) Distribution and alignment of carbon nanotubes and nanofibrils in a polymer matrix. Compos Sci Technol 62(7-8):1105–1112
18. Roberts HW, Berzins DW, Morre BK, Charlton DG (2009) Metal ceramic alloys in dentistry: a review. J Prosthodont 18(2):188–194
19. Akora T, Ucar Y, Tukay A, Balkaya MC, Brantley WA (2008) Comparison of the bond strength of laser-sintered and cast base metal dental alloys to porcelain. Dent Mater 24(10):1400–1404
20. Raj V, Mumjitha MS (2014) Formation and surface characterization of nanostructured Al_2O_3–TiO_2 coatings. Bull Mater Sci 37(6):1411–1418
21. Li CH, Hou YL, Liu ZR, Ding YC (2011) Investigation into temperature field of nano-zirconia ceramics precision grinding. Int J Abras Technol 4(1):77–89
22. Wang GK, Kong H, Bao KJ, Lv JJ, Gao F (2006) Influence on mechanical properties and microstructure of nano-zirconia toughened alumina ceramics with nano-zirconia content. West China J Stomatol 24(5)
23. Persson C, Unosson E, Ajaxon I, Engstrand J, Engqvist H, Xia W (2012) Nano grain sized zirconia-silica glass ceramics for dental applications. J Eur Ceram Soc 32(16):4105–4110
24. Thostenson ET, Ren Z, Chou T-W (2001) Advances in the science and technology of carbon nanotubes and their composites: a review. Compos Sci Technol 61(13):1899–1912
25. An JW, You DH, Lim DS (2003) Tribological properties of hot-pressed alumina-CNT composites. Wear 255(1-6):677–681
26. Choi JH, Kim MK, Woo HG, Song HJ, Park YJ (2011) Modulation of physical properties of polyvinylsiloxane impression materials by filler type combination. J Nanosci Nanotechnol 11(2):1547–1550
27. Pesqueira AA, Geiato MC, Dos Santos DM, Haddad MF, Moreno A (2012) Effect of disinfection and accelerated ageing on dimensional stability and detail reproduction of a facial silicone with nanoparticles. J Med Eng Technol 36(4):217–221
28. Habekost LV, Camacho GB, Lima GS, Oligari FA, Cubas GB, Moraes RR (2012) Nanoparticle loading level and properties of experimental hybrid resin luting agents. J. Prosthodont 21(7):540–545
29. Tang N, Li YP, Kurosu S, Matsumoto H, Koizumi Y, Chiba A (2011) Interfacial reactions between molten Al and a Co–Cr–Mo alloy with and without oxidation. Corros Sci 53(12):4324–4326

30. Tang N, Li Y, Koizumi Y, Chiba A (2014) Nitriding of Co–Cr–Mo alloy in nitrogen. Mater Chem Phys 145(3):350–356
31. Lombardo GHL, Nishioka RS, Souza ROA et al (2010) Influence of surface treatment on the shear bond strength of ceramics fused to cobalt-chromium. J Prosthodont 19(2):103–111
32. Preeti M, Hilgert E, Bottino MA, Avelar RP (2004) Evaluation of the shear bond strength of the union between two Co–Cr alloys and a detail ceramic. J Appl Oral Sci 12(4):280–284
33. Dorkhan M, Yucel- Lindberg T, Hall J, Svensater G, Davies JR (2004) Adherence of human oral keratinocytes and gingival fibroblasts to nanostructured titanium surfaces. BMC Oral Health 12(4):280–284
34. Dorkhan M, Hall J, Uvdal P, Sandell A, Svensater G, Davies JR (2014) Crystalline anatase-rich titanium can reduce adherence of oral streptococci. Biofouling
35. Sathish S, Geetha M, Aruna ST, Balaji N, Rajam KS, Asokamani R (2011) Sliding wear behaviour of plasma sprayed nanoceramic coatings for biomedical applications. Wear 271(5):934–941
36. Lan GB, Li M, Zhang Y (2013) Effects of a nano-textured titanium surface on murine pre-osteoblasts. Orthop J China 21(23)
37. Yao C, Perla V, Mckenzie JL, Slamovich EB, Webster TJ (2005) Anodized Ti and Ti_6Al_4V possessing nanometer surface features enhances osteoblast adhesion. J Biomed Nanotechnol 1(1):68–73
38. Webster TJ, Ejiofor JU (2004) Increased osteoblast adhesion on nano phase metals Ti, Ti_6Al_4V and CoCrMo. Biomaterials 25(19):4731–4739
39. Harrison A, Basker RM, Smith IS (1989) The compatibility of temporary soft materials with immersion denture cleansers. Int J Prosthodont 2:254–258
40. Nikawa H, Iwanga H, Hamada T, Yuhta S (1994) Effects of denture cleansers on direct soft tissue lining materials. J Prosthet Dent 72:657–662
41. Fu J, Ji J, Fan D, Shen J (2006) Construction of antibacterial multilayer films containing nano silver via layer-by layer assembly of heparin and chitosan-silver ions complex. J Biomed Mater Res 79:665–674
42. Nam KY (2011) In vitro antimicrobial effect of the tissue condition containing silver nanoparticles. J Adv Prosthodont 3:20–24

Chapter 8
Nanostructures in Dentistry: In Diagnosis, Drug Delivery and Oral Cancer Therapy, and their Biocompatibility

Archana Bhardwaj, Abhishek Bhardwaj and R. Nageswar Rao

8.1 Introduction

8.1.1 Nanotechnology

Nano is derived from "νανος," the Greek word for dwarf. It refers to one billionth of a unit. So, a nanometer is one billionth of a meter, roughly equivalent to the cumulative diameters of ten hydrogen atoms. The nanoscale is about a thousand times smaller than micro, which is about 1\80,000 of the diameter of a human hair [1]. Figure 8.1 depicts the various sizes of nanoscale materials for comparison.

The concept of nanotechnology was first elaborated in 1959 by Richard Feynman, a Nobel Prize-winning physicist, in a lecture titled, "There's plenty of room at the bottom." He ended the lecture concluding "this is a development which I think cannot be avoided." He described it as a technique of creating functional materials, devices and systems through control of atoms on a nanometer scale and utilization of the unique properties exhibited at that scale [2]. However, Feynman's proposal remained unrealized until the mid-1980s, when the MIT-educated engineer, K. Eric Drexler published his book, "Engines of Creation," which popularized the prospective of nanotechnology. He coined the term "nanotechnology," also known as "Molecular Nanotechnology" or "Molecular Engineering" [3]. Since then, nanotechnology has found itself useful in a multitude of applications, including dental diagnosis, material, and therapeutics.

A. Bhardwaj (✉) · A. Bhardwaj
Private Practice Abhirachna Dental, South Extension-I, E-14,
New Delhi 110049, India
e-mail: archana.sbhardwaj@gmail.com

R. N. Rao
SDM College of Dental Sciences and Hospital, Sattur,
Dharwad 580009, Karnataka, India

R. S. Chaughule (ed.), *Dental Applications of Nanotechnology*,
https://doi.org/10.1007/978-3-319-97634-1_8

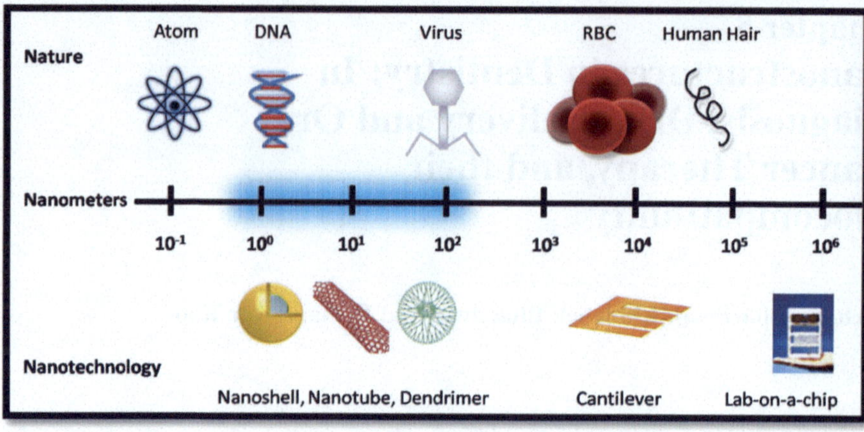

Fig. 8.1 Nanoscale

8.1.2 Principle of Nanotechnology

Nanotechnology is the production of functional materials and structures in the range of 0.1–100 nm, the nanoscale, by various physical or chemical methods. It is formulated on the basic principle of constructing functional structures by controlling atoms and molecules on an individual basis. At nanoscale, quantum physics becomes significant and the behavior of surfaces begins to govern the characteristics of bulk materials. All properties, including electrical, optical, and magnetic ones, are altered [4].

8.1.3 Advantages of Using Nanotechnology in Dentistry

Mechanical properties of dental materials deeply affect their clinical success. Dental materials should have adequate strength to withstand the static and dynamic occlusal forces for a long period of time in the complex environment of the stomatognathic system. What makes the concept of nanomaterials interesting is that their size is smaller than the critical lengths defining many physical events. The major advantage of using nanobiomaterials in dental practice is their improved mechanical properties, superior abrasion resistance, lower shrinkage, and enhanced optical and esthetic properties [5].

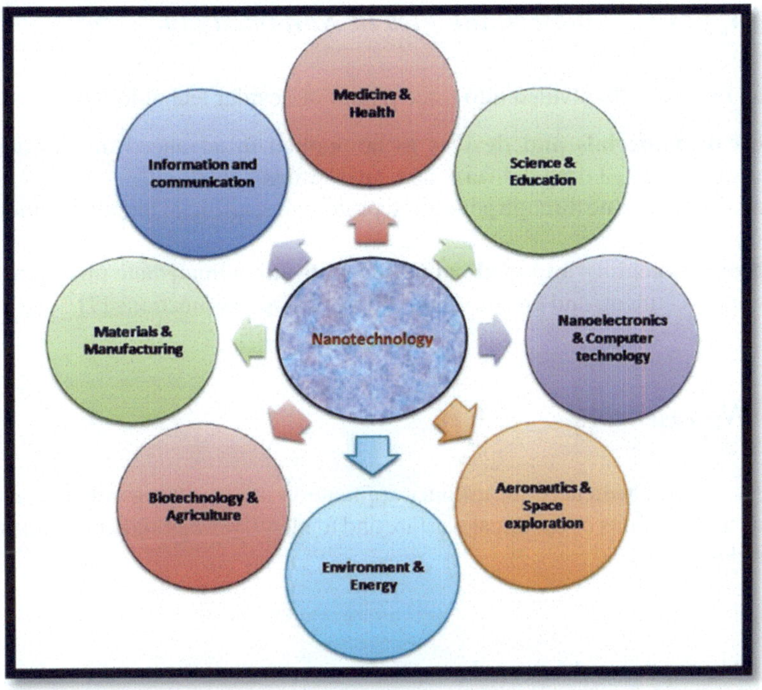

Fig. 8.2 Applications of nanotechnology

8.1.4 Applications of Nanotechnology in Dentistry

Nanotechnology has applications in multiple fields (Fig. 8.2). In the present scenario, nanobiomaterials have been used in various dental applications such as light polymerizable composites and their bonding systems, impression materials, ceramics, and dental implant coatings. Their use in the field of periodontics, orthodontics, endodontics, oral surgery, prosthodontics, oral cancer, salivary diagnostics, and nanotechnology for tooth repair and regeneration is also under research. With the application of nanotechnology in dentistry, the maintenance of ideal oral health might soon become achievable.

8.2 Nanomedicine

Nanomedicine is the science of preventing, diagnosing, and treating disease and preserving and improving human health, using nanosized particles [6].

8.2.1 *Types of Nanotechnologies in Nanomedicine*

They can be generally divided into three potent molecular technologies.

- **Nanoscale materials and devices** to be applied in advanced diagnostics and biosensors, targeted drug delivery, and smart drugs
- **Molecular medicine** through genomics, proteomics, artificial biobotics (microbial robots)
- **Molecular machines and medical nanorobots** aid in immediate microbial diagnosis and treatment, and enhancement of physiological functions [7].

8.2.2 *Nanosensors*

Nanosensors have been used for military application in identifying airborne harmful materials and weapons of chemical warfare and to identify drugs and other substances in expired air [8, 9].

8.2.3 *Nanophase Materials*

Nanophase materials are promising materials for various bio-applications as human tissues themselves are composed of nanometer size components. These include:

- **Nanophase hydroxyapatite (HA)**
- **Nanophase alumina**
- **Nanophase carbon**: Carbon nanofibers have extraordinary conjectural mechanical properties in addition to nanoscale dimensions like natural HA; these features support its proposal as a maxillofacial implant material [10].

8.2.4 *Nanorobots*

- **Fabrication**: Carbon, in the form of diamond or fullerene, is used for nanorobots fabrication due to its high strength and chemical inertness. Nanorobots are 0.5–3 μm in diameter and are constructed of parts with dimensions in the range of 1–100 nm. The external passive diamond coating provides a smooth, flawless coating making them inert to the body's immune system. Figure 8.3 shows the fabrication of nanorobots.
- **Mechanism of Action**: The powering of nanorobots can be done by metabolizing local glucose, oxygen and externally supplied acoustic energy. They can be controlled by onboard computers. A navigational network installed in the body

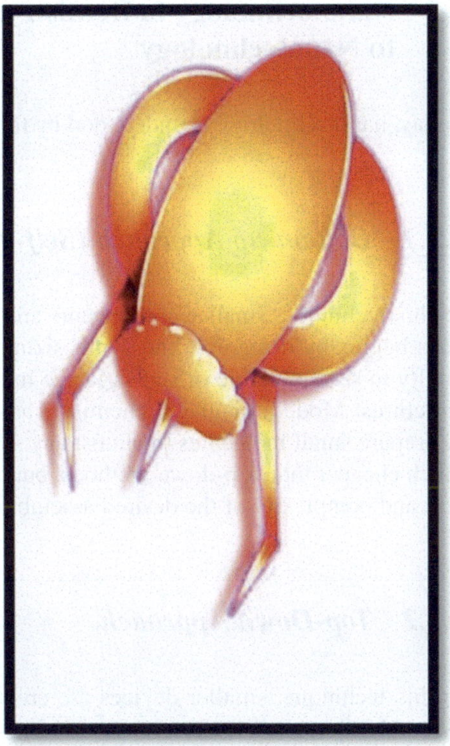

Fig. 8.3 A nanorobot

may provide high positional accuracy to all passing nanorobots and keep track of the various devices in the body. Nanorobots can distinguish between different cell types by checking their surface antigens.

- **Uses**: Nanorobots may find use in gerontology, pharmaceutics, diagnostics, in reversal of atherosclerotic damage, enhancing lung function, aiding natural immunity, repairing brain injury, modifying cellular DNA sequences, and repairing cellular damage, apart from diverse applications in dentistry.
- **Deactivation of Nanorobots**: When the task of the nanorobots is completed, they may exit via the usual human excretory channels. They can also be retrieved by active scavenger systems.
- **Nanoterminators**: Scientists are yet to build a mechanism for self-destruction of nanorobots similar to apoptosis. The question arises as what will happen if nanorobots free themselves from control. For example, those intended for anesthetizing the dental pulp may pass through the bloodstream and arrive at one of the vital centers, respiratory center at the medulla oblongata or conduction bundle in the heart, and discontinue the current of nerve impulses. In such a case, the consequences may be catastrophic. To counteract the problem, researchers advocated the development of nanoterminators or the killers of nanorobots, which will destroy the disobedient and crazy nanorobots [11].

8.3 Nanotechnology in Dentistry: Approaches to Nanotechnology

Today, nanotechnology is understood by the following approaches [12–16].

8.3.1 Bottom-up Approach (Self-assembly)

In this technique, smaller components are arranged into more complex assembly. This begins by designing and synthesizing custom-made molecules that have the ability to self-assemble or self-organize into higher-order mesoscale or macroscale structures. Modern synthetic chemistry has reached the point where it is possible to prepare small molecules to almost any structure. Such bottom-up approaches are much cheaper than top-down methods, but could potentially be overwhelmed as the size and complexity of the desired assembly increases as shown in Table 8.1.

8.3.2 Top-Down Approach

In this technique, smaller devices are created by using larger ones to direct their assembly. So, small features are made by starting with larger materials patterning and carving down to make nanoscale structures in precise patterns. Complex structures containing hundreds of millions of precisely positioned nanostructures can

Table 8.1 Approaches to nanotechnology

Bottom-up approach	Top-down approach
• Local nanoanaesthesia	• Nanocomposites
• Hypersensitivity cure	• Nanosolution
• Tooth repositioning	• Nano glass ionomer restorative
• Nanorobotic dentifrice	• Pit and fissure sealants
• Dental durability and cosmetics	• Impression materials
• Nanodiagnostics	• Nano-composite denture teeth
• Therapeutic aid in oral diseases	• Nanoencapsulation
• Nanotherapeutics/drug delivery	• Dentifrices
• Gene therapy	• Laser plasma application
• Diagnosis of oral cancer	• Bone replacement materials
• Treatment of oral cancer	• Osseo-inductive materials
• Whole tooth replacement	• Prosthetic implants
• Tooth renaturalization	• Radiopacity
• Dental biomimetics	• Orthodontic wires
• Endodontic regeneration	• Nanoneedles
• Nanoterminators	• Nanosterilizing solution

be fabricated. Materials are reduced to the nanoscale and can suddenly show very different properties, enabling unique applications (Table 8.1).

8.3.3 *The Functional Approach*

The functional approach disregards the method of production of a nanoparticle, and the objective is to produce a nanoparticle with a specific functionality. In other words, it seeks to develop components of a desired functionality without regard to how they might be assembled. Molecular scale electronics seeks to develop molecules with useful electronic properties. These could then be used as single-molecule components in a nanoelectronic device. Synthetic chemical methods can also be used to create synthetic molecular motors.

8.3.4 *The Biomimetic Approaches*

This approach seeks to apply biomolecules for applications in nanotechnology. The intersection of nanotechnology and biology is referred to by the term Bionanotechnology. It utilizes the principal of bionics and biomimicry. This fruitful collaboration between materials science, biology, and biomedicine for the advancement of biomaterials collects the most promising solutions provided by nature for the field of biomedicine, showing how to achieve the desired functionality by using biomimetics.

8.3.5 *Speculative*

This approach seeks to anticipate what inventions nanotechnology might yield, or attempt to propose an agenda along which inquiry might progress. These often take a big-picture view of nanotechnology, with more emphasis on its societal implications than the details of how such inventions could actually be created.

8.4 Therapeutic Nanostructures in Dentistry

Nanodentistry may lead to the achievement of almost perfect oral health through the use of nanomaterials, biotechnology including tissue engineering and nanorobotics (Fig. 8.4). It includes:

- Nanomaterials
- Nanodiagnostics

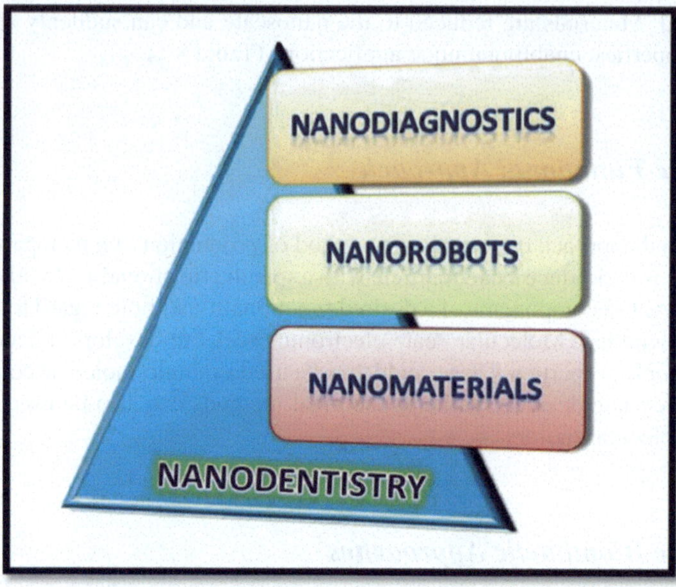

Fig. 8.4 Components of nanodentistry

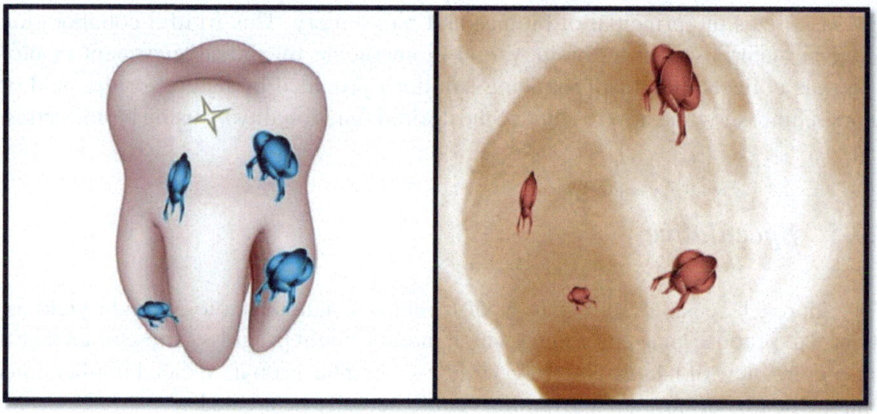

Fig. 8.5 Nanorobots on the tooth surface and within the root canal

- Nanorobotics (Fig. 8.5).

8.4.1 Nanomaterials

Nanomaterials are those materials with components less than 100 nm in at least one dimension. These may include atoms clusters, grains, fibers, films, composites, or nanoholes from these combinations and they can be constituted of metals, ceramics, polymeric materials, or composite materials.

8.4.2 Classification of Nanomaterials

A. Nanomaterials in one dimension are termed as sheets, in two dimensions as nanowires and nanotubes, and as quantum dots in three dimensions [17].
B. As defined by Richard W. Siegel, nanostructured materials can be created with various modulation dimensionalities:
    - Zero (atomic clusters, filaments, and cluster assemblies),
    - One (multilayers),
    - Two (ultrafine-grained over layers or buried layers), and
    - Three (nanophase materials consisting of equiaxed nanometer-sized grains).
C. Nanomaterials generally fall into two categories: fullerenes and inorganic nanoparticles

a. **Fullerenes**

These are a class of allotropes of carbon which are basically graphene sheets rolled into tubes or spheres. A common method used to produce fullerenes is to send a large current between two nearby graphite electrodes in an inert atmosphere. The resulting carbon plasma arc between the electrodes cools into sooty residue from which many fullerenes can be isolated. These include the carbon nanotubes which are of interest both because of their mechanical strength and also because of their electrical properties.

b. **Inorganic Nanoparticles**

Inorganic nanoparticles (e.g., quantum dots, nanowires, and nanorods), because of their interesting optical and electrical properties, could be used in optoelectronics. Furthermore, these properties, which depend on the size and shape of nanomaterials, can be adjusted during production.

Inorganic nanoparticles (either currently in use or under development) include

- Semiconductor nanoparticles
- Metal nanoparticles
- Metal oxide nanoparticles
- Silica nanoparticles
- Polyoxometalates
- Gold nanocrystals.

8.4.3 Nanostructures Used in Dentistry

Present-day nanostructures exploit carefully structured nanocarriers such as nanoparticles, nanorods, nanospheres, nanotubes, nanofibers, dendrimers, quantum dots (QDs), nanopores, nanoscale cantilevers, nanoshells, and liposomes to target-specific tissues and organs (Fig. 8.6). These nanostructures may serve in the diagnosis and cure of dental disease, as well as provide anticancer agents in dentistry.

Various types of nanostructures include:

- **Nanopores**

A nanopore (Fig. 8.6a) is simply a small hole, of the order of 1 nm in internal diameter. Nanopore sequencing is one of the most promising technologies being developed as a cheap and fast alternative to the conventional Sanger sequencing method. Protein or synthetic nanopores have been used to detect DNA or RNA molecules [18].

- **Nanotubes**

Nanotubes (Fig. 8.6h) are made up of carbon atoms bonded into hexagonal patterns and have about half the diameter of a DNA molecule. They exhibit great strength and electrical conductivity. Apart from helping in cancer diagnostics, they have been used in various dental applications. Titanium oxide nanotubes have been revealed to accelerate the kinetics of hydroxyapatite (HA) formation in vitro, mainly in the perspective of bone growth applications for dental implants [19]. Recently, modified single-walled carbon nanotubes (SWCNTs) have been shown to improve the flexural strength of composites. These SWCNTs had silicon dioxide applied to them in combination with specialized organosilane bonding agents [20]. Moreover, novel nanostructured titania tubes have been successfully introduced in the poly(methyl

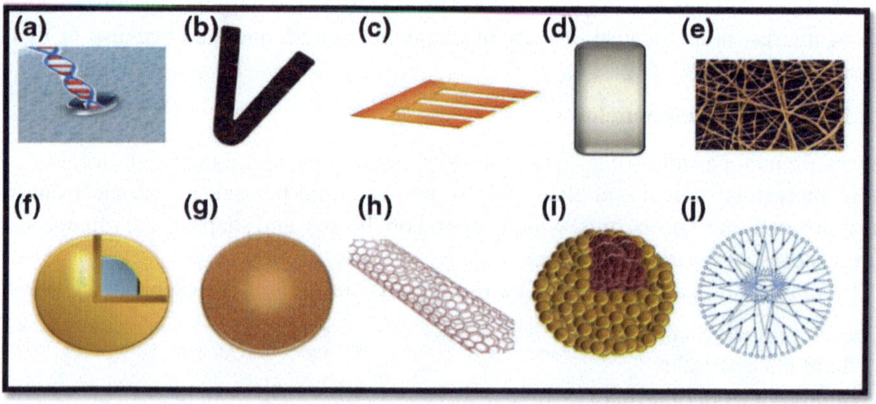

Fig. 8.6 Types of nanostructures **a** Nanopore, **b** Nanobelt, **c** Nanoscale cantilever, **d** Nanorod, **e** Nanofiber, **f** Nanoshell, **g** Nanosphere, **h** Nanotube, **i** Quantum dot, **j** Dendrimer

methacrylate) (PMMA) bone cement matrix to enhance its mechanical properties [21].

- **Quantum Dots**

Quantum dots (QDs) are typically semiconductor nanocrystals that have gained tremendous attention due to their unique electronic, magnetic, chemical, and optical properties (Fig. 8.6i). They have proven very useful in applications in many fields across the physical, engineering, chemical, biological, medical sciences, and especially in dentistry. QD conjugates with standard epifluorescence microscopy, leading to brilliant single-cell resolution of biofilms. Alves et al. studied dental resins impregnated with different concentrations of CdSe/ZnS core–shell QD and found that this modification resulted in the fabrication of restorative materials with fluorescence properties that closely match those of natural human teeth [22].

- **Nanoscale Cantilevers**

Nanoscale cantilevers (Fig. 8.6c) are flexible beams resembling a row of diving boards. Cantilever arrays with their extraordinary multiplexing capabilities could aid cancer diagnosis and could be engineered to bind to molecules associated with cancer, such as DNA sequences, single-nucleotide polymorphisms, and proteins [23].

- **Nanoshells**

A nanoshell (Fig. 8.6f) is a type of spherical nanoparticle consisting of a dielectric core that is covered by a thin metallic shell. The core is usually of silica and the metallic outer layer is composed of gold. By modifying the core-to-shell ratio, scientists can propose these shells to absorb near-infrared light creating a powerful heat that is mortal to cancer cells, without affecting adjacent normal cells. These materials have a variety of other applications in dentistry, including fluorescent diagnostic labels, catalysis, avoiding photodegradation of teeth, enhancing photoluminescence, creating photonic crystals, and preparation of bioconjugates [18].

- **Dendrimers and Dendritic Copolymers**

Dendrimers (Fig. 8.6j) are man-made molecules about the size of an average protein and have a branching shape. This shape gives those vast amounts of surface area to which scientists can attach therapeutic agents or other biologically active molecules. In addition, photopolymerized dendritic copolymers and particulate filler composites have been combined for use as dental restorative materials. The incorporation of polymerizable 2, 3-dihydroxybenzyl ether dendrimers into dental composite resins produced dental materials with enhanced physical properties [24, 25].

- **Nanobelt**

While nanotubes are a few millionths of a meter long, nanobelts (Fig. 8.6b) are millimeters long and have multiple advantages over tubes in terms of price, flexibility, and practicality.

For making nanobelts, an oxide containing zinc, tin, cadmium, gallium, or indium is evaporated for 2 h. Nanobelt is then deposited as a wool-like product. These little straps have a rectangular cross section with a width of 30–300 nm and a thickness of 10–15 nm. Each belt is a single crystal. Because the material is already an oxide, it does not undergo any further chemical reaction and has a pure, flawless surface [26].

- **Nanorods**

Nanorods (Fig. 8.6d) are nanoscale objects having dimensions ranging from 1 to 100 nm. Nanorods, having a nanosized HA adhesive system, may have practical applications in dental clinics. Chen et al. have created enamel-prism-like HA nanorods that have self-assembly properties [27]. They are analogous to the enamel rods that construct the vital crystalline construction of dental enamel; therefore, nanorods may provide a convenient artificial model of this naturally occurring structure. Shojai et al. synthesized HA nanorods which showed high dispersion stability in a dilute experimental adhesive. HA nanorods may also be regarded as an alternative to other fillers such as silicates for use in dental adhesives [28].

- **Nanospheres**

Nanospheres (Fig. 8.6g) have also been tried for their potential role in duplicating natural processes of tooth development. During the secretory stage of enamel, the organic matrix self-assembles to form nanospherical structures that align along the maturing enamel crystallites. Nanosphere assembly could be considered for a restorative purpose [29].

- **Nanofibers**

Nanofibers (Fig. 8.6e) are defined as fibers with diameters less than 1000 nm. They have a high aspect ratio and a high surface area to volume ratio, which results in greatly improved physical and mechanical properties. They have been explored for several biomedical applications. Currently, nanofibers are used to produce ceramics containing hydroxyapatite and fluorhydroxyapatite [30]. Moreover, nanofibrillar silicate crystals have been tried in strengthening dental composites. Nanofibers, added in the correct proportion and uniform distribution, were established to improve the physical properties of these composites [31]. The nanofibrous (NF)-poly(L-lactic acid) (PLLA) supported odontogenic differentiation of human dental pulp stem cells and dentin-like tissue formation, demonstrating its potential for dental tissue engineering application [32].

- **Liposomes**

Liposomes are vesicular structures which have an aqueous core surrounded by a lipid bilayer. They adsorb to hydroxyapatite, the major constituent of dental enamel. The potential of liposomes as a dental drug delivery system has also been explored recently [23].

8.5 Nanodiagnostics: Diagnosis of Oral Cancer and Other Diseases

8.5.1 Advantages of Using Nanodiagnostics

- This technology will help in the use of nanodevices and nanosystems for early disease identification at cellular and molecular level.
- The efficiency and reliability of diagnostic methods using human fluids or tissue samples could be increased further with the help of nanotechnology.
- In the case of in vivo diagnostics, it could develop devices which will be able to work inside the human body in order to identify the early presence of disease, tumor cells and also identify and quantify toxic molecules.
- In the case of cancer diagnosis, these advances will permit less invasive, less uncomfortable means of identifying and quantifying the markers of disease, thus aiding not only in diagnosis, but also monitoring recurrence or metastasis and defining the locations, biologic types, and behaviors of malignancies [33].

8.5.2 Various Techniques Used in Nanodiagnosis

- **Nanopores**

Nanopores being tiny holes that allow DNA to pass through, one strand at a time, will make DNA sequencing more efficient. The pores are so small that DNA separation is being attempted using this structure. As DNA passes through a nanopore, scientists can monitor the shape and electrical properties of each base, or letter on the strand. Since these properties are unique for each of the four bases that make up the genetic code, scientists can use the passage of DNA through a nanopore to decipher the encoded information, including errors in the code known to be associated with cancer [17, 34].

- **Nanotubes**

These carbon rods, which are about half the diameter of a molecule of DNA, can not only detect the presence of altered genes, but may also help researchers pinpoint the exact location of those changes. This may help to identify DNA changes associated with cancer [35]. A multidisciplinary team in Massachusetts Institute of Technology has developed carbon nanotubes that can be used as sensors for cancer drugs and other DNA-damaging agents inside living cells [36].

- **Quantum Dots**

These are miniscule molecule making up tiny crystals that glow when stimulated by UV light. The wavelength or color of the light depends on the size of the crystal. Latex beads filled with these crystals can be designed to bind to specific DNA sequences. By combining different sized quantum dots within a single bead, scientists can create probes that release distinct colors and intensities of light. On stimulation by UV light, each bead emits light that serves as a sort of spectral bar code, identifying a particular region of DNA. To detect cancer, scientists can design quantum dots that bind to sequences of DNA that are associated with the disease. When the quantum dots are stimulated with light, they emit their unique bar codes, or labels, making the critical, cancer-associated DNA sequences visible [1].

- **Nanoscale Cantilevers**

These flexible beams can be engineered to bind to molecules associated with cancer. They may bind to altered DNA sequences or proteins that are present in certain types of cancer and can provide rapid and sensitive detection of cancer-related molecules. As a cancer cell secretes its molecular products (DNA sequences or proteins), the antibodies coated on the cantilever fingers selectively bind to these secreted proteins, changing the physical properties of the cantilever and signaling the presence of cancer [23].

- **Nanoelectromechanical Systems (NEMS)**

Nanotechnology-based NEMS biosensors that exhibit exquisite sensitivity and specificity for analyte detection, down to single-molecule level, are being developed. They convert biochemical signals to electrical signals, thus monitoring health status, disease progression, and treatment outcome through noninvasive means [37, 38].

- **Laboratory-on-a-Chip Methods**

Laboratory on a chip (LOC) is a device which integrates several laboratory functions on a single chip. These deal with the handling of extremely small fluid volumes, down to less than picoliters. Assays are performed on chemically sensitized beads populated into etched silicon wafers with embedded fluid handling and optical detection capabilities. Complex assays can be performed with small sample volumes, short analysis times, and markedly reduced reagent costs. LOC methodologies have been used to assess the levels of interleukin-1beta, C-reactive protein, and matrix metalloproteinase-8 in whole saliva. This might prove to be the future use of these biomarkers for diagnosing and categorizing the severity and extent of periodontitis [39, 40].

- **Oral Fluid Nanosensor Test (OFNASET)**

OFNASET technology combines self-assembled monolayers (SAM), bionanotechnology, cyclic enzymatic amplification, and microfluidics for detection of salivary biomarkers for oral cancer. It was demonstrated that a combination of salivary proteomic and mRNA biomarkers detected oral cancer with high specificity and sensitivity [41].

- **Optical Nanobiosensor**

The nanobiosensor is a unique fiberoptics-based tool which allows the minimally invasive analysis of intracellular components. Components such as cytochrome C, an important protein involved in the production of cellular energy as well as in apoptosis, can be analyzed [42].

- **Nanotexturing**

Physicochemical nanoscale modification of surfaces has been proposed to analyse low-molecular-weight proteins from body fluids and other biologic samples. It will provide size segregation, selective capture, and resultant enhancement of specific regions of the proteins [38].

- **Bio-Barcode Assay**

A magnetic probe captures a target molecule using either monoclonal antibody or complementary oligonucleotide. Target-specific gold nanoparticles sandwich the target, thus distinguishing it and amplifying the signal. The barcode oligonucleotides are then released and detected using the scanometric method [38].

- **Nanowires**

Nanometer scale wires may help monitor local chemical, electrical, or physical property changes in cells or tissues [38].

- **Iodinated Nanoparticles**

Iodinated nanoparticles can be confined to the lymph nodes after bronchoscopic instillation. These may then be visualized precisely through the use of computerized tomography (CT) [38].

8.6 Nanoencapsulation and Nanotherapeutics—Nanoparticles as Dental Drug Delivery Systems

Using natural processes as a guide, substantial advances have been made at the interface of nanomaterials and biology. Nanoparticles are being developed for a host of biomedical and biotechnological applications including drug delivery, enzyme immobilization, and DNA transfection. The disadvantage that accompanies conventional drug delivery systems is short retention time in the oral cavity due to salivation, irregular swallowing, food and beverage intake, and abrasion by soft tissue movements. The use of nanostructures as a part of dental drug delivery system might overcome these problems [6, 43, 44].

8.6.1 Advantages of Nanoparticles as Dental Drug Delivery Systems

- Nanotechnologic packaging of therapeutics will provide the ability to co-localize delivery of multiple and complimentary therapeutic agents.
- Materials that now require injections potentially could be inhaled or swallowed using nanoengineered delivery devices, thus improving patient comfort and compliance.
- Vaccines may be made more comfortable, by using multitudes of micro- or nanometer-scale needles, to which human nerves are insensitive, rather than a painful injection.
- Nanoparticles formulations provide protection for agents susceptible to degradation or denaturation in regions of harsh pH and also prolong the duration of exposure of a drug by increasing retention of the formulation through bioadhesion. This may lead to increased bioavailability.
- Nanotechnology will eliminate the solubility problems, lead to a reduction in the dosage of drug, and reduce the adverse effects. This may allow the safe use of some drugs that are effective but otherwise have unacceptable toxicity profiles.
- Nanoparticulates also may allow increase in the amount of drug that reaches abnormal cells. They possess a site-specific targeting ability combined with potentially metabolism-specific targeting.

8.6.2 Drug Delivery by Liposomes

Liposomes are vesicular structures with an aqueous core surrounded by a lipid bilayer. In vitro experiments have confirmed that liposomes adsorb to HA, which is the major constituent of dental enamel. More recently, Nguyen et al. have investigated the potential of liposomes as a dental drug delivery system, specifically for teeth targeting, using in vitro adsorption of charged liposomal formulations to HA. They found that negatively charged liposomes were more effective and were the least reactive with the components of parotid saliva. The reactivity of negatively charged liposomes is based on the type of charge group bound to them. Calcium ion present in parotid saliva is a prerequisite for the interaction with negatively charged liposomes, and their affinity to the cations determines the degree of interaction [23, 45].

8.6.3 *Nanoencapsulation*

The South West Research Institute (SWRI) Texas, USA, has developed targeted release systems. These include nanocapsules in the form of new vaccines, antibiotics, and delivery of drugs with fewer adverse effects. In 2003, Osaka University in Japan made possible the targeted delivery of genes and drugs to human liver. In the time to come, dedicated nanoparticles might be developed to focus on oral tissues as well as cells derived from the periodontium.

Multilayer epoxy nanoencapsulants have a long history of medical applications in dentistry with their optimal use being implants with a medium-term (30 days to 6 months) lifespan. A typical multilayer epoxy application might be a dental implant that measures moisture in a patient's mouth and then stimulates the submandibular gland to produce more saliva [9, 46].

8.7 Nanobiomaterials in Oral Cancer Therapy

8.7.1 *Nanoshells*

Nanoshells are miniscule beads coated with gold. By manipulating the thickness of layers making up the nanoshells, the beads may absorb specific wavelengths of light. The most useful nanoshells are those that absorb near-infrared light, as it can easily penetrate several centimeters of human tissue. The light absorption results in an intense heat that is lethal to cells. Researchers can already link nanoshells to antibodies that recognize cancer cells. Scientists envision letting these nanoshells seek out their cancerous targets, then applying near-infrared light. In laboratory cultures, the heat generated by the light-absorbing nanoshells has successfully killed tumor cells while leaving neighboring cells intact [24].

8.7.2 *Dendrimers*

Dendrimers possess certain characteristics like high degree of branching, multivalency, globular structure, and well-defined molecular weight, which makes these an ideal candidate for cancer therapy. Dendrimers form nanometer by nanometer, so the number of synthetic steps or generations dictates the exact size of particles in a batch. The peripheral layer can be made to form a dense field of molecular groups that serve as hooks for attaching other useful molecules, such as DNA. Dendrimers of a certain size trigger a process called endocytosis in which the cell's outer membrane deforms into a tiny bubble or vesicle. The vesicle encloses the dendrimer which is admitted into the cell's interior. Once inside, it is released and migrates to the nucleus where it becomes part of cell's genome. It has already been tested in mammalian cell types.

Donald et al. reported using glycodendrimer "nanodecoys" to trap and deactivate influenza virus particles. In 1998, James R. Baker pursued his work to use dendrimers as a safer and more effective genetic therapy agent. According to him, these nanostructures can sneak DNA into cells while avoiding triggering an immune response [24].

8.7.3 Nanomaterials for Brachytherapy

Nanomaterials like "BrachySilTM" that delivers P^{32} are in clinical trial. Drug delivery system that can cross the blood brain barrier is vision of the future with this technology. Parkinson disease, Alzheimer disease, and brain tumors will be managed more efficiently by such techniques [33].

8.7.4 Gene Therapy

This technique prevents or treats genetic diseases, by correcting faulty genes that lead to disease development, i.e., by repairing or replacing them. Gene delivery systems may be broadly categorized into three types: viral vectors, nonviral vectors, and the direct inoculation of genes into tissues (gene guns). However, such nanovectors to correct disease at molecular aspect are yet in a developing stage [47].

8.8 Challenges Faced by Nanotechnology

Nanotechnology faces many challenges that need to be overcome such as

- Inadequate assimilation of clinical research;
- Precise positioning and assembly of nanoscale parts;
- Cost-effective nanorobot mass manufacturing methods;
- Synchronization of numerous independent nanorobots;
- Biocompatibility concern;
- Social issues of public acceptance, ethics, regulation, and human safety [37, 48].

8.9 Biocompatibility of Nanaomaterials

Nanoparticles have distinctive physicochemical properties that can be different in comparison with the same materials in macroscale size. The main difference is the disparity between the surface area and unit weight, and a much higher surface reactivity. This may lead to increased absorption through the lungs, skin, and digestive tract.

Despite efforts to improve their targeting efficiency, significant quantities of nanomaterials are cleared by the mononuclear phagocytic system before finding their targets, increasing the likelihood of unintended acute or chronic toxicity. Nanoparticles, due to their small size, can intermingle with DNA, RNA, and other intracellular apparatus, causing various adverse effects.

The main health risks related to the use of such devices consist of cytotoxicity, translocation to undesired cells, acute and chronic toxicity, and the environmental impacts [49, 50].

8.9.1 Biological Consequences of Nanomaterials

Studies indicate that the biological consequences of nanomaterials are material and formulation specific. Most of the nanomaterials are compatible with biological systems and permissive for cell growth and differentiation. Nanorobots may be pyrogenic or non-pyrogenic. The non-pyrogenic nanorobots used in vivo are bulk Teflon, carbon powder, and monocrystal sapphire, whereas pyrogenics are alumina, silica and trace elements like copper and zinc. The pyrogenic nanorobots are controlled by in vivo medical nanorobots.

Each new nanomaterial must be assessed individually, and all material properties must be taken into account. To address such concerns, the Swedish Karolinska Institute conducted a study in which various nanoparticles were introduced to human lung epithelial cells. The results, released in 2008, showed:

- Iron oxide nanoparticles caused little DNA damage and were non-toxic.
- Zinc oxide nanoparticles were slightly worse.
- Titanium dioxide caused only DNA damage.
- Copper oxide was found to be the worst offender and was the only nanomaterial identified by the researchers as a clear health risk.
- Carbon nanotubes were found to be cytotoxic and induced granulomas in the lungs of laboratory animals, as they caused DNA damage even at low levels [51].

8.9.2 Regulations for Control

Presently, there are no exact regulations for the control of nanotechnology-based materials and allied problems. Overall, there is a critical requirement to standardize these nanotechnology-based products and delivery devices. Characterization, safety, and environmental impact are the three main elements that need to be regulated. Regulatory agencies like the Food and Drug Administration (FDA), the Environment Protection Agency (EPA), and the Nuclear Protection Agency are regulating the major health risks associated with nanomaterials. Workers may be exposed to nanosized particles in the manufacturing or industrial use of nanomaterials. The National Institute for Occupational Safety and Health is also performing research on nanoparticle interaction with body systems [52].

The benefits of nanotechnology are enormous, and therefore, studies that examine the health, environmental, ethical, and safety issues should improve our understanding of how to exploit the benefits and diminish the risks.

8.10 Future Perspective of Nanotechnology in Dentistry

Nanotechnology is on the brink of initiating extraordinary advances in biomedical sciences. In the future, nanotechnology and biomimetic approach could be used for repair and restoration of damaged enamel. Based on the current knowledge, it is certain that nanotechnology has a great potential for prevention of dental caries. There are efforts to prevent the aging and degradation of resin–dentin bonding by feeding minerals back into the collagen network by guided tissue remineralization. Experimental nano-bioactive glass-ceramic composite scaffolds could be useful in periodontal regeneration. Novel properties of nanosurfaces could affect cell adhesion, proliferation, and differentiation. This would result in hastening and improving the process of osseointegration. There are already a few commercial nano-modified dental implant systems available for clinical use. Dentistry, as we know today, is ready for a paradigm shift.

The Foresight Institute has offered the $250,000 Feynman Grand Prize to the first researcher or researchers who develop two devices: a basic nanorobot and a nanocomputer. Christine Peterson, President of the Foresight Institute, estimates that the prize will be claimed between 10 and 30 years from now. Because the initial nanodevices will be basic, prototypical units, commercial applications will follow years later.

8.11 Conclusion

Nanotechnology will bring enormous changes into the field of dentistry, as it offers fresh opportunities for sensing clinically relevant markers, molecular disease imaging, and tools for therapeutic intervention. Nanotechnology in future will have a great impact on dental research, prevention, diagnostics, and treatment options. However, as with all developments, it may also pose a risk for misuse and abuse. Time, newer developments, economical and technical resources, and human needs will determine which of its applications are realized first.

References

1. Schleyer TL (2000) Nanodentistry. Fact or fiction? J Am Dent Assoc 13:1567–1568
2. Feynman RP (1961) There's plenty of room at the bottom. Reinhold, New York
3. Drexler KE (1986) Engines of creation: the coming era of nanotechnology. Anchor Press, New York
4. Drexler KE (2006) Nanosystems: molecular machinery manufacturing and computation. Wiley, New York
5. Kim JS, Cho BH, Lee IB (2005) Effect of the hydrophilic nanofiller loading on the mechanical properties and the microtensile bond strength of an ethanol based one-bottle dentin adhesive. J Biomed Mater Res B Appl Biomater 72(2):284–291
6. Kubik T, Bogunia K, Sugisaka M (2005) Nanotechnology on duty in medical applications. Curr Pharm Biotechnol 6:17–33
7. Chen Y, Jung GY, Ohlberg DA, Li X, Stewart DR, Jeppesen JO (2003) Nanoscale molecular-switch crossbar circuits. Nanotechnology 14:462–468
8. Kanaparthy R, Kanaparthy A (2011) The changing face of dentistry: nanotechnology. Int J Nanomed 6:2799–2804
9. Bhardwaj A, Bhardwaj A, Misuriya A, Maroli S, Manjula S, Singh AK (2014) Nanotechnology in dentistry: present and future. J Int Oral Health 6(1):121–126
10. Webster TJ, Ergun C, Doremus RH (2001) Enhanced osteoclast like cell functions on nanophase ceramics. Biomaterials 22:1327–1333
11. Dwivedi S, Dwivedi CD, Chandra A, Sharma N (2013) Nanotechnology boon or bane for restorative dentistry; a review. Int J Eng Sci Inv 2:1–5
12. Boncheva M, Whitesides GM (2004) Biomimetic approaches to the design of functional self-assembling systems. In: Dekkar encyclopedia of nanoscience and nanotechnology, pp 287–294
13. Ghalanbor Z, Marashi SA, Ranjbar B (2005) Nanotechnology helps medicine: nanoscale swimmers and their future applications. Med Hypoth 65(1):198–199
14. Rodgers P (2006) Nanoelectronics: single file. Nat Nanotechnol
15. Ehud G (2007) Plenty of room for biology at the bottom: an introduction to bionanotechnology. Imperial College Press, London
16. Mano JF (2012) Index in biomimetic approaches for biomaterials development. Wiley-VCH Verlag GmbH & Co, KGaA, Weinheim, Germany
17. Freitas RA Jr (1999) Nanomedicine, volume I: basic capabilities. Landes Bioscience, Georgetown, Texas
18. Burns JR, Stulz E, Howorka S (2013) Self-assembled DNA nanopores that span lipid bilayers. Nano Lett 13:2351–2356
19. Oh SH, Finones RR, Daraio C, Chen LH, Jin S (2005) Growth of nanoscale hydroxyapatite using chemically treated titanium oxide nanotubes. Biomaterials 26:4938–4943
20. Zhang F, Xia Y, Xu L, Gu N (2008) Surface modification and microstructure of single-walled carbon nanotubes for dental resin based composites. J Biomed Mater Res B Appl Biomater 86:90–97
21. Khaled SMZ, Miron RJ, Hamilton DW, Charpentier PA, Rizkalla AS (2010) Reinforcement of resin based cement with titania nanotubes. Dent Mater 26:169–178
22. Alves LP, Pilla V, Murgo DOA, Munin E (2010) Core shell quantum dots tailor the fluorescence of dental resin composites. J Dent 38:149–152
23. Nguyen S, Solheim L, Bye R, Rykke M, Hiorth M, Smistad G (2010) The influence of liposomal formulation factors on the interactions between liposomes and hydroxyapatite. Colloid Surf B 76:354–361
24. Parak WJ, Gerion D (2003) Biological applications of colloidal nanocrystals. Nanotechnology 14:R15–R27
25. Viljanen EK, Skrifvars M, Vallittu PK (2007) Dendritic copolymers and particulate filler composites for dental applications: degree of conversion and thermal properties. Dent Mater 23:1420–1427

26. Bhuvaneswarri J, Alam MN, Chandrasekaran SC, Sathya MS (2013) Future impact of nanotechnology in dentistry—a review. Int J Nanotech App 3:15–20
27. Chen H, Clarkson BH, Sun K, Mansfield JF (2005) Self-assembly of synthetic hydroxyapatite nanorods into an enamel prism-like structure. J Colloid Interfac Sci 288:97–103
28. Shojai MS, Atai M, Nodehi A, Khanlar LN (2010) Hydroxyapatite nanorods as novel fillers for improving the properties of dental adhesives: synthesis and application. Dent Mater 26:471–482
29. Fan Y, Sun Z, Wang R, Abbott C, Moradian-Oldak J (2007) Enamel inspired nanocomposite fabrication through amelogenin supramolecular assembly. Biomaterials 28(19):3034–3042
30. Kim HW, Kim HE (2006) Nanofiber generation of hydroxyapatite and fluor-hydroxyapatite bioceramics. J Biomed Mater Res B Appl Biomater 77:323–328
31. Tian M, Gao Y, Liu Y, Liao Y, Hedin NE, Fong H (2008) Fabrication and evaluation of Bis-GMA/TEGDMA dental resins/composites containing nanofibrillar silicate. Dent Mater 24:235–243
32. Wang J, Liu X, Jin X (2010) The odontogenic differentiation of human dental pulp stem cells on nanofibrous poly (L-lactic acid) scaffolds in vitro and in vivo. Acta Biomater 6:3856–3863
33. Kairemo K, Erba P, Bergstrom K, Pauwels EJ (2008) Nanoparticles in cancer. Curr Radiopharm 1:30–36
34. Jhaveri HM, Balaji PR (2005) Nanotechnology: future of dentistry. J Ind Prosth Soc 5:15–17
35. Saravana KR, Vijayalakshmi R (2006) Nanotechnology in dentistry. Ind J Dent Res 17(2):62–65
36. Sivaramakrishnan SM, Neelakantan P (2014) Nanotechnology in dentistry—what does the future hold in store? Dentistry 4:198
37. Freitas RA Jr (2000) Nanodentistry. J Am Dent Assoc 131(11):1559–1566
38. Shi W, Qi F, Wu B, Wolinsky L, Wong DT (2005) The oral fluid MEMS/NEMS Chip (OFMNC): diagnostic and translational applications. Adv Dent Res 18(1):3–5
39. Christodoulides N, Floriano PN, Miller CS, Ebersole JL (2007) Lab-on-a-chip methods for point-of-care measurements of salivary biomarkers of periodontitis. Ann NY Acad Sci 1098:411–428
40. Herr AE, Anson V (2007) Integrated microfluidic platform for oral diagnostics. Ann NY Acad Sci 1098:362–374
41. Gau V, Wong D (2007) Oral fluid nanosensor test (OFNASET) with advanced electrochemical-based molecular analysis platform. Ann NY Acad Sci 1098:401–410
42. Song JM, Kasili PM, Griffin GD, Vo-Dinh T (2004) Detection of cytochrome C in a single cell using an optical nanobiosensor. Anal Chem 76(9):2591–2594
43. El-Shabouri MH (2002) Positively charged nanoparticles for improving the oral bioavailability of cyclosporine. A Int J Pharm 249:101–108
44. Patil M, Mehta DS, Sowjanya G (2008) Future impact of nanotechnology on medicine and dentistry. J Ind Soc Periodontol 12(2):34–40
45. Nguyen S, Hiorth M, Rykke M, Smistad G (2011) The potential of liposomes as dental drug delivery systems. Eur J Pharm Biopharm 77:75–83
46. Herzog A (2002) Of genomics, cyborgs and nanotechnology: a look into the future of medicine. Conn Med 66:53–54
47. Goverdhana S, Puntel M, Xiong W, Zirger JM, Barcia C, Curtin JF (2005) Regulatable gene expression systems for gene therapy applications: progress and future challenges. Mol Ther 12(2):189–211
48. Fakruddin M, Hossain Z, Afroz H (2012) Prospects and applications of nanobiotechnology: a medical perspective. J Nanobiol 10:31–38
49. Barnard AS (2006) Nanohazards: knowledge is our first defence. Nat Mater 5:245–248
50. Sahoo SK, Parveen S, Panda JJ (2007) The present and future of nano- technology in human health care. Nanomed Nanotech Biol Med 3:20–31
51. Cobb MD, Macoubrie J (2004) Public perceptions about nanotechnology: risks, benefits and trust. J Nanoparticle Res 6:395–405
52. Roco MC (2002) Nanoscale science and engineering educational activities in the US. J Nanoparticle Res 4:271–274

Chapter 9
Nanotechnology in Orthodontics—Futuristic Approach

Dhaval Ranjitbhai Lekhadia

9.1 Introduction

The term "Nano" in Greek means "Dwarf." One nanometer is 10^{-9} m. Thus, nanotechnology refers to manipulating matter atom by atom at the nanometer scale [1–4].

9.1.1 Definition

The universal definition of nanotechnology has not been established yet, since the existing definitions vary depending on the intended use. Therefore, current attempts to define nanotechnology can be divided into the three main groups:

1. Scientific definitions,
2. Public definitions, and
3. Those that allow making certain decisions [5].

First definition of Nanotechnology by Norio Taniguchi (1974) says that "*Nanotechnology is the production technology to get the extra high accuracy and ultra fine dimensions, i.e. the preciseness and fineness of the order of 1 nm (nanometer)* 10^{-9} m *in length*" [5]. The general definition of nanomaterials is "*Nanoscale materials can be defined as those whose characteristic length scale lies within the nanometric range, i.e. in the range between one and several hundreds of nanometers (preferably between* 0 *and* 100 nm) [6]. However, the definition applicable to dentistry can be said as "*Nanodentistry is the science and technology of maintaining*

D. R. Lekhadia (✉)
Faculty of Dentistry, Unique Hospital, Surat, Gujarat, India
e-mail: drdhavallekhadia@gmail.com

R. S. Chaughule (ed.), *Dental Applications of Nanotechnology*,
https://doi.org/10.1007/978-3-319-97634-1_9

near-perfect oral health through the use of nanomaterials including tissue engineering and nanorobotics [6].

9.1.2 Classification of Nanomaterials

According to Siegel [7], nanomaterials are classified as zero-dimensional, one-dimensional, two-dimensional, and three-dimensional. Various nanostructures include nanoparticles, nanopores, nanotubes, nanorods, nanospheres, nanofibers, nanoshells, dendrimers, and dendritic copolymers. Inorganic nanoparticles include semiconductor nanoparticles, metal nanoparticles, metal oxide nanoparticles, silica nanoparticles, polyoxometalates, and gold nanocrystals.

9.1.3 Approaches Followed in Production of Nanoparticles

There are three main approaches followed in the production of nanoparticles [8].

1. Bottom-up approach. These seek to create smaller devices by using larger ones to direct their assembly.
2. Top-down approach. These seek to arrange smaller components into more complex assembly.
3. Functional approach. This approach has the objective to produce a nanoparticle with a specific functionality.

Science and technology has witnessed the fabrication of several nanoparticles that we use in our day-to-day lives, many a times not realizing it is part of the future revolution.

9.2 Nanotechnology and Orthodontic Archwires

Friction between the archwire and bracket slot have always been of concern to orthodontists when using the friction mechanics for retraction. Even leveling and aligning of teeth will require least amount of friction to be present in order to speed up tooth movement by preventing binding of the archwire and the bracket slot. Effect of different lubricants on orthodontic brackets and archwires have been evaluated and recorded in the literature. Learl et al. [9] evaluated the effect of different lubricants on friction between orthodontic brackets and archwires. They concluded that lubrication plays a role in friction forces between self-ligating brackets and CuNiTi wires, with mucin and CMC-based artificial saliva providing a reliable alternative to natural human saliva.

Friction is higher in ceramic brackets as shown in the literature. Fidalgo et al. [10] assessed the friction coefficient between brackets and wires of different materials under conditions simulating the oral environment. They found that the system formed by metallic bracket and SS wire exhibited less roughness and lower friction. The use of ceramic brackets with TMA wire should be judiciously used as this system was found to have a higher friction coefficient.

9.2.1 Nanocoatings in Archwires

Minimizing the frictional forces between the orthodontic wire and brackets has the potential to increase the desired tooth movement and therefore results in less treatment time. Nanoparticles have been used as a component of dry lubricants in recent years. Dry lubricants are solid phase materials capable to reduce friction between two surfaces sliding against each other without the need for a liquid media.

Materials used as potent dry lubricants include inorganic fullerene-like nanoparticles of tungsten sulfide (IF-WS2). These can be used as self-lubricating coatings for orthodontic stainless steel wires. The coating consists of electrodeposited Ni film impregnated with inorganic fullerene-like nanospheres of tungsten disulfide.

Redlich et al. [10] coated stainless steel wire with nickel–phosphorus electroless film impregnated with inorganic fullerene-like nanoparticles of tungsten disulfide (IF-WS2) by inserting stainless steel (SS) wires into electroless solutions of nickel–phosphorus (Ni–P) and IF-WS2. Friction tests simulating archwire functioning of the coated and uncoated wires were carried out by an Instron machine and SEM/EDS analysis of the coated wires showed clear impregnation of the IF-WS2 nanoparticles in the Ni–P matrix. The friction forces measured on the coated wire were reduced by up to 54%.

Katz et al. [11] have shown reduction in friction between archwires and self-ligated brackets. The friction value was reduced and this reduced value was maintained throughout the tests.

9.2.2 Shape-Memory esthetic Polymer Wires

Over the past decade, there has been an increased interest in producing esthetic orthodontic wires to complement tooth-colored brackets. Shape-memory esthetic polymer is an area of potential research. These are a class of stimuli-responsive materials, which have the capacity to remember a preprogrammed shape imprinted during the synthesis; can be reformed at a higher temperature to impart a desired temporary shape; and recover their original shape when influenced by a stimulus, such as heat, light, or magnetic field [12].

These wires can also be made with clinically relevant levels of elastic stiffness. Once placed in the mouth, these polymers can be activated by the body temperature or photoactive nanoparticles activated by light and thus influence tooth movement. Future research directions in shape-memory nanocomposite polymers to produce esthetic orthodontic wires can be of interesting potential in orthodontic biomaterial research [13].

Jung et al. [12] tested the use of shape-memory polymer wires in orthodontic application, and it was found that this shape recovery force was sufficient to correct misaligned teeth in the orthodontic test. The shape-memory PU wire possesses strong potential as a novel orthodontic appliance with esthetically appealing appearance.

Shape-memory esthetic polymer is an area of potential research. Shape-memory polymers (SMPs) are materials that have the ability to memorize a macroscopic or equilibrium shape and then be manipulated and fixed to a temporary or dormant shape under specific conditions of temperature and stress. They can later relax to the original, stress-free condition under thermal, electrical, or environmental condition. This relaxation is associated with elastic deformation stored during prior manipulation. The return of the SMP toward its equilibrium shape can be accompanied by an adequate and prescribed force, useful for an orthodontic tooth movement, or macroscopic shape change, which is useful for ligation mechanisms. Due to the SMP's ability to have two shapes, these devices meet needs unattainable with current orthodontic materials allowing for easier and more comfortable procedure for the orthodontist to insert into the mouth of the patient [14, 15].

These SMPs orthodontic wires can provide improvements over traditional orthodontic materials as they will provide lighter, more constant forces which in turn may cause less pain for the patients. In addition, the SMP materials are clear, colorable, and stain resistant, providing the patient a more esthetically appealing appliance during treatment. The high percent elongation of the SMP appliance (up to about 300%) allows for the application of continuous forces over a long range of tooth movement and, hence, results in fewer visits for the patient [16, 17].

Future research directions in shape-memory nanocomposite polymers to produce esthetic orthodontic wires can be of interesting potential in orthodontic biomaterial research.

9.2.3 Hollow NiTi Wires

To improve the handling of orthodontic tools, a project involving nickel titanium nanoparticles for orthodontics was proposed. A replacement for the bulk nickel titanium wire with a textile or polymer fiber coated with NiTi nanoparticles via electrospinning and then removing the fiber could produce a hollow wire for orthodontic purposes. This wire could potentially have the same shape-memory and superelasticity properties, while possibly reducing the material needed for the wire production. After producing such a hollow NiTi wire, its properties would have to be compared with the original NiTi orthodontic wire already in use [18].

9.3 Nanotechnology and Orthodontic Adhesives

Polymer nanocomposites are a new class of materials that contain nanofillers that are 0.005–0.01 microns in size. To make filler particles of the mechanically strong composites of today (such as macrofills, hybrids, and microhybrids), one starts from dense, large particles like mined quartz, melt glasses, ceramics and comminute them to small particle size [19].

9.3.1 Reduced Polymerization Shrinkage and Increased Mechanical Properties

Due to the reduced dimension of the particles and a wide size distribution, an increased filler load can be achieved that reduces polymerization shrinkage [20] and also increases mechanical properties such as tensile and compressive strength and resistance to fracture.

Geraldeli and Perdigao [21] compared nanocomposites with that of total-etch adhesives and reported that nanocomposites had a good marginal seal to enamel and dentine compared with total-etch adhesives.

9.3.2 Literature Review of Nanocomposites

A study undertaken by Chalipa et al. in 2013 [22] reveals that nanocomposites, while maintaining the shear bond strength value of regular composites, can be successfully used for bonding orthodontic brackets.

Hosseinzadeh-Nik et al. in 2013 [23] conducted a study to determine the bond strength and the location of bond failure in the brackets bonded to the tooth with a nanocomposite. The value of the tensile bond strength for the bracket bonded with the nanocomposite was found to be more than the minimum value of tensile bond strength recommended for successful composite adhesive and nanocomposite groups.

In recent times, a nanoionomer which is resin-modified GIC (Ketac TM N1OO Light Curing Nanoionomer) has been introduced to operative dentistry [24, 25]. This light curing nanoionomer is composed of nanofillers fluoroaluminosilicate glass, and nanofiller "clusters" combined to improve mechanical properties and high fluoride release.

Uysal et al. [26] tested nanocomposite (Filtek Supreme Plus Universal) and a nanoionomer (Ketac™ N100 Light Curing Nanoionomer) restorative to determine their shear bond strength (SBS) and failure site locations in comparison with a conventional light-cure orthodontic bonding adhesive (Transbond XT). The results suggest that nanocomposites and nanoionomers may be suitable for bonding since they achieve the previously suggested SBS ranges for clinical acceptability. But they are inferior to a conventional orthodontic composite.

Bishara and Ajlouni [27] compared the shear bond strength of a nanohybrid restorative material, Grandio (Voco, Germany), and traditional adhesive material (Transbond XT; 3 M Unitek) when bonding orthodontic brackets and concluded that nanofilled composite materials can potentially be used to bond orthodontic brackets to teeth if its consistency can be made more flowable to readily adhere to the bracket base.

9.3.3 What Are Nanocomposites?

Non-agglomerated discrete nanoparticles are homogeneously distributed in resins or coatings to produce nanocomposites. The nanofiller used includes an aluminosilicate powder having a mean particle size of 80 mm and a 1:4 M ratio of alumina to silica and a refractive index of 1.508 [28].

9.3.4 Advantages of Nanocomposites

The advantages of nanocomposite materials include excellent optical properties, easy handling characteristics, and superior polishability [29]. Also, nanofillers can decrease surface roughness of orthodontic adhesives, which is one of the most significant factors for bacterial adhesion [30, 31].

Other advantages include superior hardness, flexible strength, modulus of elasticity, translucency, esthetic appeal, excellent color density, high polish and polish retention, and excellent handling properties.

9.4 Nanotechnology and Elastomeric Ligatures

Orthodontic appliances predispose to the accumulation of bacterial plaque and increase the risk of developing periodontal disease [32] and carious lesions that may rapidly progress because of the intense demineralization around the brackets and under orthodontic bands.

Studies have shown that most patients that undergo orthodontic treatment have some type of enamel lesion after the removal of the appliance [33, 34]. During orthodontic treatment, there is a significant increase in the number of *Streptococcus mutans* CFUs [33–38] and lactobacilli [39] which is associated with the development of white spots and later development of carious lesions [40].

Successful attempts to reduce susceptibility to enamel decalcification have been made with the use of fluoride agents [41–43] that enhance remineralization in the presence of critical pH, inhibit plaque glycolysis, and promote the formation of high-quality fluorapatite. Fluoride has bactericide and bacteriostatic properties that act on microorganisms in the oral cavity, such as *S. mutans* [44].

Fluoride-releasing elastomers can continuously release these agents in areas close to the areas more susceptible to demineralization [45–47]. In vitro studies have demonstrated that the initial amount of fluoride released by fluoride-releasing elastomeric ligature ties is high, but levels decrease logarithmically after one week [48–51]. The use of these materials also temporarily reduces the levels of *S. mutans* in saliva and increases enamel resistance at a 20-mm depth after one month in the oral cavity [52].

Elastomeric ligatures can serve as a carrier scaffold for delivery of nanoparticles that can be anticariogenic, anti-inflammatory, and antibiotic drug molecules embedded in the elastomeric matrix. The release of anticariogenic fluoride from elastomeric ligatures has been reported in the literature previously [53, 54]. The studies conclude that the fluoride release is characterized by an initial burst of fluoride during the first few days followed by a logarithmic decrease. For optimum clinical benefit, the fluoride ties should be replaced monthly [55].

Banks et al. conducted a prospective controlled clinical trial to evaluate the effectiveness of stannous fluoride-releasing elastomeric modules (Fluor I Ties) and chain (Fluor I Chain) in prevention of decalcification of enamel during fixed orthodontic therapy [45]. It was concluded that the use of fluoride-releasing elastomeric modules and chains reduced post fixed appliance treatment enamel decalcification scores per tooth by 49 percent. Significant differences in decalcification between the two groups were seen in all but the occlusal enamel zones. Fluoride-releasing elastomerics appear to provide a clinically worthwhile reduction in enamel decalcification during fixed appliance therapy when they are changed at each treatment visit.

9.5 Nanoenamel Remineralization Agents in Orthodontics

Control of oral biofilms during orthodontic treatment is very important in order to prevent dental caries and maintain a healthy to tooth structure. There are number of published papers that have studied possibilities for dental caries prevention of using the benefits of nanotechnology [56–59].

9.5.1 Antibacterial Nanoparticles

NPs (nanoparticles) present a greater surface-to-volume ratio (per unit mass) when compared with non-nanoscale particles, interacting more closely with microbial membranes and provide considerably larger surface area for antimicrobial activity. Metal NPs in the size range of 1–10 nm have particularly shown the greatest biocidal activity against bacteria [60]. Silver has a long history of use in medicine as an antimicrobial agent [61]. The antibacterial properties of NPs have been used through the mechanism of combining dental materials with NPs or coating surfaces with NPs to prevent microbial adhesion, with the aim of reducing biofilm formation [62, 63]. Resin composites containing silver ion-implanted fillers that release silver ions have been found to have antibacterial effects on oral streptococci [64].

9.5.2 Review of Nanoenamel Remineralization Agents and Antibacterial Properties

Ahn et al. [65] compared an experimental composite adhesives (ECAs) containing silica nanofillers and silver nanoparticles with two conventional composite adhesives and resin-modified glass ionomer cement(RMGIC) to study surface characteristics, physical properties, and antibacterial activities against cariogenic streptococci. The results suggest that ECAs had rougher surfaces than conventional adhesives due to the addition of silver nanoparticles. Bacterial adhesion to ECAs was less than to conventional adhesives, which was not influenced by saliva coating. Bacterial suspension containing ECAs showed slower bacterial growth than those containing conventional adhesives. There was no significant difference in shear bond strength and bond failure interface between ECAs and conventional adhesives. This study suggested that ECAs can help prevent enamel demineralization around brackets without compromising physical properties.

Lin et al. (2011) showed that incorporation of nanofluorapatite in resin-modified glass ionomer cement lead to improvement in the fluoride release properties and reducing decalcification during fixed appliance treatment than other preventive modalities [66].

Liu et al. [67] performed a study to access the antimicrobial and osteoinductive properties of stainless steel wire coated with silver nanoparticles. The results show that stainless steel wire coated with silver nanoparticles exhibit strong bactericidal and osteoinductive properties that make it a promising pharmaceutical material in orthopedic surgery [67].

On examining the antimicrobial activity of supra-gingival plaque on nanoscale HA-coated Ti against early and late colonizers, Murakami et al. [68] observed that HA-coated Ti may have potential for use as implants [68].

Zhao et al. [69] investigated nanohardness, wear resistance, and pseudoelasticity of hafnium implanted NiTi shape-memory alloy and found it to exhibit better wear resistance than the untreated NiTi in the aspects of a lower initial friction coefficient with a much longer fretting time. Also, the Hf-NiTi exhibited improved pseudoelastic behavior and retained their surface integrity even after being strained in tension to 10% [69].

The study performed by Borzabadi-Farahani et al. [22] shows that although the use of nanoparticles in orthodontics can offer new possibilities, information on the long-term performance of orthodontic material using nanotechnology is lacking and necessitates further investigation and so do possible safety issues (toxicity) , which can be related to the size of the nanoparticles [22].

A study undertaken by Chalipa et al. [70] reveals that nanocomposites, while maintaining the SBS value of regular composites, can be successfully used for bonding orthodontic brackets [70].

Hosseinzadeh-Nik et al. [71] conducted a study to determine the bond strength and the location of bond failure in the brackets bonded to the tooth with a nanocomposite. The value of tensile bond strength for the bracket bonded with the nanocomposite was found to be more than the minimum value of tensile bond strength, recommended for successful composite adhesive and nanocomposite groups [71].

Zhang et al. [72] undertook a study to investigate the effects of dentine primer containing dual antibacterial agents, namely 12-methacryloyloxy dodecyl pyridinium bromide (MDPB) and nanoparticles of silver (NAg), on dentine bond strength, dental plaque microcosm biofilm response, and fibroblast cytotoxicity. The results indicate that the method of using dual agents MDPB + NAg in the primer yielded potent antibacterial properties and, thus, this method may be promising to combat residual bacteria in tooth cavity and invading bacteria at the margins [72].

Argueta-Figueroa et al. [73] developed a process to prepare bimetallic Cu–Ni nanoparticles as well as tested their antibacterial activity against standard human pathogens. While the experimental results were found to be promising for potential use in dental materials science, a further investigation was deemed to be necessary [73].

Gao et al. [74] studied the effect of titanium nanotubes with embedded silver oxide nanoparticles on bacteria and osteoblasts and observed that the said arrays can effectively kill Escherichia coli and Staphylococcus aureus even after immersion for 28 days, demonstrating the long-lasting antibacterial ability [74].

Mhaske et al. [75] performed a study to access the antiadherent and antibacterial properties of surface-modified stainless steel and NiTi orthodontic wires with silver nanoparticles against *Lactobacillus acidophilus*. The authors observed that the wires coated with silver showed an antiadherent effect against *L. acidophilus* leading to minimization of accumulation of dental plaque and, thus, the minimization of dental caries during orthodontic treatment [75].

Gao et al. [76] presented a novel strategy to control plaque biofilms using catalytic nanoparticles (CAT-NP) with peroxidase-like activity that trigger extracellular matrix degradation and cause bacterial death within acidic niches of caries-causing biofilm [76].

Medeiros et al. [77] undertook a study to analyze the effect of calcium nanophosphate paste, fluoride gel, and varnish to protect against enamel erosion. Results showed that the calcium nanophosphate paste showed similar protection against enamel erosion compared with high-concentrated fluoride agents, even containing lower fluoride concentration [77].

An experimental study was done in 2008 by Lackovic et al. in which it was concluded that silver nanoparticles reduce the attachment of *S. mutans* and thus reduced biofilm formation [78].

Elsaka et al. in their study in 2011 concluded that addition of titanium dioxide nanoparticles to glass ionomer cement improved its mechanical and antibacterial properties [79].

9.6 Nanotechnology and Orthodontic Miniscrews

Orthodontic microimplants are manufactured from materials such as commercially pure titanium, surgical stainless steel, and titanium alloys (Ti6Al4V). Several factors including osseointegration at the bone-implant interface and the amount of bacterial colonization around the implants influence the final characteristics and outcomes of dental and orthodontic microimplants [80]. According to previous studies, the long-term survival or stability of implants primarily depends on the inflammatory condition that causes peri-implantitis and loss of supporting bone [81]. Inflammation leads to progressively escalating damage to the cortical bone, especially around the neck of the implants [82, 83]. Peri-implant mucositis, an inflammatory lesion, not only affects the soft tissues but also damages the supporting bone [84]. According to a previous study, 43% of individuals with implants suffer from peri-implantitis, and peri-implant mucositis was seen in up to 50% of the cases [85]. Thus, limiting the amount of inflammation is vital for ensuring long-term stability and success of the implant. Infection of the implants and their subsequent loosening is most commonly seen in titanium-based prosthesis and can largely be avoided by inhibiting the adhesion of microbes on the surface of the implanted devices [86]. The initial colonizers that adhere to tooth and implant surfaces include *Streptococcus oralis*, *Streptococcus sanguinis*, and *S. mutans*. In addition, Aggregatibacter actinomycetemcomitans, a gram-negative bacterium, is also essentially known to be responsible for many periodontal and peri-implant diseases [87].

9.6.1 Nanoparticles in Miniscrews to Prevent Infection of Implants

Many methods including surface treatments such as polishing and modification of surface free energy [88–91] have been employed to reduce or nullify the bacterial aggregation around titanium-based prosthesis. Elemental silver has been specifically known for its array of antimicrobial properties for many years [92].

Many previous studies have demonstrated the excellent antibacterial properties of silver-impregnated hydroxylapatite films as well as silver zeolite against different cultured bacteria [93–95].

It has been speculated that several mechanisms are involved in the antibacterial potency of nanosilver. Due to an extraordinarily large surface area, silver nanoparticles (AgNPs) exhibit better antibacterial properties than metallic silver [96]. Therefore, in the recent past, the amalgamation of AgNPs with different biomaterials to improve the biocompatibility and antibacterial capability has been intensely studied, and much incorporation such as silver-doped hydroxylapatite, polymer AgNPs, and AgNPs on TiO_2 have been developed [97–99].

Vanugopal et al. surface-treated titanium microimplants with AgNPs to achieve antibacterial properties and concluded that Ti-BP-AgNPs surface had remarkable antibacterial activity toward the oral bacteria S. mutans, S. sanguinis, and A. actinomycetemcomitans. These data suggest that Ti-BP-AgNPs has excellent antibacterial properties, making it a promising implantable biomaterial [100].

9.6.2 Ideal Mini-Implant Surface Using Nanoparticles

It is postulated that the balance lies in the fabrication of an ideal surface that could stimulate initial osseointegration and facilitate its removal once the TAD is no longer needed. Biocompatible coatings like titanium nanotubes can be studied to evaluate if the nanotubular layer can enhance initial osseointegration and can serve as an interfacial layer between the newly formed bone and the TAD [101].

Recently, three nanostructured implant coatings are developed.

1. Nanostructured diamond: They have ultrahigh hardness, improved toughness over conventional microcrystalline diamond, low friction, and good adhesion to titanium alloys [102].
2. Nanostructured processing applied to hydroxyapatite coatings: This is used to achieve the desired mechanical characteristics and enhanced surface reactivity and has been found to increase osteoblast adhesion, proliferation, and mineralization [102].
3. Nanostructured metalloceramic coatings: These provide continuous variation from a nanocrystalline metallic bond at the interface to the hard ceramic bond on the surface [102]. Nanostructured ceramics, carbon fibers, polymers, metals, and composites enhance osteoblast adhesion and calcium/phosphate mineral deposition.

Studies have suggested that nanophase ZnO and TiO_2 may reduce S. epidermidis adhesion and increase osteoblast functions necessary to promote the efficacy of orthopedic implants [103].

9.7 Nanotechnology for Rapid Tooth Movement in Orthodontics

9.7.1 Terminologies

1. Bio-MEMS (Biomedical Microclectromechanical systems): It is defined as the science and technology of operating at the microscale for biological and biomedical applications, which may or may not include any electronic or mechanical functions. They are made up of micromachined elements usually on silicon substrates, including gears, motors, and actuators with linear and rotary motion for applications to biological systems.
2. NEMS (Nanoelectromechanical systems): These are devices integrating electrical and mechanical functionality on the nanoscale level.

9.7.2 Application of Bio-MEMS and NEMS

Evidence suggests that orthodontic tooth movement can be enhanced by supplementing the mechanical forces with electricity [104, 105].

Animal experiments indicated that when 15–20 μA of low direct current (DC) was applied to the alveolar bone by modifying the bioelectric potential osteoblasts and periodontal ligament cells demonstrated increased concentrations of the second messengers, cAMP and cGMP. These findings suggest that electric stimulation enhanced cellular enzymatic phosphorylation activities, leading to synthetic and secretory processes associated with accelerated bone remodeling.

However, the intraoral source of electricity is a major problem that has to be addressed. It has been proposed that microfabricated biocatalytic fuel cells (enzyme batteries) can be used to generate electricity to aid orthodontic tooth movement. An enzymatic microbattery when placed on the gingiva near the alveolar bone might be a possible electrical power source for accelerating orthodontic tooth movement. It is proposed that this device uses organic compound (glucose) as the fuel and is noninvasive and nonosseointegrated. The enzyme battery can be fabricated with the combination of two enzyme electrodes and biocatalysts such as glucose oxidase or formate dehydrogenase to generate electricity. However, there are several issues like soft tissue biocompatibility, effect of food with different temperature and pH range on the output of such microfabricated enzyme battery that need to be addressed. The use of microenzyme batteries has issues like enzyme stability, electron transfer rate, and enzyme loading which result in shorter lifetime and poor power density. Many nanostructured materials, such as mesoporous media, nanoparticles, nanofibers, and nanotubes, have been demonstrated as efficient hosts of enzyme immobilization. When nanostructure of conductive materials is used, the large surface area of these nanomaterials can increase the enzyme loading and facilitate reaction kinetics, and thus improve the power density of the biofuel cells [106].

It is expected that the MEMS-/NEMS-based system will be applied over the next few years to develop biocompatible powerful biofuel cells, which can be safely implanted in the alveolus of the maxilla or mandible to enhance orthodontic tooth movement.

9.8 Nanotechnology for Bone Growth and Prevention of Root Resorption in Orthodontics

9.8.1 NanoLIPUS Devices for Bone Growth

Ultrasound (US) is a form of mechanical energy that is transmitted through and into biological tissues as an acoustic pressure wave at frequencies above the limit of human hearing, is used widely in medicine as a therapeutic, operative, and diagnostic tool [107, 108].

Low-intensity pulsed US (LIPUS) has been reported to be effective in liberating preformed fibroblast growth factors from a macrophage-like cell line (U937), and it enhances angiogenesis during wound healing [109].

Also, LIPUS has been reported to enhance bone growth into titanium porous-coated implants [110] and bone healing after fracture [111, 112] and after mandibular distraction osteogenesis [113]. The specific mechanisms by which US stimulation works on bone cell activities are still unknown.

El-Bialy et al. [114] applied LIPUS on the temporomandibular joint (TMJ) region of growing rabbits and baboon monkeys for 20 min daily. Their results show a significant increase in mandibular cartilaginous growth, especially under chronic mandibular advancement [115].

In another study by Oyonartea et al. [116] experimental rats were stimulated with LIPUS in the TMJ region unilaterally, for 10 or 20 min for 20 days. The results showed that LIPUS application may affect mandibular growth pattern in rats acting at the cartilage and bone level.

9.8.2 NanoLIPUS Devices to Prevent Root Resorption in Orthodontics

El-Bialy et al. [117] observed that LIPUS can promote dental tissue formation in rabbits and concluded that it may be used to treat root resorption. Similar results were found by Liu et al. [118].

The initial devices were bulky but with nanotechnology nanoLIPUS device can be made with system on a chip design. The wireless design of the ultrasound transducer means the miniscule device will be able to fit comfortably inside a patient's mouth while packed with biocompatible materials. The unit will be easily mounted on a bracket or even a plastic removable crown. An energy sensor can also be used that will ensure the LIPUS power is reaching the target area of the teeth roots within the bone.

9.9 Nanotechnology and Orthodontic Brackets

9.9.1 Smart Brackets with Nanomechanical Sensors

Quantitative knowledge of the three-dimensional (3D) forcemoment systems applied for orthodontic tooth movement is of utmost importance for the predictability of the course of tooth movement as well as the reduction of traumatic side effects. The concept of a smart bracket with integrated sensor system for 3D force and moment measurement has recently been published. Nanomechanical sensors can be fabricated and be incorporated into the base of orthodontic brackets in order to provide real-time feedback about the applied orthodontic forces. This real-time feedback allows the orthodontist to adjust the applied force to be within a biological range to efficiently move teeth with minimal side effects.

Lapaki et al. [118, 119] reported on the introduction of a "smart" bracket for multidimensional force and moment control. They reported on a large-scale prototype bracket that utilized microsystem chip encapsulated into small low profile contemporary bracket systems with reduced dimensions to allow clinical testing of this technology. They concluded that their methodological approach is generally suitable for monitoring the relatively low forces and moments exerted on individual teeth with fixed orthodontic appliances.

9.9.2 Nanorobotics Coupled with Smart Brackets—Futuristic Approach

Nanorobotics centers are self-sufficient machines which are functional at the nanoscale. The nanorobot design consists of a biocompatible glycocalyx-coated diamondoid material with molecular sorting rotors and a robot arm (telescoping manipulator) [120]. Different nanorobot molecule types are distinguished by a series of chemotactic sensors, and their functioning is controlled by a stimulator.

Nanorobots may be used for manipulation of tissues directly at nanolevel and research has begun on the use of nanorobotics for medical applications like drug delivery, management of aneurysms, and tumors.

The theory of use of such nanorobots could be extended to dentistry and orthodontics in distant future, where nanorobots with specific motility mechanisms would navigate through periodontium to remodel it directly allowing accelerated orthodontic tooth movement. Nanorobotics when coupled with smart brackets could bring about accelerated tooth movement with relatively low forces and moments exerted on individual teeth with fixed orthodontic appliances.

9.10 Probable Risks of Nanotechnology

Postorthodontic treatment, infection based on the growth of bacteria remains one of the most common complications. Nanomaterials can be used to load and deliver antibacterial agents to targeted spaces inside the oral cavity. However, scientists [121, 122] have concerns about the unknown harmful impacts of these materials on the human body by inhalation into the lungs.

Yao et al. [123] and Zhang et al. [124] mention that there are certain negative impacts of nanotechnology on environment in many ways, such as increased toxicological pollution on the environment due to the uncertain shape, size, and chemical compositions of some of the nanotechnology products (or nanomaterials). The authors opine that it is necessary to establish environmental risk assessment upon nanomaterials, which will finally promote the developments in the fields of nanotechnology.

The amount of free nanoparticles in nature depends on various factors such as their physicochemical properties, quantity, and time of exposure. Nanomaterials released in the environment can be further modified by temperature, pH, different biological conditions, and presence of other pollutants.

In this interaction nanomaterials can alter atmosphere, soil, and water. These transformations and interactions can adversely affect the current state of the environment and be harmful to human health and balance of the ecosystems [125, 126].

Nanoparticles have unique physicochemical properties that can be different in comparison with the same material in macroscale size. The main characteristic of these particles is greater difference between the surface area and unit weight, and higher surface reactivity [127–129]. It might lead to increased absorption through the lungs, skin, digestive tract, and it might cause side effect to the lungs and other organs [130, 131]. Nanoparticles are so small that they can interact with DNA, RNA, and other intracellular components.

The question is what kind of immune response could be activated by these small particles, and what will be the metabolism, absorption, and elimination path of these products? There is a need for systemic solutions, monitoring, and recording of potential hazard as well as finding timely responses in order to achieve safety for human health and environment.

9.11 Conclusion

There is a huge potential in research in this area including nanodesigned orthodontic bonding material, possible nanovector for gene delivery for mandibular growth stimulation and nanoLIPUS devices. A successful future for nanotechnology will only be achieved through open sharing of ideas and research. Although biosafety of nanoparticles and materials is a subject of concern, the future in orthodontic treatment will benefit enormously through nanotechnology should all the current attempts succeed to its clinical application at a reasonable cost to the orthodontist and patients.

References

1. Taniguchi N (1974) On the basic concept of nano technology. Proc ICPE Tokyo 21:8–23
2. Schleyer TL (2000) Nanodentistry. Fact or fiction? J Am Dent Assoc 131:1567–1568
3. Whitesides GM, Love JC (2001) The art of building small. Sci Am 285:33–41
4. Kaehler T (1994) Nanotechnology: Basic concepts and definitions. Clin Chem 40:1797–1799
5. Balogh L (2010) Why do we have so many definitions for nanoscience and nanotechnology? Nanomed Nanotechnol Biol Med 6:397–398
6. Govindankutty D (2015) Applications of nanotechnology in orthodontics and its future implications: a review. Int J Appl Dent Sci 1(4):166–171
7. "Nanomaterial"s—From Wikipedia, the free encyclopedia en.wikipedia.org/wiki/Nanomaterials
8. Silva GA (2004) Introduction to nanotechnology and its applications to medicine. Surg Neurol 61:216–220
9. Leal RC, Amaral FLB, França FMG, Basting RT, Turssi CP (2014) Role of lubricants on friction between self-ligating brackets and archwires. Angle Orthod 84(6):1049–1053
10. Redlich M, Katz A, Rapoport L, Wagnerb HD, Feldmanb Y (2008) Improved orthodontic stainless steel wires coated with inorganic fullerene-like nanoparticles of WS2 impregnated in electroless nickel–phosphorous film. Dent Mater 24:1640–1646
11. Katz A, Redlich M, Rapoport L, Wagner HD, Tenne R (2006) Self-lubricating coatings containing fullerene-like WS2 nanoparticles for orthodontic wires and other possible medical applications. Tribol Lett 21:135–139
12. Jung YC, Cho JW (2010) Application of shape memory polyurethane in orthodontic. J Mater Sci Mater Med 21(10):2881–2886
13. Urban M, Strankowski M (2017) Shape memory polyurethane materials containing ferromagnetic iron oxide and graphene nanoplatelets. Materials 10:1083
14. Gunes S, Jana C (2008) Shape memory polymers and their nanocomposites: a review of science and technology of new multifunctional materials. J Nanosci Nanotechnol 8(4):1616–1637
15. Stuart MA, Huck WT, Genzer J, Muller M, Ober C, Stanm M (2010) Emerging applications of stimuli-responsive polymer materials. Nat Mater 9:110–113
16. Meng Q, Hu J (2009) A review of shape memory polymer composites and blends. Compos Appl Sci Manuf 40(11):1661–1672
17. Leng J, Lan X, Liu Y, Du S (2011) Shape-memory polymers and their composites: stimulus methods and applications. Prog Mater Sci 56(7):1077–1135
18. Majeric P, Rudolf R, Anzel I, Bogovic J, Stopic S et al (2015) Synthesis of NiTi/Ni-TiO_2 composite nanoparticles via ultrasonic spray pyrolysis. Mater Technol 49(1):75–80
19. Mitra SB, Dong WU, Holmes BN (2003) An application of nanotechnology in advanced dental materials. J Am Dent Assoc 134:1382–1390
20. Moszner N, Salz U (2001) New developments of polymeric dental composites. Program Polymer Sci 26:535–576
21. Geraldeli S, Perdigao J (2003) Microleakage of a new restorative system in posterior teeth. J Dent Res 81(special issue A):126 (Abstract)
22. Chalipa J et al (2013) Evaluation of shear bond strength of orthodontic brackets bonded with nano filled composites. J Dent (Tehran) 10(5):461–465

23. Hosseinzadeh-Nik T, Karimzadeh A, Ayatollahi MR (2013) Bond strength of nano composites used for bonding ceramic orthodontic brackets. Mater Des 51:902–906
24. Killian CM, Croll TP (2010) Nano-ionomer tooth repair in pediatric dentistry. Pediatr Dent 32(7):330–535
25. Kotmaj V (2010) Influence of different conditioning methods on shear bond strength of novel light curing nano-ionomer restorative to enamel and dentin. Lawers Med Sci 25:861–866
26. Uysal T, Yagci A, Uysal B, Akdogan G (2010) Are the nano-composites and nano-ionomers suitable for orthodontic bracket bonding? Eur J Orthod 32:78–82
27. Bishara SE, Ajlouni R, Soliman MM, Oonsombat C, Laffoon JF, Warren J (2007) Evaluation of a new nano- filled restorative material for bonding orthodontic brackets. World J Orthod 8:8–12
28. Saravana Kumar R, Vijayalakshmi R (2006) Nanotechnology in dentistry. Ind J Dent Res 17(2):62–65
29. Lee YK, Lim BS, Rhee SH, Yang HC, Powers JM (2004) Changes of optical properties of dental nano-filled resin composites after curing and thermocycling. J Biomed Mater Res 71B:16–21
30. Quirynen M, Bollen CM (1995) The influence of surface roughness and surface free-energy on supra- and subgingival plaque formation in man. J Clin Periodontol 22:1–14
31. Quirynen M, Marechal M, Busscher HJ, Weerkamp AH, Darius PL, van Steenberghe D (1990) The influence of surface free energy and surface roughness on early plaque formation, an in vivo study in man. J Clin Periodontol 17:138–144
32. Diamanti-Kipioti A, Gusberti FA, Lang NP (1987) Clinical and microbiological effects of fixed orthodontic appliances. J Clin Periodontol 14(6):326–333
33. Gorelick L, Geiger AM, Gwinnett AJ (1982) Incidence of white spot formation after bonding and banding. Am J Orthod Dentofacial Orthop. 81(2):93–98
34. Ogaard B (1989) Prevalence of white spot lesions in 19-year-olds: a study on untreated and orthodontically treated persons 5 years after treatment. Am J Orthod Dentofacial Orthop. 96(5):423–427
35. Balenseifen JW, Madonia JV (1970) Study of dental plaque in orthodontic patients. J Dent Res 49(2):320–324
36. Fournier A, Payant L, Bouclin R (1998) Adherence of Streptococcus mutans to orthodontic brackets. Am J Orthod Dentofacial Orthop 114(4):414–417
37. Rosenbloom RG, Tinanoff N (1991) Salivary Streptococcus mutans levels in patients before, during, and after orthodontic treatment. Am J Orthod Dentofacial Orthop. 100(1):35–37
38. Scheie AA, Arneberg P, Krogstad O (1984) Effect of orthodontic treatment on prevalence of Streptococcus mutans in plaque and saliva. Scand J Dent Res 92(3):211–217
39. Chang HS, Walsh LJ, Freer TJ (1999) The effect of orthodontic treatment on salivary flow, pH, buffer capacity and levels of mutans streptococci and lactobacilli. Aust Orthod J 15(4):229–234
40. Ikeda T, Sandham HJ, Bradley EL (1973) Changes in Streptococcus mutans and lactobacilli in plaque in relation to the initiation of dental caries in Negro children. Arch Oral Biol 18(4):555–566
41. Mitchell L (1992) Decalcification during orthodontic treatment with fixed appliances—an overview. Br J Orthod 19(3):199–205
42. Todd MA, Staley RN, Kanellis MJ, Donly KJ, Wefel JS (1999) Effect of a fluoride varnish on demineralization adjacent to orthodontic brackets. Am J Orthod Dentofacial Orthop 116(2):159–167
43. Zachrisson BU (1975) Fluoride application procedures in orthodontic practice, current concepts. Angle Orthod 45(1):72–81
44. Lundström F, Krasse B (1987) Streptococcus mutans and lactobacilli frequency in orthodontic patients; the effect of chlorhexidine treatments. Eur J Orthod 9(2):109–116
45. Banks PA, Chadwick SM, Asher-McDade C, Wright JL (2000) Fluoride-releasing elastomerics—a prospective controlled clinical trial. Eur J Orthod 22(4):401–407

46. Benson PE, Douglas CW, Martin MV (2004) Fluoridated elastomers: effect on the microbiology of plaque. Am J Orthod Dentofacial Orthop. 126(3):325–330
47. Mattick CR, Mitchell L, Chadwick SM, Wright J (2001) Fluoride-releasing elastomeric modules reduce decalcification: a randomized controlled trial. J Orthod 28(3):217–219
48. Doherty UB, Benson PE, Highan SM (2002) Fluoride-releasing elastomeric ligatures assessed with the in situ caries model. Eur Orthod Soc 24(4):371–378
49. Joseph VP, Grobler SR, Rossouw PE (1993) Fluoride release from orthodontic elastic chain. J Clin Orthod 27(2):101–105
50. Storie DJ, Regennitter F (1994) Characteristics of fluoride-releasing elastomeric chain. Angle Orthod 64(3):199–210
51. Wiltshire WA (1996) Determination of fluoride from fluoride-releasing elastomeric ligature ties. Am J Orthod Dentofacial Orthop 110(4):383–387
52. Wilson TG, Love B (1995) Clinical effectiveness of fluoride-releasing elastomers. II: Enamel microhardness levels. Am J Orthod Dentofacial Orthop 107(4):379–381
53. Wiltshire WA (1996) Determination of fluoride from fluoridereleasing elastomeric ligature ties. Am J Orthod Deniofacial Thop 110(4):383–387
54. Wiltshire WA (1999) In vitro and in vivo fluride release from orthodonic elastoomerc ligature ties. Am J Orthod Dentofac Orthop 115(3):288–292
55. Miura KK (2007) Anitcariogenic effect of fluoride-releasing elastomers in orthodontic patients. Braz Oral Res 21(3):228–233
56. Hannig M, Hannig C (2010) Nanomaterials in preventive dentistry. Nat Nanotech 5:565–569
57. Azami M, Jalilifiroozinezhad S, Mozafari M, Rabiee M (2011) Synthesis and solubility of calcium fluoride/hydroxy-fluorapatite nanocrystals for dental applications. Ceram Int 37:2007–2014
58. Tschoppe P, Zandim D, Martus P, Kielbassa A (2011) Enamel and den- tine remineralization by nano-hydroxyapatite toothpastes. J Dent 39:430–437
59. Mendonça G, Mendonça D, Aragão F, Cooper L (2008) Advancing dental implant surface technology—from micron- to nanotopography. Biomaterials 29:3822–3835
60. Allaker RP (2010) The use of nanoparticles to control oral biofilm formation. J Dent Res 89:1175–1186
61. Hill WR, Pillsobury DM (1939) The pharmacology of silver. The Williams & Wilkins Co, Baltimore
62. Hannig M, Kriener L, Hoth-Hannig W, Becker-Willinger C, Schmidt H (2007) Influence of nanocomposite surface coating on biofilm formation in situ. J Nanosci Nanotechnol 7:4642–4648
63. Monteiro DR, Gorup LF, Takamiya AS, Ruvollo-Filho AC, De Camargo ER, Barbosa DB (2009) The growing importance of materials that prevent microbial adhesion: antimicrobial effect of medical devices containing silver. Int J Antimicrob Agents 34:103–110
64. Yamamoto K, Ohashi S, Aono M, Kokubo T, Yamada I, Yamauchi J (1996) Antibacterial activity of silver ions implanted in SiO_2 filler on oral streptococci. Dent Mater 12:227–229
65. Lee SJ, Kook JK, Lim BS (2009) Experimental antimicrobial orthodontic adhesives using nanofillers and silver nanoparticles. Dent Mater 25(2):206–213
66. Lin J, Zhu J, Gu X, Wen W, Li Q, Fischer Brandies H et al (2011) Effects of incorporation of nano-fluorapatite or nano-fluoro hydroxyapatite on a resin modified glass ionomer cement. Acta Biomater 7:1346–1353
67. Liu Y, Zheng H, Zara JN, Hsu C, Soofer DE, Lee KS et al (2012) The antimicrobial and osteoinductive properties of silver nanoparticle/poly (dl-lactic-co-glycolic acid)-coated stainless steel. Biomaterials 33:8745–8756
68. Murakami A, Arimoto T, Suzuki D, Iwai Yoshida M, Otsuka F, Shibata Y et al (2012) Antimicrobial and osteogenic properties of a hydrophilic modified nanoscale hydroxyapetite coating on titanium. Nanomed Nanotech Biol Med 8:374–382
69. Zhao T, Li Y, Liu Y, Zhao X (2012) Nanohardness, wear resistance and pseudo elasticity of hafnium implanted Ni-Ti shape memory alloy. J Mech Behav Biomed Mater 13:174–184

70. Borzabad Farahani A, Borzabad E, Lynch E (2013) Nanoparticles in orthodontics, A review of antimicrobial and anticaries application. Acta Odondol Scand 72:1–5
71. Hossaeinzadeh-Nik T, Karimzadeh A, Ayatollahi MR (2013) Bond strength of nanocomposites used for bonding ceramic orthodontic brackets. Mater Des 51:902–906
72. Zhang K, Cheng L, Emazato S, Antonucci JM, Lin NJ, Lin-Gibson S et al (2013) Effects of dual antibacterial agents MDPB and nanosilver in primer on micro caused biofilm, cytotoxicity and dentine bond properties. J Dent 41:464–474
73. Argueta Figuerao L, Morales-Luckie RA (2014) Scougall characterization and antibacterial activity of copper, nickel and bimetallic Cu-Ni for potential use in dental materials. Prog Nat Sci 24(321–328):22
74. Gao A, Hang R, Huang X, Zhao L, Zhang X, Wang L et al (2014) The effects of titania nanotubes with embedded silver oxide nanoparticles on bacteria and osteoblast. Biomaterials 35:4223–4235
75. Mhaske AR, Shetty PC, Bhatt NS, Ramachandra CS, Laxmikanth SM, Nagarahalli K et al (2015) Antiadherant and antibacterial properties of stainless steel and Ni-TI orthodontic wires coated with silver against Lactobacillus acidophilus-an in vitro study. Prog Orthod 16(40):1–6
76. Gao L, Liu Y, Kim D, Li Y, Hwang G, Naha PC et al (2016) Nanocatalyst promote streptococcus mutans biofilm matrix degradation and enhance bacterial killing to suppress dental caries in vivo. Biomaterials 101:272–284
77. Medeiros IC, Brasil VLM, Carlo HL, Santo RLD, De Lima B, De Carvalho FB (2014) In vitro effect of calcium nanophosphate and high concentrated fluoride agents on enamel erosion: An AFM study. Int J Paediatr Dent. 24:168–174
78. Lackovic KP, Furman B, Hatch JP, Whang K, Chu L, Rawls HR. Silver nanoparticles to Improve Antimicrobial Properties of Orthodontic Bracket-Bonding Cement.iadr. 2008. Toronto techprogram.abstract _109194
79. Elsaka SE, Hamouda IM, Swain MV (2011) Titanium dioxide nanoparticles addition to a conventional glass-ionomer restorative: influence on physical and antibacterial properties. J Dent 39:589–598
80. Schierholz JM, Beuth J (2001) Implant infections: a haven for opportunistic bacteria. J Hosp Infect 49:87–93
81. Quirynen M, De Soete M, van Steenberghe D (2002) Infectious risks for oral implants: a review of the literature. Clin Oral Implants Res 13:1–19
82. Zitzmann NU, Berglundh T, Ericsson I, Lindhe J (2004) Spontaneous progression of experimentally induced periimplantitis. J Clin Periodontol 31:845–849
83. Ericsson I, Berglundh T, Marinello C, Liljenberg B, Lindhe J (1992) Long-standing plaque and gingivitis at implants and teeth in the dog. Clin Oral Implants Res 3:99–103
84. Lindhe J, Meyle J (2008) Peri-implant diseases: consensus report of the Sixth European workshop on periodontology. J Clin Periodontol 35(8 Suppl):282–285
85. Zitzmann NU, Berglundh T (2008) Definition and prevalence of peri-implant diseases. J Clin Periodontol 35(8 Suppl):286–291
86. Wadström T (1989) Molecular aspects of bacterial adhesion, colonization, and development of infections associated with biomaterials. J Invest Surg 2:353–360
87. Liao J, Anchun M, Zhu Z, Quan Y (2010) Antibacterial titanium plate deposited by silver nanoparticles exhibits cell compatibility. Int J Nanomedicine 5:337–342
88. Meredith DO, Eschbach L, Riehle MO, Curtis AS, Richards RG (2007) Microtopography of metal surfaces influence fibroblast growth by modifying cell shape, cytoskeleton, and adhesion. J Orthop Res 25:1523–1533
89. Chang YY, Huang HL, Lai CH, Hsu JT, Shieh TM, Wu AY et al (2013) Analyses of antibacterial activity and cell compatibility of titanium coated with a Zr-C-N film. PLoS One 8:e56771
90. Oh EJ, Nguyen TDT, Lee SY, Jeon YM, Bae TS, Kim JG (2014) Enhanced compatibility and initial stability of Ti6Al4V alloy orthodontic miniscrews subjected to anodization, cyclic recalcification, and heat treatment. Korean J Orthod 44:246–253
91. Cho YC, Cha JY, Hwang CJ, Park YC, Jung HS, Yu HS (2013) Biologic stability of plasma ion-implanted miniscrews. Korean J Orthod 43:120–126

92. Crede CSF (1884) Die verhutung der augenentzundung der neugeborenen (Ophthalmoblenn-orrhoea neonatorum) der haufigsten und wuchtigsten ursache der blindheit. A. Hirschwald, Berlin
93. Buckley JJ, Lee AF, Olivic L, Wilsonb K (2010) Hydroxyapatite supported antibacterial Ag_3PO_4 nanoparticles. J Mater Chem 20:8056–8063
94. Ciobanu CS, Massuyeau F, Constantin LV, Predoi D (2011) Structural and physical properties of antibacterial Ag-doped nano-hydroxyapatite synthesized at 100 °C. Nanoscale Res Lett 6:613
95. Hotta M, Nakajima H, Yamamoto K, Aono M (1998) Antibacterial temporary filling materials: the effect of adding various ratios of Ag-Zn-Zeolite. J Oral Rehabil 25:485–489
96. Kvítek L, Panáćek A, Soukupová J, Kolář M, Večeřová R, Prucek R et al (2008) Effect of surfactants and polymers on stability and antibacterial activity of silver nanoparticles (NPs). J Phys Chem C 112:5825–5834
97. Rusu VM, Ng CH, Wilke M, Tiersch B, Fratzl P, Peter MG (2005) Size-controlled hydroxyapatite nanoparticles as self-organized organic-inorganic composite materials. Biomaterials 26:5414–5426
98. Lim SI, Zhong CJ (2009) Molecularly mediated processing and assembly of nanoparticles: exploring the interparticle interactions and structures. Acc Chem Res 42:798–808
99. Zheng J, Yu H, Li X, Zhang S (2008) Enhanced photocatalytic activity of TiO_2 nano-structured thin film with a silver hierarchical configuration. Appl Surf Sci 254:1630–1635
100. Muthuchamy AVN, Anantha-Iyengar-Gopalanc HT, Leeb KP, Leee Hee Moon Kyunga HJ (2017) Incorporation of silver nanoparticles on the surface of orthodontic microimplants to achieve antimicrobial properties. Korean J Orthod 47(1):3–10
101. Govindankutty D (2015) Applications of nanotechnology in orthodontics and its future implications: a review. Int J Appl Dent Sci 1(4):166–171
102. Colon G, Ward BC (2006) Webster TJ Increased osteoblast and decreased Staphylococcus epidermidis functions on nanophase ZnO and TiO_2. J Biomedical Mater Res 78(3):595–604
103. Meyer U, Bühner M (2006) Fast element mapping of titanium wear around implants of different surface structures. Clini Oral Implant Res 17(2):206–211
104. Davidovitch Z, Finkelson MD, Steigman S, Shanfrid JL, Montgomery PC, Korostoff E et al (1980) Electric currents, bone remodeling and orthodontic tooth movement. II. Increase in rate of tooth movement and periodontal cyclic nucleotide level by combined force and electric current. Am J Orthod 77:33–47
105. Davidovitch Z, Finkelson MD, Steigman S, Shanfrid JL, Montgomery PC, Korostoff E et al (1980) Electric currents, bone remodeling, and orthodoniic tooth movement I. The effect of electric currents on periodontal cyclic nucleotides. Am J Orthod 77:14–32
106. Kolahi J, Abrishami M, Davidovitch Z (2009) Microfabricated biocatalytic fuel cells: a new approach to accelerating the orthodontic tooth movement. Med Hypotheses 73:340–341
107. Maylia E, Nokes LD (1999) The use of ultrasonics in orthopaedics-a review. Technol Health Care 7:1–28
108. Ziskin MC (1987) Applications of ultrasound in medicine—comparison with other modalities. In: Rapacholi MH, Grandolfo M, Rindi A (eds) Ultrasound: medical applications, biological effects, and hazard potential. Plenum Press, New York, NY, pp 49–59
109. Young SR, Dyson M (1990) The effect of therapeutic ultrasound on angiogenesis. Ultrasound Med Biol 16:261–269
110. Tanzer M, Harvey E, Kay A, Morton P, Bobyn JD (1996) Effect of noninvasive low intensity ultrasound on bone growth into porouscoated implants. J Bone Joint Surg 14:901–906
111. Abramovich A (1970) Effect of ultrasound on the tibia of the young rat. J Dent Res 49:1182
112. Heckman JD, Ryaby JB, McCabe J, Frey JJ, Kilcoyne RF (1994) Acceleration of tibial fracture-healing by non-invasive, low-intensity pulsed ultrasound. J Bone Joint Surg. 76:26–34
113. El-Bialy TH, Royston TJ, Magin RL, Evans CA, Zaki AM, Frizzell LA (2002) The effect of pulsed ultrasound on mandibular distraction. Ann Biomed Eng 30(10):1251–1261
114. El-Bialy T, El-Shamy I, Graber TM (2003) Growth modification of the rabbit mandible using therapeutic ultrasound: is it possible to enhance functional appliance results? Angle Orthod 73:631–639

115. El-Bialy T, Hassan A, Albaghdadi T, Fouad HA, Maimani AR (2006) Growth modification of the mandible with ultrasound in baboons: a preliminary report. Orthod Dentofacial Orthop 130(4):e7–e14
116. Oyonarte R (2009) MZa´rate, odriguez FR. Low-intensity pulsed ultrasound stimulation of condylar growth in rats. Angle Orthod 79(5):964–970
117. Liu Z, Xu J, Eb L, Wang D (2012) Ultrasound enhances the healing of orthodontically induced root resorption in rats. Angle Orthod 82(1):48–55
118. Lapatki BG, Paul O (2007) Smart backets for 3D-force-moment measurement in orthodontic research and therapydevelopmental status and prospects. J Orofac Orthop 68(5):377–396
119. Lapaiki HG, Bartholomeyczik J, Ruther P, Jonas IE, Paul O (2007) Smart bracket for multi-dimensional force and moment measurement. J Dent Res 86(1):73–78
120. Cavalcanti A, Rosen L, Kretly LC, Rosenfeld M, Einav S (2004) Nanorobotic challenges in biomedical applications, design and control IEEE ICECS international conference on electronics, circuits and systems, Tel-Aviv, Israel
121. Lam CW, James JT, McCluskey R, Hunter RL (2004) Pulmonary toxicity of single wall nanotubes in mice 7 and 90 days after intratracheal instillation. Toxicol Sci 77:126–134
122. Warheit DB, Laurence BR, Reed KL, Roach DH, Reynolds GAM, Webb TR (2004) Comparative pulmonary toxicity assessment of single wall carbon nano tubes in rats. Toxicol Sci 77:117–125
123. Yao D, Chen Z, Zhao K, Wang Q, Zhang W (2013) Limitation and challenge faced to the researchers on environmental risk of nanotechnology. Procedia Environ Sci. 18:149–156
124. Zheng B, Misakh, Dhanasekaran PS, Kallad D, Asmatulu R (2011) Environmental impact of nanotechnology and its product. In: Proceedings of the 2011 Midwest Section Conference of the American Engineering Education
125. Barnard AS (2006) Nanohazards: knowledge is our first defense. Nature Mater 5:245–248
126. Haramina T (2008) Rizici uporabe nanočestica. Polimeri 29:115
127. Jandt KD, Sigusch BW (2009) Future perspectives of resin-based dental materials. Dent Mater 25:1001–1006
128. Ryan V, Hart TR, Schiller R (1980) Size determination of streptococcus mutans 10449 by laser light scattering. Biophys J 31:313–324
129. Kanaparthy R, Kanaparthy A (2011) The changing face of dentistry: nano-technology. Int J Nanomed 6:2799–2804
130. Sahoo SK, Parveen S, Panda JJ (2007) The present and future of nano- technology in human health care. Nanomed Nanotechnol Biol Med 3:20–31
131. Freitas RA Jr (2005) Nanotechnology, nanomedicine and nanosurgery. Int J Surg 3:243–246

Chapter 10
Nanobiomaterials and Their Application in Prosthodontics

Channamsetty Praveena, Prakash Manne, Lohitha Kalluri and Ravikanth Anne

10.1 Introduction

With the improvement of people's living standards and the promotion of oral health knowledge, prosthodontics increasingly received widespread attention [1]. Prosthodontics is a highly specialized branch of dentistry concerned with the making of artificial replacements for missing parts of the mouth and jaw.

Restorative materials used in Prosthodontics can mainly be divided into three categories: polymers/resins, ceramics, and metals [2]. Metals are inherently strong but conduct heat rapidly and are opaque. Ceramics and polymers are thermally insulating and tend to be more translucent but lack toughness. Polymers are so much weaker and more flexible than metals and ceramics. No one class of materials possesses all the desired properties [2]. The pursuit for ideal restorative materials has lead to the exploration of newer technologies into dentistry.

Nanotechnology has revolutionized the field of dentistry. Bulk material when reduced to nanoscale there is significant change in the optical, thermal, and antimicrobial properties [1]. This alteration of the desired physicochemical properties of nanomaterials has led to the conceptual development of "**Nanodentistry.**"

C. Praveena (✉)
Malla Reddy Institute of Dental Sciences,
Hyderabad, Telangana, India
e-mail: praveena.channamsetty@gmail.com

P. Manne
Sibar Institute of Dental Sciences, Guntur, Andhra Pradesh, India

L. Kalluri
Safa Multispeciality Dental Hospital, Guntur, Andhra Pradesh, India

R. Anne
Pinnamaneni Siddhartha Institute of Dental Sciences,
Gannavaram, Andhra Pradesh, India

R. S. Chaughule (ed.), *Dental Applications of Nanotechnology*,
https://doi.org/10.1007/978-3-319-97634-1_10

The potential benefit of nanotechnology is its ability to exploit the atomic and molecular properties of materials and to develop newer materials with better properties. Nanomaterials are mainly used in ceramic, resin, and metal, providing a huge space for the improvement and innovation of dental material [1].

Nanotechnology and nanoscience have developed new dental composite material and bonding agents with superior properties. Various nanoparticles such as zirconium oxide, titanium oxide and carbon nanotubes have been incorporated to improve the performance of polymethylmethacrylate [1]. Nanoceramics are superplastic materials designed to meet need for translucency and strength of dental restoration [1]. The tissues to replace implanted biodegradable material will provide new vistas in the field of tissue regeneration. The basic idea of nanotechnology, used in narrow sense of the word, is to employ individual atoms and molecules to construct functional structures and maintenance of near-perfect oral health through the use of nanomaterials, biotechnology, tissue engineering nanorobotics, etc. [3].

The objective of this chapter is to describe the principles and need for nanotechnology in dentistry emphasizing their applications in the branch of regeneration of hard and soft tissues around a solid implant, or development of new prosthodontics.

10.1.1 Nanotechnology Definition

While many definitions of nanotechnology exist, the most widely used is from the US Government's National Nanotechnology Initiative (NNI). According to the NNI, nanotechnology is defined as "Research and technology development at the atomic, molecular, and macromolecular levels in the length scale of approximately 1–100 nm range, to provide a fundamental understanding of phenomena and materials at the nanoscale and to create and use structures, devices and systems that have novel properties and functions because of their small and/or intermediate size" [3].

Nanotechnology or molecular engineering is the production of functional materials and structures in the range of 0.1–00 nanometers–nanoscales by various physical or chemical methods [4]. The term "nano" is derived from "***nanos***," the Greek word for "***dwarf***." A nanometer is 10^{-9} a meter or one-billionth of a meter [5]. In simple terms, it is engineering at the atomic and molecular scale. It is a highly multidisciplinary field and cuts across many disciplines, including colloidal science, chemistry, applied physics, and biology.

10.1.2 Historical Review

Nanotechnology is not a new term. This phenomenon is seen in the nature in the locomotor organs of various organisms and in our biological system. Although nanotechnology has been around since the beginning of time, the discovery of nanotechnology is widely attributed to the American Physicist and Nobel Lau-

Table 10.1 Classification of nanomaterials based on dimensions

Dimension	Characteristics	Examples
Zero	Clusters/powders	Atomic clusters, filaments, and cluster assemblies
One	Multilayers	Nano-thin film
Two	Ultrafine grained over-layers or buried layers	Nanotubes, nanofibers, nanowires
Three	Nanophase materials consisting of equiaxed nanometer-sized grains	Nanoparticles, nanopowders, dendrimers, fullerenes, quantum dots, nanostructures, nanocapsules, nanopores

reate, Dr. Richard Phillips Feynman [3]. The first use of the word "nanotechnology" has been attributed to Taniguchi in 1974. In 1986, Eric Drexler introduced and popularized the term "nanotechnology" in his book "Engines of Creation" [3]. Dr. Robert A. Fretias Jr is one among the pioneer scientists who has written about nanomedicine, nanodentistry, and its future changes [3]. It was introduced into dentistry as nanocomposites in the year 2002 by Filtek Supreme [6].

10.1.3 Classification of Nanomaterials

Siegel has classified nanomaterials [6, 7] based on dimension as shown in Table 10.1.

10.1.4 Approaches in Nanotechnology

The fabrication techniques of the nanoscale materials can be divided into the following three approaches [5, 6].

a. Larger to smaller (top-down approach)

Top-down fabrication reduces large pieces of materials all the way down to the nanoscale (Fig. 10.1). This approach requires larger amounts of materials and can lead to waste if excess material is discarded. Here, larger materials are patterned and carved down to make nanoscale structures in precise patterns. Materials reduced to the nanoscale can suddenly show very different properties, enabling unique applications (Table 10.2).

b. Simple to complex (bottom-up approach)
   The bottom-up approach to nano-manufacturing creates products by building them up from atomic- and molecular-scale components, which can be time-consuming (Fig. 10.1). This begins by designing and synthesizing custom-made

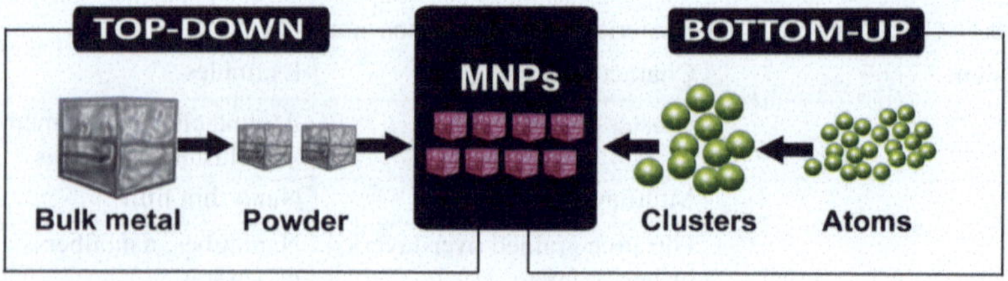

Fig. 10.1 Approaches in nanotechnology

Table 10.2 Examples of top-down and bottom-up approach

Top-down approach examples	Bottom-up approach examples
1. Salivary diagnostics	1. Inducing local anesthesia
2. Laser plasma application for periodontia	2. Hypersensitivity cure
3. Nanotechnology-based root-end sealant	3. Tooth repair
4. Nano-needles	4. Nanorobotic dentifrice (dentifrobots)
5. Nano-bone fibers	5. Orthodontic nanorobots
6. Nanocomposites	6. Dental durability and cosmetics
7. Nanotechnology for GIC	7. Nanotech floss
8. Nanoceramic technology	8. Photosensitizers and carriers
9. Nanobond	9. Diagnosis and treatment of oral cancer
10. Nanosolutions	
11. Coating agents	
12. Nanotechnology for impression materials	
13. Nanocomposite denture teeth	
14. Nanoparticles as antimicrobial agents	
15. Implants surface coatings	
16. Nano-bone replacement materials	

molecules that have the ability to self-assemble or self-organize into higher order structures.

c. Functional Approach

In this approach, components of a desired functionality are developed without regard to how they might be assembled.

10.2 Need for Nanotechnology in Dentistry

Nanomaterials differ from other materials in two principle factors [1]: ***relative surface area and quantum effect***.

a. ***Relative Surface Area***: There is marked increase in surface area when reduced to nanoscale. For example, a particle of 30 nm has 5% of its atoms on its surface, at 10 nm 20% of its atoms, and at 3 nm 50% of its atoms as seen from Fig. 10.2.
b. ***Quantum Effects***: Bulk material when reduced to nanoscale there is significant change in the optical, thermal, and antimicrobial properties (Fig. 10.3).

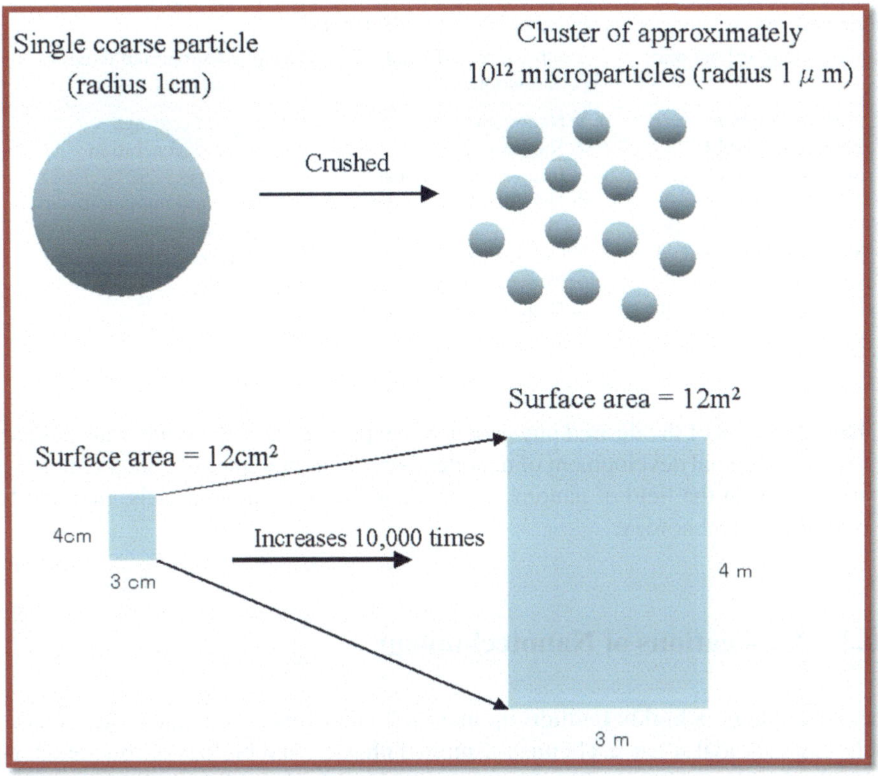

Fig. 10.2 Increased surface area of nanomaterials

Fig. 10.3 Reduction of bulk material to nanoscale. **a** Opaque to transparent, **b** platinum as catalyst, **c** silicone turns as conductor

- An opaque material turns translucent.
- An inert platinum material acts as catalyst.
- Silicone insulator becomes conductor.
- Solids at room temperature turn to liquids.
- Increased fillers increase mechanical strength.

Table 10.3 Applications of nanobiomaterials in prosthodontics

Removable prosthodontics	Tooth-supported fixed prosthodontics	Implant-supported fixed prosthodontics
1. Carbon nanotubes reinforced PMMA 2. Metal oxide reinforced PMMA 3. POSSs reinforced silicones 4. Metal oxide reinforced silicones 5. Nanocomposite denture teeth	1. Nanocomposites 2. Nanosolutions 3. Impression materials 4. Nanofilled resin-modified GIC 5. Coating agents 6. Nanocare gold 7. Nano-optimized ceramics	1. Implant coatings 2. Tissue regeneration scaffolds 3. Structural implant materials and bone repair 4. Bioresorbable materials

- This alteration of the desired physicochemical properties of nanomaterials has led to the conceptual development of nanodentistry. The principles of nanotechnology are applied in the field of nanomedicine, dentifrobots, nanomaterials, implantology, and biotechnology.

10.3 Applications of Nanotechnology

Nanotechnology is highly multidisciplinary field and cuts across many disciplines, including colloidal science, chemistry, applied physics, and biology. Growing interest in the future medical applications of nanotechnology is leading to the emergence of a new field called "**nanomedicine**" preserving and improving human health using nanoscale-structured materials, biotechnology, and genetic engineering and eventually complex molecular machine systems and nanorobots [8].

The application of the principles of nanotechnology in the field of dentistry has led to the conceptual development of "**nanodentistry**" [9]. It is foreseen as a means of making possible the maintenance of near-perfect oral health through the use of nanomaterials, biotechnology including tissue engineering and nanorobotics.

10.4 Nanobiomaterials in Prosthodontics

The branch of prosthodontics in dentistry is a unique blend of science, art, and creativity which aims at rehabilitation of oral health of people with missing teeth and facial structures, providing various treatment modalities ranging from removable partial and complete dentures to fixed tooth-supported and implant-supported prostheses. Nanomaterials in prosthodontics are mainly used in ceramic, resin, and metal, providing a huge space for the improvement and innovation of dental material. Applications of nanobiomaterials in prosthodontics can be grouped as given in Table 10.3.

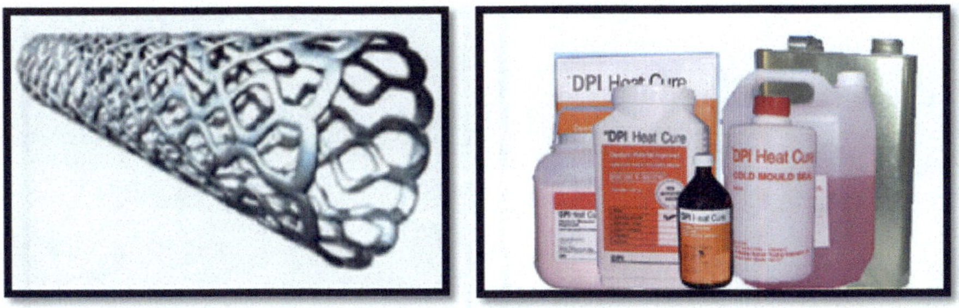

Fig. 10.4 Carbon nanotubes reinforced to PMMA

10.4.1 Removable Prosthodontics

10.4.1.1 Carbon Nanotubes Reinforced Polymethylmethacrylate (PMMA)

Heat-activated acrylic denture base resins are the best materials currently used for denture prostheses. In acrylic resin, the two main unavoidable dimensional changes are expansion and shrinkage. Dimensional changes occur as a result of shrinkage of monomer and release of stress during polymerization [1].

According to carbon nanotubes were regarded as the best reinforcement materials in resins [10]. Micro-additions of 0.125–0.5% of carbon nanotubes are added to monomer by ultrasonic agitation. Augmentation of polymethylmethacrylate (PMMA) resins with carbon nanotubes improves the strength of the prostheses to better withstand the forces of mastication (Fig. 10.4). Carbon nanotubes have tensile strengths up to 4000 times stronger than steel and as much as 200 times stronger than carbon fibers [11].

Carbon nanotubes bond to matrices like PMMA polymer by comparatively weak van der Waals forces. The carbon nanotube/PMMA matrix adhesion strength is very large, such that the bond strength is greater, and the mechanical fatigue strength and compression strength are enhanced. These carbon nanotubes present in PMMA prevent shrinkage and dimensional changes in the resin during and after polymerization, helping in better denture fit or bone/implant interface eliminating the need for metal reinforcement in stress-bearing areas [11].

10.4.1.2 Metal Oxide Reinforced Polymethylmethacrylate (PMMA)

To date, up to 95% dental prostheses are composed of Polymethylmethacrylate (PMMA), due to its advantages, including its optical properties, biocompatibility, and esthetics. Microbial adhesion onto PMMA, lack of strength, and toxicity are inherent drawbacks of acrylic polymeric materials [1].

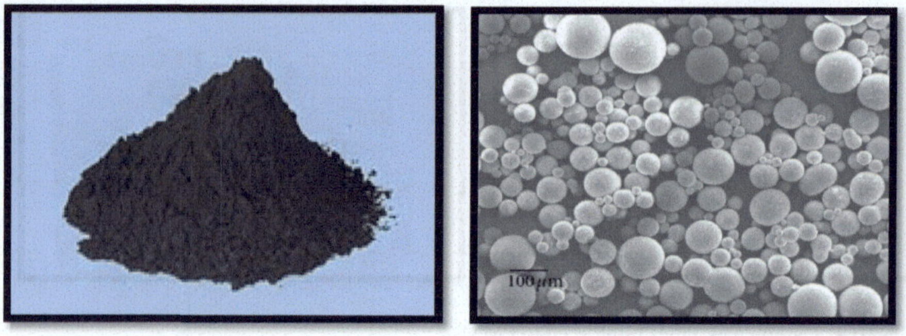

Fig. 10.5 Metal oxide particles reinforced to PMMA

Nanosized TiO_2, Fe_2O_3, and silver particles (150–350 nm in diameter) can be reinforced during synthesis of PMMA (Fig. 10.5). These metal oxide particles possess high photocatalytic activity by increasing their disinfection effect. The introduction of nanosized metal oxide materials for preparing acrylic resins allows the production of polymer with both color and surface modifications. Addition of metal oxides particles improves the flexural strength, tensile strength, antimicrobial property and reduces the occurrence of porosity [12, 13].

10.4.1.3 Polyhedral Oligomeric Silsesquioxanes (POSSs) Reinforced Silicone Elastomers

Silicone elastomers are used in maxillofacial restorations due to their ease of fabrication and realistic appearance. They have properties that are important for human maxillofacial prosthetics such as elasticity, esthetics, heat, and chemical stability. Main reasons for mechanical failure in maxillofacial prostheses include tensile and tearing loads.

A new approach that has the potential to improve polymeric materials is the use of polyhedral oligomeric silsesquioxanes (POSSs) as a reinforcing agent. POSSs are a nanoscale organic–inorganic hybrid containing a 1.5 nm silica cage with eight pendant organic groups [14].

The use of POSSs as a reinforcing agent has enhanced the tensile and tearing strengths of conventional materials (Fig. 10.6). There is no statistically significant decrease in the extension at failure. Interestingly, the extension is increased with POSS loading suggesting that POSS may have a plasticizing effect on polymer matrixes.

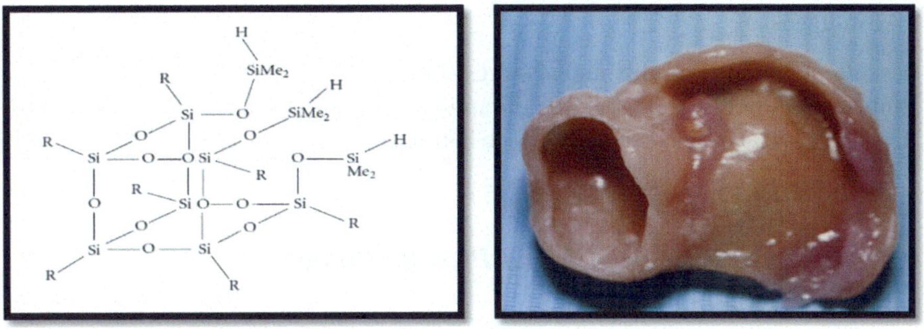

Fig. 10.6 POSSs reinforced to PMMA

10.4.1.4 Nano-Metal Oxides to Silicone Elastomer

Silicone elastomer is the most preferred material for maxillofacial prosthesis. But the advent disadvantage is the aging process of the material over a period of time. Addition of nano-titanium particles to silicones improves the physical properties of the materials. Even the anti-thermal aging properties are markedly increased [15].

10.4.1.5 Nanocomposite Denture Teeth

Conventional denture teeth have their own inherent disadvantage. Porcelain is highly wear resistant, but is brittle, lacks bonding ability to the denture base, and is not easy to polish. Acrylic on the other hand is to adjust, but undergo undue wear. Nanocomposite denture teeth (Fig. 10.7) are made of polymethylmethacrylate (PMMA) and homogeneously distributed nanofillers [16].

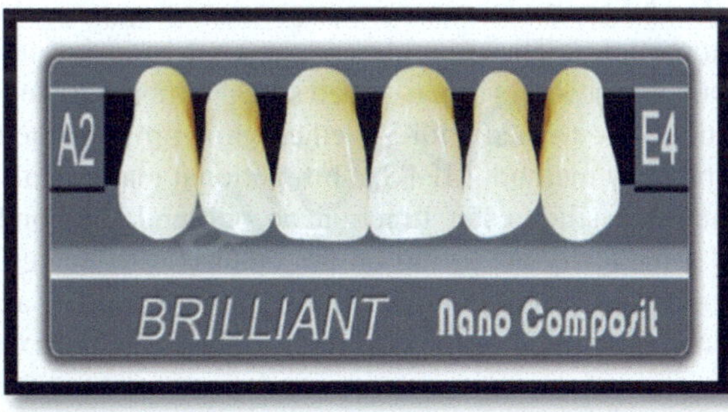

Fig. 10.7 Nanocomposite teeth

Advantages are

- Excellent polishing ability and stain resistant.
- Improved esthetics and increased durability.
- Enhanced wear resistance and surface hardness.

10.4.2 Tooth-Supported Fixed Prosthodontics

10.4.2.1 Nanocomposites

The development of polymeric dental composites has revolutionized the field of dentistry over the past 30 years. This development has been achieved mainly through organic monomer discovery, modifications in formulation and filler technology, advances in light curing equipment and efficient photoinitiators. Despite these developmental advances, dental composites are still limited by problems such as polymerization shrinkage, and wear resistance. The post-gel polymerization shrinkage causes significant stresses in the surrounding tooth structure and composite tooth bonding leading to premature restoration failure. Other problems such as uncured organic monomers leaching from the dental composites into the surrounding gum tissue have caused cytotoxic effects [3]. One of the recent advances in restorative dentistry has been the introduction of nanofiller particles (nanoparticles) into the resin matrix (Fig. 10.8) thereby producing a newer light-cured composite resin nanocomposite [3].

None of the present-day composite resin satisfy all the requirements, and hence, the search for developing a composite dental filling material that could be used in all areas of the mouth with high initial polish and superior polish retention (typical of microfills), as well as excellent mechanical properties suitable for high stress-bearing restorations (typical of hybrid composites), is still on. Since nanocomposites fulfill most of these requirements, they can be considered to be the ideal restorative material of the present era [17].

Nanofiller Particles

Two new types of nanofiller particles–nanomeric particles (NM) and nanoclusters (NC) have been synthesized and incorporated into the resin matrix [3].

The NM particles are monodisperse, non-aggregated, and non-agglomerated silica nanoparticles. These can be synthesized by using aqueous colloidal silica sols to obtain dry powders of nanosized silica particles 20 and 75 nm in diameter. The silica particles are treated with 3-methacryloxypropyltrimethoxysilane or MPTS, using a proprietary method. MPTS, a bifunctional material also known as a coupling agent, contains a silica ester function on one end for bonding to the inorganic surface and a methacrylate group on the other end to make the filler compatible with the resin before curing to prevent any agglomeration or aggregation. MPTS also allows chemical bonding of the NM filler to the resin matrix during curing.

Fig. 10.8 Nanofillers

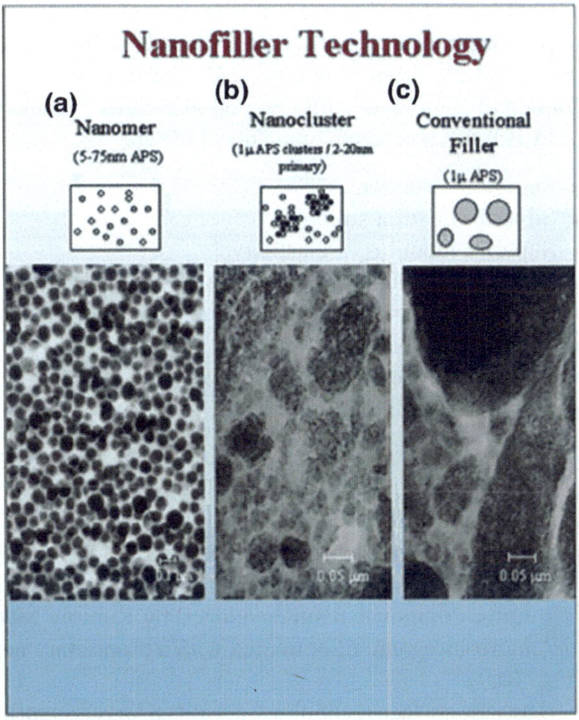

The nanoclusters particles can be of two types:

(1) The first type consists of zirconia–silica particles synthesized from a colloidal solution of silica and a zirconyl salt. The primary particle size of this NC filler ranges from 2 to 20 nm, while the spheroidal agglomerated particles have a broad size distribution, with an average particle size of 0.6 μm. This formulates the dentin, body, and enamel shades of Filtek Supreme standard paste.

(2) The second type of NC filler is synthesized from 75-nm primary particles of silica and has a broad secondary particle size distribution with a 0.6-μm average. This formulates the Filtek Supreme transparent shades.

The surfaces of both types of nanocluster filler particles are treated with an MPTS coupling agent to provide compatibility and chemical bonding with the organic resin.

Resin system: The resin system used in the nanocomposites contains bisphenol A glycidyl dimethacrylate, ethoxylated bisphenol A dimethacrylate, triethylene glycol dimethacrylate,1,6-bis(2-methacryloyloxycarbonylamino)-2,4,4-trimethylhexane, photoinitiator, and stabilizer.

To make filler particles of the mechanically strong composites of today (such as macrofills, hybrids, and microhybrids) the dense, large particles (mined quartz, melt glasses, ceramics) are comminuted to small-particle size. However, these milling procedures usually cannot reduce the filler particle size below 100 nm (1 nm=

1/1000 μm). To circumvent this roadblock, synthetic chemical processes are used to produce building blocks on a molecular scale. These materials are assembled into progressively larger structures and then transformed into nanosized fillers suitable for dental composite with low polymerization shrinkage [3].

Advantages of nanocomposites [18] are

- Superior hardness.
- Superior flexural strength.
- Superior modulus of elasticity.
- Superior translucency and esthetic appeal.
- Excellent color density.
- High polish and polish retention.
- About 50% decrease in filling shrinkage.
- Excellent handling properties.

Commercial products are Filtek Supreme, Clearfil majesty posterior, PureNano, and Premise.

Filtek Supreme Universal Restorative. It consists of nanofillers and a resin matrix introduced by 3 M ESPE Dental Products, St. Paul, (2002).

Clearfil Majesty Posterior. It is a light-cure, nano-superfilled, radiopaque restorative composite resin developed by Kuraray Medical Inc. composed of nano- and micro-inorganic filler treated with a proprietary new surface coating technology (Fig. 10.9).

The new surface technology permits a larger quantity of nanofiller to be dispersed in the resin matrix which consists of monomer and microfiller. The resulting matrix is reinforced with the filler loading of 92 wt% (82 vol.%).

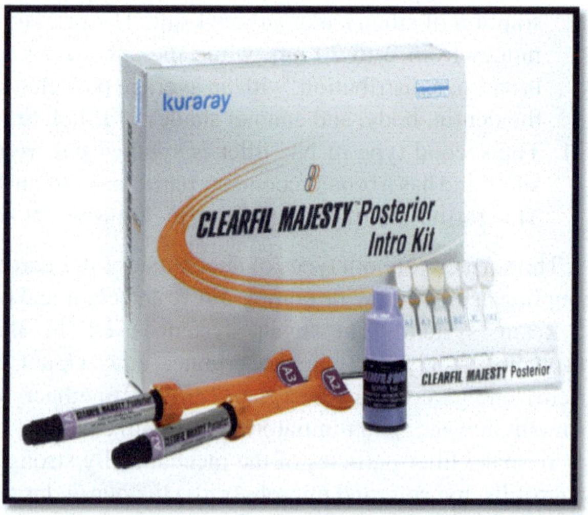

Fig. 10.9 Clearfil majesty nanocomposites

In spite of the extremely high filler loading, Clearfil majesty posterior also has a very shapeable consistency and the high refractive matrix provides benefit of a very minor transparency shift after light curing.

Composition:

Bisphenol A diglycidyl methacrylate

Hydrophobic aromatic dimethacrylate

Triethylene glycol dimethacrylate

Nano glass ceramics

Surface treated alumina microfiller and silica nanoparticles filler

Camphorquinone and pigments

Camphorquinone and pigments.

Premise (Kerr Corporation). It is a dental composite restorative having significant technological advancements over competing products combining esthetics, ease of placement, durability, and long-term polish (Fig. 10.10). Its a single combination of three types of loads (0.02, 0.4 μm and charged pre-polymerized) makes it possible to reach a load factor of 84% which contributes to the improved properties of the material.

As visible light hits the 0.02 μm filler particles, the light is scattered much more than a larger particle composite. More scattering allows blending in of the restoration (the "chameleon effect" and also gives life-like esthetics).

PureNano™ has been developed by NanoProducts Corporation and contains non-agglomerated discrete nanoparticles that are homogeneously distributed in resins or coatings. The nanofiller used include an aluminosilicate powder having a mean particle size of about 80 nm and a 1:4 M ratio of alumina to silica. The nanofiller has a refractive index of 1.508 and offers similar properties and advantages as Filtek supreme universal restorative nanocomposites. PureNanoTM nanoparticles enable

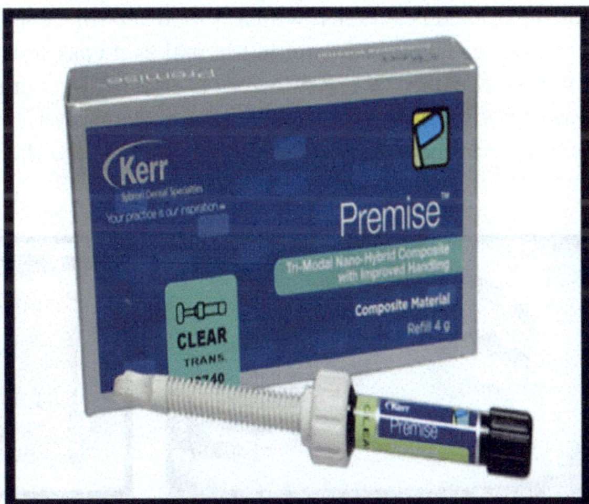

Fig. 10.10 Premise nanocomposites

nearly 50% reduction in filling shrinkage and are particularly useful for fabricating load bearing and cosmetic restorations.

10.4.2.2 Nanosolutions

Nanosolutions produce unique and dispersible nanoparticles, which can be used in bonding agents [6].

Trade names. Adapter Single Bond, Prime & Bond NT Adhesive, GC G-Bond and Excite Bond.

Adper Single Bond Plus. The adhesive is a fast, easy, and convenient nano-based bonding agent offering exceptional bond strength (Fig. 10.11). The adhesive has been recognized as a new option in total-etch visible light activated dental adhesives. It contains BisGMA, HEMA, dimethacrylates, ethanol, water, a novel photoinitiator system acid, and a methacrylate functional copolymer of polyacrylic and polyitaconic acids. It incorporates 10% by weight of 5-nm-diameter spherical silane-treated silica particles through a process that prevents agglomeration. As discrete particles, their extremely small size keeps them in colloidal suspension [19]. The use of silica nanofiller nanotechnology contributes to higher bond strength performance and provides a stable, filled adhesive. Since the particles do not cluster together or settle out of dispersion, the adhesive does not need to be shaken prior to use.

It is available in two delivery options: a unit dose system and in a single bottle system. The unit dose delivery system requires no measuring, mixing or cleanup. It is disposable and hence offers hygienic dispensing. It contains the adhesive and brush-tip applicator in a self-contained system. The large blister is squeezed to transfer the adhesive into the chamber enclosing the applicator. The applicator is briefly spun to fully saturate with adhesive and is then ready to use.

The single bottle use is a convenient squeeze bottle delivery system meant for multiple uses. It is attached with a "pinch and flip" cap which helps to reduce spillage and wastage. The translucent bottle makes it easy to see how much adhesive is left.

Prime & Bond NT Adhesive. It is a fifth-generation (i.e., "one-component") dentin bonding agent marketed by the Dentsply/Caulk Company (Fig. 10.11). It contains extremely small 7-nm fillers which are the basis for including "NT" in

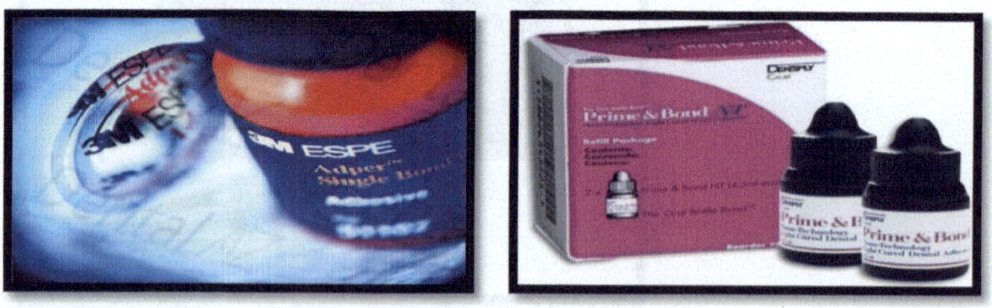

Fig. 10.11 Nanosolutions

the product's name: "NT" for "nanotechnology." Although Prime & Bond NT Dual Cure is described as a "single-component" product, it consists of a 34% phosphoric acid etchant, an acetone-based, nanofilled adhesive resin, and an acetone/ethanol-based, sulfinate self-cure activator. Other improvements to the product's composition include incorporation of a small resin to better infiltrate the dentin surface, a crosslinking agent to strengthen the adhesive, and a higher overall resin concentration which makes it possible to achieve adequate coverage with only one application.

For doing direct bonding of resin composites, compomers, and veneers, only the etchant and bonding resin is used. For bonding indirect restorations such as inlays, onlays, and crowns and the bonding resin is combined 1:1 with the self-cure activator. This converts the material into a dual-cure adhesive. Prime & Bond NT Dual Cure is also recommended for amalgam bonding and as the preparatory step before the cementation of endodontic posts with a resin cement.

GC G-Bond. It is a seventh-generation one-component bonding agent. Etching, priming, and bonding are done in a single step with a single coat application. It produces high bond strength with dry, moist or wet, dentin or enamel. It is proposed that it is advanced 4-MET and Phosphoric Acid Ester Monomer Adhesive Technology combined with nanofilled particles forms a unique nanofilled interface with the dentin a "Nano-Interaction Zone" (NIZ). This "nano"-level reaction produces insoluble calcium compound for a better bond less likely to deteriorate from enzymes contained in the mouth.

Excite Bond. Ivoclar Vivadent Exite contains 12-nm fillers and is packaged in a single-use capsule. It is very fast to apply in one coat and comes with a graphics-only instruction card.

10.4.2.3 Impression Materials

Impression materials are available with nanotechnology application. Nanofillers are integrated into the vinyl polysiloxanes, producing a unique addition siloxane impression material [6, 19, 20].

Advantages:

- Better flow.
- Improved hydrophilic properties, hence fewer voids at margin and better model pouring.
- Enhanced detail precision.

Trade name: Nanotech elite H-D+ and Imprint 2 Penta H

Nanotech Elite H-D+

In the evolution of polyvinyl siloxanes, Zhermack, Italy, has developed a new and more effective platinum-based system elite HD+ to obtain a complete and stable reticulation (Fig. 10.12). It is an insoluble, odorless, two-part system with a relative density of 1.50 g/cc. The insertion of particles with a nanometric size and structure

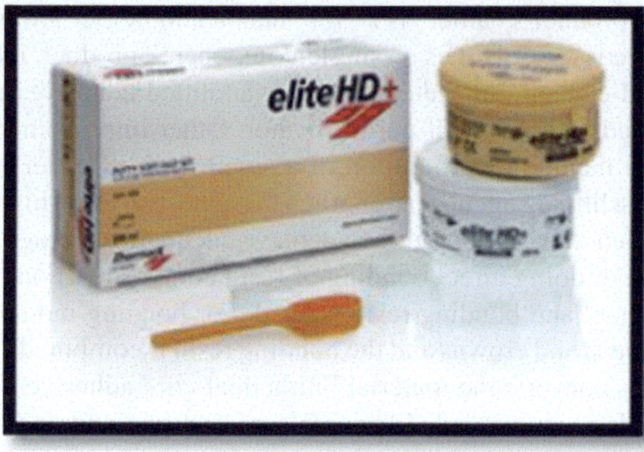

Fig. 10.12 Impression materials

causes atomic and molecular interaction that macroscopically influences the properties of the impression material. The nanometric particles and the other components of the material create a nanostructure which enables a degree of fluidity completely different from the initial viscosity to be obtained. This improves the flow, tear resistance, hydrophilic properties, resistance to distortion, and heat resistance. When pressure is exerted in taking the impression, an excellent reproduction of infinitely small details is obtained [6].

Elite also has a snap set with a total time in the mouth for 2.5 min. Its rigid final set minimizes the possibility of distortion and offers excellent tear resistance and adhesion with the wash. No specific hazards are encountered under normal use of the product. Contact with the eyes may, however, cause slight irritation and reddening of the conjunctiva.

10.4.2.4 Nanofilled Resin-Modified Glass Ionomer

A new nanofilled RMGI restorative material has been introduced. It is based on a prior RMGI with a simplified dispensing and mixing system (paste/paste) that requires the use of a priming step, but no separate conditioning step. Its primary curing mechanism is by light activation, and no redox or self-curing occurs during setting (Fig. 10.13).

Apart from the user-friendliness, the major innovation of this material involves the incorporation of nanotechnology, which allows a highly packed filler composition (69%), of which approximately two-thirds are nanofillers.

Composition: Chemistry of nanoionomer is based on the methacrylate modified polyalkenoic acid, which is capable of both crosslinking via pendate methacrylate groups as well as the acid–base reaction between the fluoroaluminosilicate glass (FAS) and the acrylic and itaconic acid copolymer groups. It contains surface treated nanofillers (approx 5–25 nm) and nanoclusters (approx 1 to 1.6 microns). Filler loading is approx. 69% by weight of which the relative proportion of two

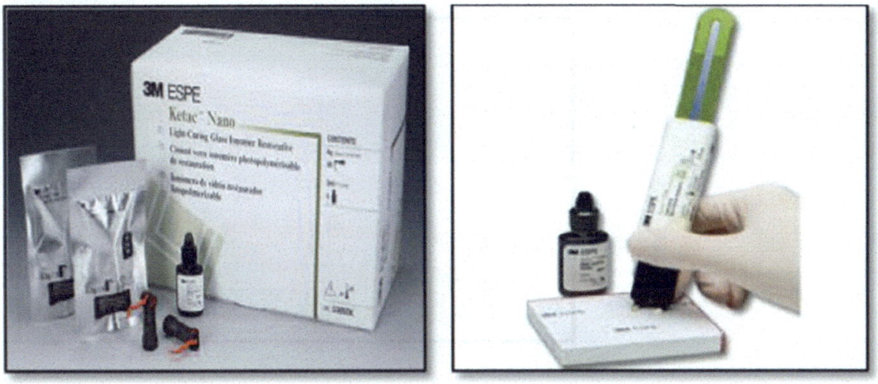

Fig. 10.13 Nanofilled resin-modified GIC

filler types (FAS and combination of nanofillers) are approx 2/5 and 3/5, respectively. All nanofillers are further surface modified with methacrylate silane coupling agents to provide covalent bond formation into free radically polymerized matrix.

Advantages: Good luster and enhanced mechanical properties [9].

10.4.2.5 Coating Agents

These are used as final coating over esthetic restorations (Fig. 10.14). Nanotechnology uniformly disperses nanofillers [6].

Advantages:

- Higher wear resistance.

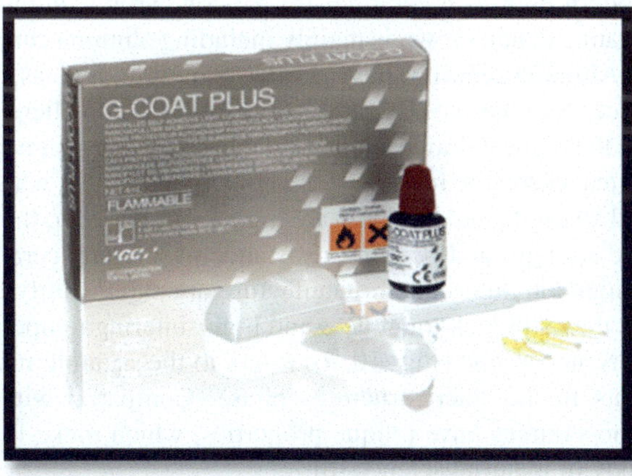

Fig. 10.14 G-coat plus

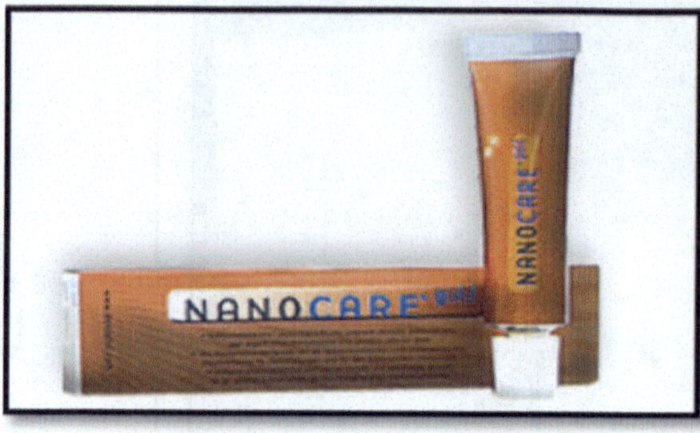

Fig. 10.15 Nanocare gold

- Prevents abrasion and discoloration.
- Smooth high luster finish retained over time.

10.4.2.6 Nanocare Gold

Nanocare gold dental material is used directly before embedding dental films, ceramic or porcelain crowns, bridges, porcelain veneers (Fig. 10.15), inlays or onlays.

Advantages:

- Provides antimicrobial effect.
- Improves adhesion to tooth tissues which greatly increases durability of dental restorations.

10.4.2.7 Nano-Optimized Moldable Ceramics

Traditional ceramics are made of clay and other natural occurring materials, while modern high-tech ceramics use silicon carbide, alumina, and zirconia. At present, ceramic dental crown is mainly including alumina ceramic and zirconia ceramic [1].

Alumina ceramics have good esthetics, high gloss, chemical stability, wear resistance, high hardness, good biocompatibility, no allergies, and no effect on the MRI, but the biggest drawback is brittleness. ZrO_2 has a good abrasion resistance, physiological corrosion resistance, and biocompatibility, whose modulus of elasticity, flexural strength, and hardness are higher, compared to those of HA and titanium alloys. The strength and bending resistance of zirconia ceramics through computer-aided design/computer-aided manufacture are significantly higher than alumina ceramic, but they still lack toughness and high sintering temperature [1].

Nanoceramic (Fig. 10.16) refers to the ceramic material with nanoscale dimensions in the microstructures phase. Compared with the conventional ceramics, nanoceramics have unique properties, which make it become the hot topics in the study of material science [6].

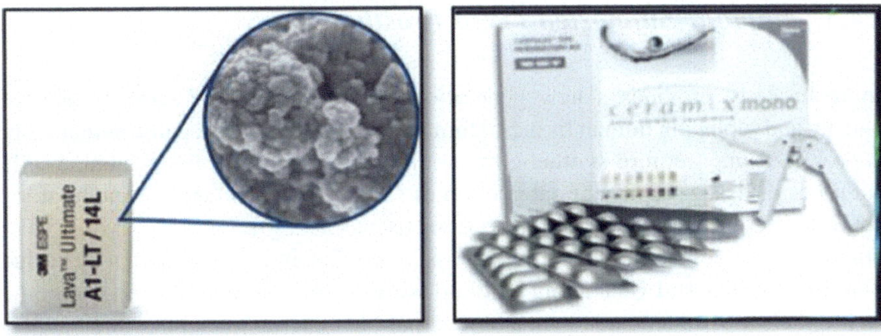

Fig. 10.16 Nano-optimized moldable ceramics

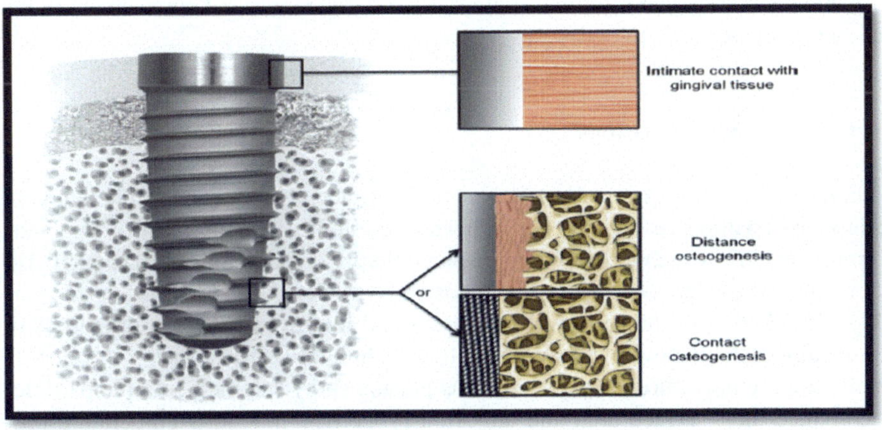

Fig. 10.17 Distance and contact osteogenesis

Composition:

- **Nanofillers**: Enhance polish ability and reduce wear.
- **Nanopigments**: Adjust the shade of the restoration to the surrounding teeth (chameleon effect).
- **Nanomodifiers**: Increase the stability (non-slump) of the material and prevent sticking to instruments.

Advantages:

- Nanoceramics have superplasticity.
- Show good toughness and ductility.
- Superior mechanical properties: The hardness and strength of many nanoceramics are 4–5 times higher than those of the traditional materials.
- Improved translucency and corrosion resistance.

10.4.3 Implant-Supported Fixed Prosthodontics

Nanotechnology provides a new generation of biocompatible nanomaterials for repairing and replacing human tissues. Human tissue that is diseased or traumatically compromised may require synthetic materials for its repair or replacement. While most types of tissues repair the interaction of stem cells with chemical modulators, there are differences in the ways that various tissues heal [3].

In many cases, the failure occurs at the tissue–implant interface, which may be due to the implant material weakening its bond with the natural material. To overcome this, implants are often coated with a biocompatible material to increase their adherence properties and produce a greater surface area to volume ratio for the highest possible contact area between the implant and natural tissue.

Nanotechnology can offer new solutions for tissue repair and replacement in the following areas.

10.4.3.1 Implant Coatings

Nanotechnology brings a variety of new high surface area biocompatible nanomaterials and coatings to increase the adhesion, durability, and lifespan of implants. Ceramic materials such as calcium phosphate (hydroxyapatite or HAP) (Fig. 10.18) are made into implant coatings using nanosized particles instead of micro-sized particles. In addition to the higher surface areas and improved adhesion properties of the nanoparticle coatings, improved coating techniques are also being developed. While high-temperature processes such as plasma spray can melt ceramic particles and reduce their surface area and adhesion properties, new low-temperature processes with electromagnetic fields can maintain the nanomaterial properties. This provides the maximum possible contact area between the implants and bone surface to improve the potential for in-growth in the host bone (Fig. 10.17).

New types of nanomaterials are being evaluated as implant coatings to improve interface properties [21, 22]. For example, nanopolymers (Table 10.4) such as polyvinyl alcohol (PVA) can be used to coat implantable devices that are in contact with blood (e.g., artificial hearts, vascular grafts, catheters) for dispersing clots or preventing their formation [13, 14].

10.4.3.2 Tissue Regeneration Scaffolds

Nanostructures arc being researched for the preparation and improvement of tissue regeneration scaffolds. Research areas include the ability to develop molecular sensitive polymers using the optical properties of nanoparticles as control system, manipulating the stiffness, and strength of scaffolds using hybrid nanostructures and

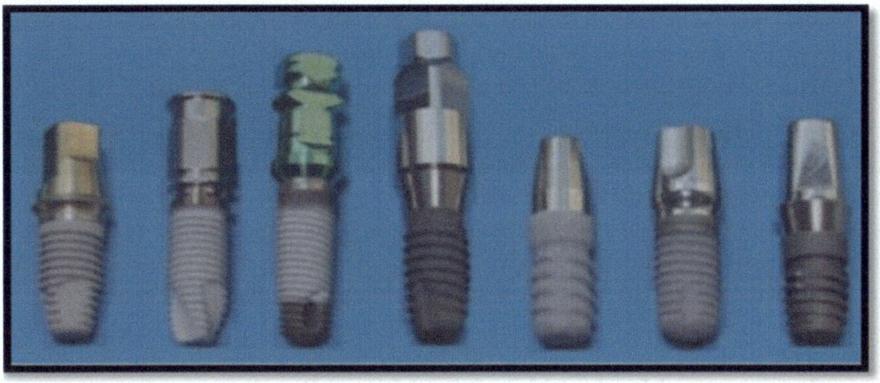

Fig. 10.18 New generation HAp-coated implants

Table 10.4 Range of materials added to calcium phosphate for the production of nanocomposite coatings and their advantages and concerns

Material	Advantages	Concerns
CNT	• Excellent mechanical properties • Used to reinforce the HAp	• Non-biodegradable • Possible transfer of internal organs after the degradation of the matrix • Free graphite is difficult to disperse homogeneously
Bioactive glass	• Increased surface bioactivity • Improves the bioactivity of metallic dental implants	• Special attention is needed to control the surface reactivity rates
TiO_2	• Capable of inducing cell growth and enhances osteoblastic activity	• Crystallinity control is required
Silica	• Alters the surface and interfacial properties of Hap composites	• Long-term adhesion and reliability
Collagen	• Improved osteogenic effects • Formation of new bone tissue without encapsulation	

the use of nanotechnology to prepare molecular imprints to maximize long-term viability and function of cells on scaffold surfaces [3].

With the ultimate objective of growing large complex organs, a variety of nanomaterials and nanotechnology fabrication techniques are being investigated as tissue regeneration scaffolds that provide improved structural requirements and guide the activity of seeded cells [14]. Some examples are as follows:

a. Nanoscale polymers such as polyvinyl alcohol (PVA) are being molded into heart valves and seeded with fibroblasts and endothelial cells.
b. Polymer nanocomposites are being researched for bone scaffolds.

Scientific challenges related to a better understanding of molecular/cell biology and fabrication methods for producing large three-dimensional scaffolds are among the many obstacles yet to be overcome.

Nanostructures are also being, used to study the fundamental properties of implanted tissues. In areas of in vivo analysis, nanostructures are used as tracers for implanted cells and to study the response of host to implanted tissues.

10.4.3.3 Structural Implant Materials and Bone Repair

Nanotechnology provides a new generation of biocompatible materials that can be used as implants or temporary bioresorbable structures [23]. Bone is a high strength material that is used as both weight bearing and non-weight bearing. Bones are more than just structural materials as they also contain interconnected pores that allow body fluids to carry nutrients and permit interfacial reactions between hard and soft tissues. In the case of bone fractures, grafts disorders and dental applications, bones may require repair or replacement [3].

A variety of natural materials are used as bone substitutes. These include autografts from the patient's pelvis, allograft from another human, bovine material or coral blocks. Natural materials are tend to be brittle and can lose mechanical strength during sterilization. They can also cause inflammation, pain at the pelvis graft site and potentially transmit disease. Bone cavities can also be filled with synthetic bone cement. Current bone cements containing PMMA act as filler or grout, which is injected as flowable paste and then hardens vivo. While PMMA cement can offer adequate mechanical properties and bonding, it is typically recommended only for non-weight-bearing bones. PMMA has also been linked to tissue damage, nerve root pain, and other side effects.

High strength nanoceramic materials, such as calcium phosphate appetite (CPA) and hydroxyapatite (HAP), can be made into flowable, moldable nanoparticle paste that can confirm to and interdigitate with bone. As natural bone is approximately 70% by weight CPA including hydroxyapatite (HAP), biocompatibility is thought to be extremely high with minimal side effects. As its dense surface and tight three-dimensional crystalline structures will allow for superior compressive strength to PMMA, nanoceramics may be suitable for both weight-bearing and non-weight-bearing bones [3].

10.4.3.4 Bioresorbable Materials

Bioresorbable polymers are currently being used in degradable medical applications such as sutures and orthopedic fixation devices. With new production methods, nanostructures are being fabricated which could be used as temporary implants. Bioresorbable implants will biodegradable and do not have to be removed in subsequent operation. Nanostructured implants are being designed to degrade at a rate that will slowly transfer load to a healing bone that it is supporting. Thus, it is possible in near future to have nanomaterials used in the body to exceed current human performance on multiple dimensions and beyond current human limitations [6].

10.5 Social Implications

Potential risks of nanotechnology can broadly be grouped into three areas [24].

(a) The risk to health from nanoparticles and nanomaterials.
(b) The risk to environment from nanoparticles and nanomaterials.
(c) The risk posed by molecular manufacturing (or advanced nanotechnology).

Studies of the health impact of airborne particles are the closest thing, and we have to use as a tool for assessing potential health risks from free nanoparticles. These studies have generally shown that the smaller the particles get the more toxic they become. This is due in part to the fact that, given the same mass per volume, the close in terms of particle numbers increases as the particle size decreases. Regulatory bodies have concluded that nanoparticles form the potential for an entirely new risk, and that it is necessary to carry out an extensive analysis of the risk.

The concerns expressed by national and international groups are also constantly being considered and monitored as the potential usages, risks, and applications of nanotechnology are recognized [25].

10.6 Conclusion

The futuristic view of multiple treatment opportunities offered by nanotechnological advances may sound unlikely, implausible, or even heretic. Yet, the theoretical and applied research to turn them into reality is progressing rapidly.

As claimed by the National council of science and technology (USA), "Molecular technology is destined to become the core technology changing the nature of every human-made object in the next century."

Dentistry is also facing a major revolution in the wake of this technology having already been targeted with novel "nanomaterials." The development of nano-based materials such as nanocomposites, bonding agents, impression materials, and implant surface modifications has influenced the clinical dental practice to a great extent. This is particularly true for purpose-designed nano- and microstructures. Biologic approaches such as tissue and genetic engineering will yield new diagnostic and therapeutic approaches.

However, the search for ideal materials still continues and may lead to the development of newer materials with superior properties and improvement in the properties of the currently available dental materials with the advances in nanotechnological research.

References

1. Wang W et al (2015) Recent applications of nanomaterials in prosthodontics. J Nanomater 3:3. https://doi.org/10.1155/2015/408643
2. O'Brien WJ (2002) A comparison of metals, ceramics, and polymers: dental materials and their selection, 3rd edn. Quintessence Publishing Co, Batavia, pp 1–11
3. Subramani K, Ahmed W (2012) Emerging nanotechnologies in dentistry. Elesevier, New York, pp 1–08, 78–97, 322–389
4. Mitra SB, Holmes BN (2003) An application of nanotechnology in advanced dental materials. J Am Dent Assoc 134:1382–1390
5. Lakshmi Sree, Balasubramanian B, Deepa D (2013) Nanotechnology in dentistry-a review. Int J Dent Sci Res 2:40–44
6. Satyanarayana TSV, Rai R (2011) Nanotechnology: the future. J Interdisc Dent 1(2):93–100
7. Hofmann H (2011) Introduction and classification of nanomaterials, pp 5–12
8. Freitas RA Jr (2000) Nanodentistry. J Am Dent Assoc 131:1559–1566
9. Shilpa SS, Sathyajith NN, Shashibhushan KK, Poornima P, Shivayogi MH, Roshan NM (2014) Nanodentistry: the next big thing is small. Int J Contemp Dent Med Rev 2:1–6
10. Bal S, Samal SS (2007) Carbon nanotube reinforced polymer composites—a state of the art. Bull Mater Sci 30:379–388
11. Turagam N, Mudrakola DP (2012) Effect of micro-additions of carbon nanotubes to polymethylmethacrylate on reduction in polymerization shrinkage. J Prosthodont 22:1–7
12. Laura S, Elvira N, Genoveva H, Victor M (2011) Biocompatible metal-oxide nanoparticles: nanotechnology improvement of conventional prosthetic acrylic resins. J Nanomater 2011:1–8
13. Shirkavand S, Moslehifard E (2014) Effect of TiO_2 nanoparticles on tensile strength of dental acrylic resins. J Dent Res Dent Clin Dent Prospect 8(4):197–203
14. Mohammad SA, Wee AG, Deborah J, Rumsey DJ, Scott R, Schricker SR (2010) Maxillofacial materials reinforced with various concentrations of polyhedral silsesquioxanes. J Dent Biomech 2010:1–6
15. Wang L, Liu Q, Jing D, Zhou S, Shao L (2014) Biomechanical properties of nano-TiO_2 addition to a medical silicone elastomer: the effect of artificial ageing. J Dent 42:475–483
16. Freitas RA Jr (2005) Nanotechnology, nanomedicine and nanosurgery. Int J Surg 3(4):243–246
17. Ozak ST, Ozkan P (2013) Nanotechnology and dentistry. Eur J Dent 7:145–151
18. van Dijken JW, Jan WV, Pallesen U (2014) A randamized 10 year prospective follow- up of class II nanohybrid and conventional hybrid resin composite restorations. J Adhes Dent 16(6):585–592
19. Jhaveri HM, Balaji PR (2005) Nanotechnology: furure of dentistry. J Indian Prosthodont Soc 5:15–17
20. Saravana Kumar R, Vijayalakshmi R (2006) Nanotechnology in dentistry. Indian J Dent Res 17:62–65
21. Choi AH, Ben-Nissan B, Matinlinna JP, Conway RC (2013) Current perspectives: calcium phosphate nanocoatings and nanocomposite coatings in dentistry. J Dent Res 92(10):853–859
22. Lavenus S, Louarn G, Layrolle P (2010) Nanotechnology and dental implants. Int J Biomater 2010(10):1–9
23. Jurczyk M (2012) Bionanomaterials for dental applications. Taylor & Francis Group, Milton Park, pp 29–180

24. Hoet PTM, Brüske-Hohlfeld I, Salata OV (2004) Nanoparticles—known and unknown health risks. J Nanobiotechnol 2(12):1–15
25. Schulte P, Geraci C, Zumwalde R, Hoover M, Kuempel E (2008) Occupational risk management of engineered nanoparticles. J Occup Environ Hyg 5:239–249

Chapter 11
Nanomaterials for the Management of Periodontal Diseases

Radhika Arjunkumar

11.1 Introduction

Periodontitis is a complex disease and has a multifactorial etiology. It can occur as the less severe gingivitis, chronic and aggressive periodontitis and as a manifestation of numerous systemic diseases like diabetes and hematological disorders. Periodontal disease is characterized by infection and inflammation, which leads to progressive attachment loss and bone loss around teeth, which if left untreated can lead to premature loss of tooth [1].

Dental plaque is the main causative factor of periodontal disease, and plaque control forms the mainstay of periodontal therapy. Mechanical and chemical plaque control methods have been employed to treat the disease with success. However, these treatment modalities are not without shortcomings. In the quest for better treatment options for patients with a greater success rate and lesser side effects, nanotechnology has played a significant role.

11.2 Nanorobotic Dentifrices

When properly designed, dentifrobots could identify and destroy pathogenic bacteria residing in dental plaque and elsewhere, while sparing the 500 or so species of harmless oral microflora and allowing them to flourish in a healthy ecosystem. Dentifrobots would also provide a continuous barrier to halitosis, since bacterial putrefaction is the central metabolic process involved in oral malodor. Subocclusal dwelling

R. Arjunkumar (✉)
Department of Periodontics, Saveetha Dental College,
Saveetha Institute of Medical and Technical Sciences,
Chennai, India
e-mail: radhikaarjunkumar@gmail.com

R. S. Chaughule (ed.), *Dental Applications of Nanotechnology*,
https://doi.org/10.1007/978-3-319-97634-1_11

nanorobotic dentifrice which can be delivered by mouthwash or toothpaste could patrol all supragingival and subgingival surfaces at least once a day, metabolising trapped organic matter into harmless and odorless vapors concomitantly performing continuous calculus debridement [2].

The invisibly small dentifrobots with purely mechanical devices [1–10 micron in size], crawling at 1–10 μ/s, would be inexpensive and would safely deactivate themselves if swallowed and would be programmed with strict occlusal avoidance protocol [3].

11.3 Hypersensitivity Management

Tooth hypersensitivity is one of the most common problems encountered in clinical practice. Unlike dental caries and periodontal disease, it is one of the most painful and least successfully treated chronic problems of the teeth. It has been reported that 8–30% of the adult are affected by dentin hypersensitivity [4].

Numerous theories have been put forth to explain the mechanism of hypersensitivity, which includes the direct stimulation theory, odontoblastic transducer mechanism, gate control theory, and the hydrodynamic theory. The hydrodynamic theory is the most accepted, and according to this, a rapid shift in the fluid flow within the dentinal tubules appears to be responsible for causing the odontoblastic pain [5]. Several treatment methods have been tried to reduce dentinal hypersensitivity, ranging from home-use, over the counter products such as desensitizing mouthwashes, dentifrices, or tray application forms to in-office application products such as varnishes, liners, restorative material, dentinal adhesives, iontophoresis procedures, and more recently, laser [6].

Among the various treatment options mentioned above, dentifrices have emerged as an important means of treating hypersensitivity. Many of these dentifrices contain potassium nitrate, stannous fluoride, sodium fluoride, sodium mono fluorophosphate, and strontium chloride as an active ingredient and have proven to be effective, leading it to be a frequent choice among both patients and dentists for the treatment of sensitive teeth [7].

Research into the ultrastructure of teeth has led to a renewed interest in hydroxyapatite (HAP), which is a building block for the mineralized dental tissues. This has led to the preparation of HAP for the occlusion of dentinal tubules. Among all the formulations of HAP, nano-hydroxyapatite (n-HAP) is considered one of the most biocompatible and bioactive materials and is widely accepted in medicine and dentistry in recent years [8]. Nanosized particles have a similarity to the apatite crystals of tooth enamel in morphology and crystal structure [9]. Different formulations of n-HAP have been developed, and early data has suggested remineralization of subsurface lesions [10]. The n-HAP-containing toothpaste was effective in reducing dentin hypersensitivity with standard dentifrices tested and could also be advocated in the management of hypersensitivity [7, 11]. Apart from reduction of hypersensitivity by tubule occlusion, n-HAP dentifrices also have shown to improve the microhardness of dentin in in vitro settings (Unpublished data by the author).

Reconstructive dental nanorobots could selectively and precisely occlude specific tubules within minutes, offering patients a quick and permanent cure. Nanorobots pass through the journey of enamel, dentin, and reach into the pulp. Once they are installed in the pulp, nanorobots establish control over nerve impulse traffic. The analgesic dental nanorobots may be commanded by the dentist to shut down all sensitivity in selected tooth that requires treatment. When the dentist passes the icon for the desired tooth on the handheld controlled display monitor, the nerve is immediately anesthetized. After the oral treatment procedures are completed, the dentist orders the nanorobots via the same acoustic data links to restore all sensation, to relinquish control the nerve traffic and to retrieve from the tooth via similar path. This analgesic technique is a patient-friendly one as it reduces anxiety, needle phobia, is quick, and has a completely reversible action [12, 13].

11.4 Drug Delivery

Chemotherapy in the management of periodontal diseases has a significant albeit adjunctive effect. Systemic antimicrobials are found to be beneficial in the treatment of chronic and aggressive forms of periodontal disease. Among all groups of antibiotics, a combination of amoxicillin and metronidazole was found to be most effective [14, 15]; use of systemic antibiotics, however, leads to high propensity for antibiotic resistance, non-compliance to the regimens, possibility of adverse drug reactions, and a very high dosage.

Periodontitis being a site-specific disease can be treated by local drug delivery systems as well. This approach is more favorable as compared to systemic approach because it mainly focuses on improving the therapeutic outcomes by achieving factors like site-specific delivery, low-dose requirement, bypass of first-pass metabolism, reduction in gastrointestinal side effects, and decrease in dosing frequency. Overall, it provides a safe and effective mode of treatment, which enhances patient compliance [16].

Poor oral bioavailability of conventional oral formulations has paved the way for new therapeutic options, like the use of nanotechnology for drug delivery. A number of polymer-based delivery systems like films, chips, strips, fibers, microparticles, nanofibers, and nanoparticles made from a variety of natural and synthetic materials have been successfully tested to deliver a variety of drugs. These systems are biocompatible and biodegradable, completely fill the pockets, and have strong retention on the target site due to excellent mucoadhesion properties [17].

11.4.1 Nanoparticles

An emerging therapeutic strategy is the use of nanoparticles as drug carriers [18, 19]. The factors that make nanoparticles more advantageous than microparticles, microspheres, and emulsion- based delivery systems are [20].

- They are particulate dispersions or solid particles with a size range from 10–1000 nm.
- The drug is dissolved, entrapped, encapsulated, or attached to a nanoparticle matrix [21].
- They are highly dispersible in aqueous medium, offer controlled release rate and enhanced stability.
- Because of their small size, nanoparticles can access sites unreachable for other devices, like the periodontal pocket regions below the gingival margin.
- A uniform drug distribution for prolonged time period is obtained, thus decreasing the dosage frequency.

Studies show that the immobilization/encapsulation process improves the overall biologic effectiveness of active agents including pharmacokinetic parameters (bioavailability, clearance) and reduces the side effects [22, 23]. In addition, it has been noted that the antibiotic resistance of drug-resistant pathogens may be overcome by using nanoparticles as antibiotic carriers [24].

11.4.1.1 Magnetic Nanoparticles

Among the many types of nanostructures, core–shell magnetic iron oxide nanoparticles are the most commonly used drug carriers in biomedical studies [19, 25]. Chlorhexidine (CHX) is a second-generation bis-biguanide antiseptic. It acts on the inner cytoplasmic membrane; hence, it is a membrane active type of substance. It is dicationic at pH levels above 3.5. It prevents plaque accumulation; hence, it is a antiplaque and antigingivitis agent [26] and reduces the adherence of *Porphyromonas gingivalis* to epithelial cells [27]. It can be bacteriostatic or bactericidal depending on the dose. It acts against a wide array of bacteria including Gram-positive and Gram-negative bacteria, dermatophytes, and lipolytic viruses and fungi.

However, CHX application is limited due to inactivation in body fluid and cytotoxicity toward human cells, especially at high concentrations. To overcome the limitations of CHX, Tokajuc et al. [28] synthesized nanosystems composed of aminosilane-coated magnetic nanoparticles functionalized with chlorhexidine (MNP@CHX). In the presence of human saliva, MNPs@ CHX displayed significantly greater bactericidal and fungicidal activity against planktonic and biofilm-forming microorganisms than free CHX. In addition, CHX attached to MNPs has an increased ability to restrict the growth of mixed-species biofilms compared to free CHX. The observed depolarization of mitochondria in fungal cells treated with MNP@CHX suggests that induction of oxidative stress and oxidation of fungal structures may be a part of the mechanism responsible for pathogen killing. Nanoparticles functionalized by CHX did not affect host cell proliferation or their ability to release the proinflammatory cytokine, IL-8. The use of MNPs as a carrier of CHX has great potential for the development of antiseptic nanosystems.

Apart from CHX numerous drugs like tetracycline, doxycycline, metronidazole, silver, and leaf extracts can also be used as nanoparticle drug delivery systems to treat periodontal diseases (Table 11.1).

11.4.1.2 Nanofibers

Polymeric fibers having diameters in submicron or nanometer range (e.g., 10×10^3 100×10^3 mm) are called nanofibers [33]. They provide numerous remarkable characteristics such as

- Very large surface area to volume ratio (this ratio can be as large as 103 times that of microfiber),
- Suppleness in surface functionalities, and
- Superior mechanical performance (e.g., stiffness and tensile strength).

Nanofibres can be prepared by

- Drawing,
- Template synthesis,
- Phase separation,
- Self- assembly,
- Electrospinning.

Among these electrospinning has been accepted as an efficient technique for the fabrication of continuous nanofibers from a variety of polymers. These polymers can either be first dissolved in suitable solvent and then electrospun or can be directly electrospun in molten form [34]. Both hydrophobic and hydrophilic drugs can be incorporated.

Polymers can be fabricated to form nanofibrous scaffolds and seeded with various cells. Due to high surface area, porosity, and resemblance of this 3D structure to natural extracellular matrix, they enable cell adhesion and provide a nano-environment for cellular growth and function. This has led to their wide application in tissue engineering [35]. Few drug-loaded nanofibers used for the treatment of periodontitis are listed in Table 11.2.

11.5 Anti-biofilm Approaches

Rapid advances in nanoscale engineering provide opportunities to develop new nanomaterials against virulent biofilms. Nanoparticles can carry and selectively release

Table 11.1 Nanoparticle drug delivery systems to treat periodontal diseases

Polymer	Drug	Inferences	Reference
2-Hydroxyethyl methacrylate (HEMA)	Tetracycline nanoparticles in calcium sulfate beads	• Cytocompatible $CaSO_4$-Tet nanoparticle composite beads could be valuable in declining bacterial count at the infection site • The size range of nanoparticles was 130 ± 20 nm with an entrapment efficiency of 89% • The sustained release followed Higuchi model • The composite beads were found to have antibacterial activity against Staphylococcus aureus and Escherichia coli	Sindhura Reddy et al. [29]
Polymersomes	Metronidazole, doxycycline	• Amphiphilic block copolymer vesicles having ability to encapsulate drugs • Can act intracellularly • They enter early endosomes of cells by the process of endocytosis and release drug by disassembling due to low pH • decreased count of *P. gingivalis* • This can be an efficient delivery device for antibiotics which cannot normally enter host cells	Wayakanon et al. [30]

(continued)

Table 11.1 (continued)

Polymer	Drug	Inferences	Reference
Poly(DL-lactide-co-glycolide) (PDLGA)	Silver	• Silver possesses antibacterial activity with no acute toxicity to human cells • More effective against aerobic bacteria than the anaerobic oral pathogenic bacteria	Lu et al. [31]
PDLGA	Harungana madagascariensis leaf extract	• PLG nanoparticles enhanced the bactericidal effect of the extract	Moulari et al. [32]

antimicrobial agents after attachment or within oral biofilms, resulting in their disruption. The mechanism involves "smart release" of agents when triggered by pathogenic microenvironments (e.g., acidic pH or low oxygen levels) for localized and controlled drug delivery to simultaneously kill bacteria and dismantle the biofilm matrix.

11.5.1 Zinc Oxide (ZnO) Nanoparticles

- Zinc oxide (ZnO) nanoparticle-coated titanium disks have demonstrated anti-adhesion properties. These ZnO-modified surfaces reduce viable bacteria like *S. aureus* and streptococci without cytotoxic effect on osteoblasts and human mesenchymal cells [39].
- The use of metal oxide nanoparticles to coat implants could provide osteoconductive and antimicrobial functionalities to prevent failure.
- Nevertheless, it may be difficult preventing bacterial adhesion altogether in the complex oral environment
- The benefits of ZnO nanoparticles in the presence of saliva and other host-derived components as well as a complex microbiota are yet to be proven clinically [40].

11.5.2 Silver Nanoparticles (AgNPs)

The use of AgNPs for surface modification has recently emerged as a promising approach.

Table 11.2 Drug-loaded nanofibers used for the treatment of periodontitis

Polymer	Drug	Inferences	Reference
Electrospun PLGA membranes	Human periodontal ligament (hPDL) cells	• Multilayered hybrid arrangements had developed by deposition of supplementary membrane layers on the existing ones • Cell adhesion, viability, and osteogenic differentiation were obtained indicating the feasibility of use in periodontal tissue regeneration	Inanc et al. [36]
Polytetrafluoro–thylene (PTFE) nanofibers coated on smooth glass and rough titanium substrates	Cultured human gingival fibroblasts	• Cell and collagen alignment persuaded by PTFE nanofibers and controlled by implant surface may enable improvement in gingival support	Kearns et al. [37]
Polydioxanone (PDS) biodegradable nanofiber matrices	Metronidazole and ciprofloxacin	• The nanofiber matrices did not impair the growth of periodontal beneficial bacteria • Growth inhibition of Fusobacterium nucleatum and Aggregatibacter actinomycetem-comitans was obtained from ciprofloxacin-containing nanofibers	Bottino et al. [38]

- AgNPs exhibited both high anti-adhesion and antibacterial actions against *Streptococcus mutans* when they were coated on the surface of methacrylate resins in vitro [41].
- The AgNPs coatings can induce death of bacteria either through cell membrane–nanoparticle interactions or through secondary effects induced by long-term released AgNPs or Ag^+ ions, by the production of reactive oxygen species, by alteration of the membrane integrity, and by protein-site-specific interactions that could prevent DNA replication [42].
- The concentration of dissolved oxygen has a direct implication on the AgNPs dissolution via the Ag^0 oxidation, and a low (or absence of) antimicrobial activity has been reported in anaerobic conditions [43].

11.5.3 Nano-hydroxyapatite-Modified Surfaces

- Surfaces modified by hydroxyapatite nanorod arrays exhibited antibacterial activity against *S. aureus* and *Escherichia coli* (significantly superior to titanium surfaces) and induced no cytotoxic effect toward human bone marrow stromal cells [44].

11.5.4 Other Coatings with Anti-adhesion and Antibacterial Properties

1. Hexametaphosphate associated with chlorhexidine also displayed antibacterial effects when coating pure grade II titanium. The antimicrobial efficacy of the coated surfaces appears to be associated with long-term release of chlorhexidine, which was capable of inhibiting the growth of *Streptococcus gordonii*, an early colonizer [45].
2. Graphene oxide (GO) has shown to have anti-adhesion and antibacterial properties [46]. The mechanism in which GO induces antimicrobial effects is not yet fully understood, although the size of the GO sheet appears to play an important role [47].

As these coatings are assembled in very small-length scales (nanometers), there is little influence on the mechanical properties of the bulk upon which they are assembled (e.g., hardness, strength, flexibility) and therefore exert limited action against mechanically induced failures [48]. However, antimicrobial coatings can help prevent interface (e.g., adhesive) failure from the deleterious effects of biofilms.

11.6 Role of Nano-proresolving Medicines in Management of Periodontal Diseases

Nonsoluble communication between cells have emerged as an efficient means of long-range communication beyond soluble cytokines and autacoids [49]. Membrane-shed vesicles, termed *microparticles*, can be employed to construct nano-proresolving medicines (NPRM) incorporating proresolving lipid mediators (LMs) that can target specific tissues without dilution or inactivation of the mediators [50]. Incorporation of proresolving mediators into NPRM has proven to be an effective approach for promoting survival in animal models of sepsis [51], reducing inflammation and stimulating resolution [52]. Both lipoxin A4 and resolvin E1 (RvE1) facilitate bone regeneration in small animal models of periodontitis [53–55].

T.E. Van Dyke et al in a proof-of-principle experiment in 2014 used nano-proresolving medicines (NPRM) containing a novel lipoxin analog (benzo-lipoxin A4, bLXA4) to promote regeneration of hard and soft tissues irreversibly lost to periodontitis in the Hanford miniature pig. In this experiment, NPRM-bLXA4 dramatically reduced inflammatory cell infiltrate into chronic periodontal disease sites treated surgically and dramatically increased new bone formation and regeneration of the periodontal organ. These findings indicate that NPRM-bLXA4 is a mimetic of endogenous resolving mechanisms with potent bioactions that offers a new therapeutic tissue engineering approach for the treatment of chronic osteolytic inflammatory diseases [56].

11.7 Conclusion

Recent developments in nanomaterials and nanotechnology have provided a promising insight into the commercial applications of nanomaterials in the management of periodontal diseases. Nanotechnology has the potential to provide controlled release devices with autonomous operation guided by need. It is envisaged that this trend will be further improved in the future as more and more nanotechnologies are commercially explored.

References

1. Flemmig TF (1999) Periodontitis. Ann Periodontol 4(1):32–37
2. Kleinberg I, Codipilly M (1999) Modeling of the oral malodor system and methods of analysis. Quintessence Int Berl Ger 1985 30(5):357–369
3. Saravana KR, Vijayalakshmi R (2006) Nanotechnology in dentistry. Indian J Dent Res Off Publ Indian Soc Dent Res 17(2):62–65
4. Rees JS, Addy M (2002) A cross-sectional study of dentine hypersensitivity. J Clin Periodontol 29(11):997–1003
5. Krauser JT (1986) Hypersensitive teeth. Part II Treatment. J Prosthet Dent 56(3):307–311

6. Blatz MB (2012) Laser therapy may be better than topical desensitizing agents for treating dentin hypersensitivity. J Evid-Based Dent Pract 12(3 Suppl):229–230
7. Gopinath NM, John J, Nagappan N, Prabhu S, Kumar ES (2015) Evaluation of dentifrice containing nano-hydroxyapatite for dentinal hypersensitivity: a randomized controlled trial. J Int Oral Health (JIOH) 7(8):118–122
8. Tschoppe P, Zandim DL, Martus P, Kielbassa AM (2011) Enamel and dentine remineralization by nano-hydroxyapatite toothpastes. J Dent 39(6):430–437
9. Vandiver J, Dean D, Patel N, Bonfield W, Ortiz C (2005) Nanoscale variation in surface charge of synthetic hydroxyapatite detected by chemically and spatially specific high-resolution force spectroscopy. Biomaterials 26(3):271–283
10. Hannig M, Hannig C (2010) Nanomaterials in preventive dentistry. Nat Nanotechnol 5(8):565–569
11. Vano M, Derchi G, Barone A, Covani U (2014) Effectiveness of nano-hydroxyapatite toothpaste in reducing dentin hypersensitivity: a double-blind randomized controlled trial. Quintessence Int Berl Ger 1985 45(8):703–711
12. Freitas RA (2000) Nanodentistry. J Am Dent Assoc 1939 131(11):1559–1565
13. Chandki R, Kala M, Kumar KN, Brigit B, Banthia P, Banthia R (2012) "Nanodentistry": exploring the beauty of miniature. J Clin Exp Dent 4(2):e119–e124
14. Keestra JAJ, Grosjean I, Coucke W, Quirynen M, Teughels W (2015) Non-surgical periodontal therapy with systemic antibiotics in patients with untreated chronic periodontitis: a systematic review and meta-analysis. J Periodontal Res 50(3):294–314
15. Keestra JAJ, Grosjean I, Coucke W, Quirynen M, Teughels W (2015) Non-surgical periodontal therapy with systemic antibiotics in patients with untreated aggressive periodontitis: a systematic review and meta-analysis. J Periodontal Res 50(6):689–706
16. Greenstein G, Polson A (1998) The role of local drug delivery in the management of periodontal diseases: a comprehensive review. J Periodontol 69(5):507–520
17. Joshi D, Garg T, Goyal AK, Rath G (2016) Advanced drug delivery approaches against periodontitis. Drug Deliv 23(2):363–377
18. Huh AJ, Kwon YJ (2011) "Nanoantibiotics": a new paradigm for treating infectious diseases using nanomaterials in the antibiotics resistant era. J Control Release Off J Control Release Soc 156(2):128–145
19. Piktel E, Niemirowicz K, Wątek M, Wollny T, Deptuła P, Bucki R (2016) Recent insights in nanotechnology-based drugs and formulations designed for effective anti-cancer therapy. J Nanobiotechnology 14(1):39
20. Kong LX, Peng Z, Li S-D, Bartold PM (2000) Nanotechnology and its role in the management of periodontal diseases. Periodontol 2006(40):184–196
21. Garg T, Singh S, Goyal AK (2013) Stimuli-sensitive hydrogels: an excellent carrier for drug and cell delivery. Crit Rev Ther Drug Carrier Syst 30(5):369–409
22. Niemirowicz K, Prokop I, Wilczewska AZ, Wnorowska U, Piktel E, Wątek M et al (2015) Magnetic nanoparticles enhance the anticancer activity of cathelicidin LL-37 peptide against colon cancer cells. Int J Nanomedicine 10:3843–3853
23. Michalak G, Głuszek K, Piktel E, Deptuła P, Puszkarz I, Niemirowicz K et al (2016) Polymeric nanoparticles—a novel solution for delivery of antimicrobial agents. Med Stud Med 32(1):56–62
24. Gu H, Ho PL, Tong E, Wang L, Xu B (2003) Presenting vancomycin on nanoparticles to enhance antimicrobial activities. Nano Lett 3(9):1261–1263
25. Niemirowicz K, Markiewicz KH, Wilczewska AZ, Car H (2012) Magnetic nanoparticles as new diagnostic tools in medicine. Adv Med Sci 57(2):196–207
26. Corbet E, Tam J, Zee K, Wong M, Lo E, Mombelli A et al (1997) Therapeutic effects of supervised chlorhexidine mouthrinses on untreated gingivitis. Oral Dis 3(1):9–18
27. Grenier D (1996) Effect of chlorhexidine on the adherence properties of Porphyromonas gingivalis. J Clin Periodontol 23(2):140–142
28. Tokajuk G, Niemirowicz K, Deptuła P, Piktel E, Cieśluk M, Wilczewska AZ et al (2017) Use of magnetic nanoparticles as a drug delivery system to improve chlorhexidine antimicrobial activity. Int J Nanomedicine 12:7833–7846

29. Sindhura Reddy N, Sowmya S, Bumgardner JD, Chennazhi KP, Biswas R, Jayakumar R (2014) Tetracycline nanoparticles loaded calcium sulfate composite beads for periodontal management. Biochim Biophys Acta 1840(6):2080–2090
30. Wayakanon K, Thornhill MH, Douglas CWI, Lewis AL, Warren NJ, Pinnock A et al (2013) Polymersome-mediated intracellular delivery of antibiotics to treat Porphyromonas gingivalis-infected oral epithelial cells. FASEB J Off Publ Fed Am Soc Exp Biol 27(11):4455–4465
31. Lu Z, Rong K, Li J, Yang H, Chen R (2013) Size-dependent antibacterial activities of silver nanoparticles against oral anaerobic pathogenic bacteria. J Mater Sci Mater Med 24(6):1465–1471
32. Moulari B, Lboutounne H, Chaumont J-P, Guillaume Y, Millet J, Pellequer Y (2006) Potentiation of the bactericidal activity of Harungana madagascariensis Lam. ex Poir. (Hypericaceae) leaf extract against oral bacteria using poly (D, L-lactide-co-glycolide) nanoparticles: in vitro study. Acta Odontol Scand 64(3):153–158
33. Gagandeep, Garg T, Malik B, Rath G, Goyal AK (2014) Development and characterization of nano-fiber patch for the treatment of glaucoma. Eur J Pharm Sci Off J Eur Fed Pharm Sci 53:10–16
34. Huang Z-M, Zhang Y-Z, Kotaki M, Ramakrishna S (2003) A review on polymer nanofibers by electrospinning and their applications in nanocomposites. Compos Sci Technol 63(15):2223–2253
35. Vasita R, Katti DS (2006) Nanofibers and their applications in tissue engineering. Int J Nanomedicine 1(1):15–30
36. Inanç B, Arslan YE, Seker S, Elçin AE, Elçin YM (2009) Periodontal ligament cellular structures engineered with electrospun poly(DL-lactide-co-glycolide) nanofibrous membrane scaffolds. J Biomed Mater Res A 90(1):186–195
37. Kearns VR, Williams RL, Mirvakily F, Doherty PJ, Martin N (2013) Guided gingival fibroblast attachment to titanium surfaces: an in vitro study. J Clin Periodontol 40(1):99–108
38. Bottino MC, Arthur RA, Waeiss RA, Kamocki K, Gregson KS, Gregory RL (2014) Biodegradable nanofibrous drug delivery systems: effects of metronidazole and ciprofloxacin on periodontopathogens and commensal oral bacteria. Clin Oral Invest 18(9):2151–2158
39. Abdulkareem EH, Memarzadeh K, Allaker RP, Huang J, Pratten J, Spratt D (2015) Anti-biofilm activity of zinc oxide and hydroxyapatite nanoparticles as dental implant coating materials. J Dent 43(12):1462–1469
40. Song F, Koo H, Ren D (2015) Effects of material properties on bacterial adhesion and biofilm formation. J Dent Res 94(8):1027–1034
41. Ionescu AC, Brambilla E, Travan A, Marsich E, Donati I, Gobbi P et al (2015) Silver-polysaccharide antimicrobial nanocomposite coating for methacrylic surfaces reduces Streptococcus mutans biofilm formation in vitro. J Dent 43(12):1483–1490
42. de Lima R, Seabra AB, Durán N (2012) Silver nanoparticles: a brief review of cytotoxicity and genotoxicity of chemically and biogenically synthesized nanoparticles. J Appl Toxicol JAT 32(11):867–879
43. Loza K, Diendorf J, Sengstock C, Ruiz-Gonzalez L, Gonzalez-Calbet JM, Vallet-Regi M et al (2014) The dissolution and biological effects of silver nanoparticles in biological media. J Mater Chem B 2(12):1634–1643
44. Tian B, Chen W, Dong Y, Marymont JV, Lei Y, Ke Q et al (2016) Silver nanoparticle-loaded hydroxyapatite coating: structure, antibacterial properties, and capacity for osteogenic induction in vitro. RSC Adv 6(11):8549–8562
45. Wood NJ, Jenkinson HF, Davis SA, Mann S, O'Sullivan DJ, Barbour ME (2015) Chlorhexidine hexametaphosphate nanoparticles as a novel antimicrobial coating for dental implants. J Mater Sci Mater Med 26(6):201
46. Zeng Z, Yu D, He Z, Liu J, Xiao F-X, Zhang Y et al (2016) Graphene oxide quantum dots covalently functionalized PVDF membrane with significantly-enhanced bactericidal and antibiofouling performances. Sci Rep 2(6):20142
47. Perreault F, de Faria AF, Nejati S, Elimelech M (2015) Antimicrobial properties of graphene oxide nanosheets: why size matters. ACS Nano 9(7):7226–7236

48. Spencer P, Ye Q, Misra A, Goncalves SEP, Laurence JS (2014) Proteins, pathogens, and failure at the composite-tooth interface. J Dent Res 93(12):1243–1249
49. Distler JHW, Distler O (2010) Inflammation: microparticles and their roles in inflammatory arthritides. Nat Rev Rheumatol 6(7):385–386
50. Dalli J, Serhan CN (2012) Specific lipid mediator signatures of human phagocytes: microparticles stimulate macrophage efferocytosis and pro-resolving mediators. Blood 120(15):e60–e72
51. Dalli J, Norling LV, Montero-Melendez T, Federici Canova D, Lashin H, Pavlov AM et al (2014) Microparticle alpha-2-macroglobulin enhances pro-resolving responses and promotes survival in sepsis. EMBO Mol Med 6(1):27–42
52. Norling LV, Spite M, Yang R, Flower RJ, Perretti M, Serhan CN (2011) Cutting edge: humanized nano-proresolving medicines mimic inflammation-resolution and enhance wound healing. J Immunol Baltim Md 1950 186(10):5543–5547
53. Hasturk H, Kantarci A, Goguet-Surmenian E, Blackwood A, Andry C, Serhan CN et al (2007) Resolvin E1 regulates inflammation at the cellular and tissue level and restores tissue homeostasis in vivo. J Immunol Baltim Md 1950 179(10):7021–7029
54. Herrera BS, Ohira T, Gao L, Omori K, Yang R, Zhu M et al (2008) An endogenous regulator of inflammation, resolvin E1, modulates osteoclast differentiation and bone resorption. Br J Pharmacol 155(8):1214–1223
55. Van Dyke TE (2011) Proresolving lipid mediators: potential for prevention and treatment of periodontitis. J Clin Periodontol 38(Suppl 11):119–125
56. Van Dyke TE, Hasturk H, Kantarci A, Freire MO, Nguyen D, Dalli J et al (2015) Proresolving nanomedicines activate bone regeneration in periodontitis. J Dent Res 94(1):148–156

Chapter 12
Oral Biofilms: From Development to Assessment and Treatment

Heeresh Shetty and Pankaj Gupta

12.1 Introduction

Rodney M. Donlan has defined biofilms as "An assemblage of microbial cells that is irreversibly associated with a surface and enclosed in a matrix of primarily polysaccharide material" [1]. Although James et al. is credited with describing the detailed structure of biofilms in trickling filters of water treatment plants using scanning and transmission electron microscopy in 1969 [2], it was Antonie van Leeuwenhoek who first observed microbes with his crude microscope on the tooth structure and thus is credited with its discovery.

Though the structure and composition of biofilms were correctly theorized by Costerton [3] half a century ago, it was not till the last two decades that the ultrastructure of biofilms and the genes involved in cell adhesion have been characterized by the use of confocal laser scanning electron microscope. Biofilms can form on any fluid—solid interface like indwelling medical devices, domestic and industrial water lines, natural aquatic systems like ponds and lakes, and living tissues. The general cycle of biofilm formation is illustrated in Fig. 12.1. Though the structure of biofilms varies greatly in different places, some characteristics related to the structure, composition, and the environmental variables can be generalized.

1. Variables related to cell adhesion and biofilm formation
2. Biofilm architecture

H. Shetty (✉) · P. Gupta
Department of Conservative Dentistry and Endodontics,
Nair Hospital Dental College, Mumbai, India
e-mail: heereshshetty@yahoo.com

R. S. Chaughule (ed.), *Dental Applications of Nanotechnology*,
https://doi.org/10.1007/978-3-319-97634-1_12

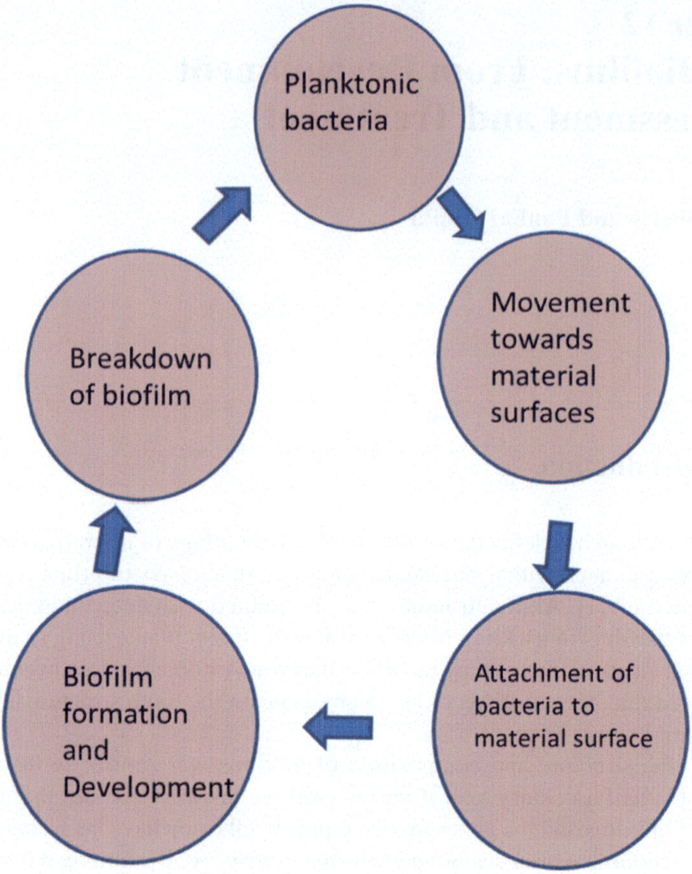

Fig. 12.1 Cycle of biofilm formation

12.1.1 Variables Related to Cell Adhesion and Biofilm Formation

There are multitudes of factors which affect the attachment of the cells to the surface (Fig. 12.2). These include surface characteristics of the solid, the fluid dynamics of the aqueous medium, and the properties of the attaching cells.

A rough surface attracts higher bacterial colonization because of increased surface area and reduced shear stress [4]. The surface characteristics of the solid also have effect on microbial colonization; for example, hydrophilic surfaces like glass attract less bacterial colonization than hydrophobic substances like Teflon [5, 6].

The aqueous medium to which the surface is exposed will affect the rate and extent of microbial attachment by coating the surface with polymers from the aqueous

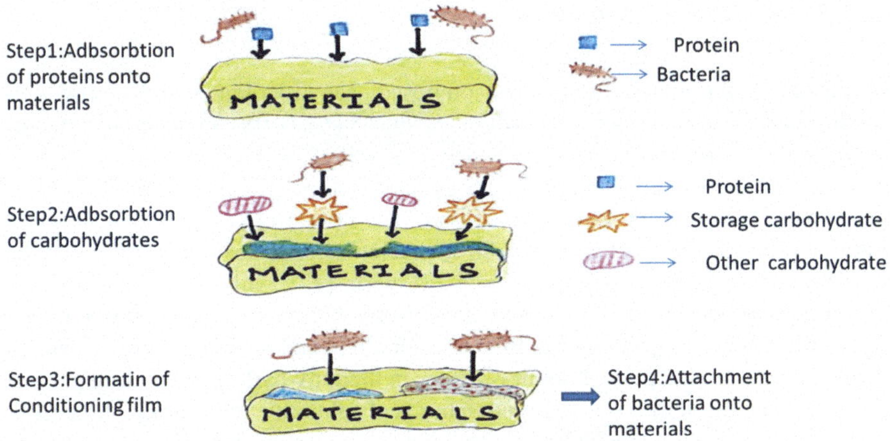

Fig. 12.2 Concept of conditioning film formation on materials

medium. This phenomenon was demonstrated for the first time by Loeb and Neihof on surfaces exposed to seawaters [7].

Hydrodynamic boundary layer referes to the layer near the solid–liquid interface where fluid velocity is negligible, the thickness of which is inversely proportional to the fluid velocity. When the fluid velocity is low, the thickness of this layer is more and the cell depends on its own motility and size for association with the surface. Therefore, high velocity of the fluid will cause rapid association with the surface until the velocities are high enough to dislodge the cells from the surface [1, 8].

Many other factors like the temperature of the aqueous medium, nutritional levels, and its acidity also affect the attachment of the microbes to the solid surface. The characteristics of the attaching cells like the presence of fimbriae and flagella, cell surface hydrophobicity, and the production of extracellular polymeric substances also influence the association of the microbial cell to the surface but the details of which are beyond the scope of this chapter, for a detailed review of all the factors, the reader is referred to an excellent article by Donlan [1].

Not only the characteristics of the solid surface and the aqueous medium affect the attachment of the microbes to the surface, but also the microbes themselves show an up- or downregulation in genetic expressions upon initial attachment which leads to a favorable colonization of the surface; for example, it was demonstrated that Staphylococcus aureus in biofilms shows an upregulation of the glycolysis genes probably because of deficiency of oxygen in the developed biofilms [9].

12.1.2 Biofilm Architecture

Biofilms are heterogenous aggregates composed of microcolonies of bacterial cells which are encased in extracellular polymeric substances (EPS) separated by interstitial voids [10]. EPS are the primary matrix material of the biofilms and can constitute anywhere between 50 and 90% of the total organic carbon of the biofilms [11]. The composition of EPS varies greatly, but it is primarily composed of polysaccharides. The EPS vary greatly in their characteristics; for example, they may be polyanionic for biofilms of gram-negative bacteria [12], while it may be cationic for gram-positive ones [13]. EPS may be hydrophobic in some cases, though most EPS have hydrophobic and hydrophilic areas in them and most are highly hydrated because they absorb a large amount of water in them by hydrogen bonding [12]. A high degree of variability is also found in terms of solubility and rigidity of the EPS.

Biofilms can be influenced by non-microbial particles from the host or the environment, and an example of this can be found in the calcification of plaque from the saliva which protects the microbes from mechanical removal as well as the action of antibacterial like chlorhexidine. A micro-colony is the basic structural unit of established biofilms. Various interactions within and between them bestow upon them characteristics which are not available to the planktonic cells. This can range from multidrug resistance as a result of gene transfers to attachment of different species of bacteria to each other as a result of quorum sensing (cell to cell signalling).

It is important to appreciate the heterogeneity of biofilms in all aspects. They are almost ever-changing in terms of space and time caused by various external and internal factors which provide a number of tactical advantages to the colonizing microbes over free microbes, namely, the resistance to removal because of the tough intercellular matrix, the tight adhesion to the underlying surface, which makes its complete removal by mechanical and chemical means difficult [14]. It has been reported that bacteria in biofilms are up to 1000 times more resistant to antimicrobial agents when compared to planktonic bacteria [15]. In the branch of dental health care, biofilms can be found not only in the oral cavity but also in the waterlines of the dental chair, both of which can have an effect on the treatment outcomes.

12.2 Biofilms in the Oral Cavity

Dental plaque was the first thing to be ever observed under a microscope, and research on it has been going on for more than a century. Though the view of plaque as a biofilm was proposed for the first time around 25 years ago by Marsh, he proposed that a change in environmental variables can trigger a change in the residential microflora to a disease-associated composition [16].

It is difficult to define particular species of microbes as being part of oral biofilms because the composition and structure of biofilms vary significantly between healthy individuals and it also changes as the person grows old. That's said, some species like

Neisseria, Streptococcus, Veillonella, Actinomyces, and Granulicatella have been found to be associated with biofilms in health with increasing consensus [17].

The oral biofilm associated with health consists of multiple species of microbes in close proximity of each other, which allows for a lot of interactions between species. Saliva seems to be the primary nutrient source for the microbial population of the oral biofilms which is broken down using proteases and glycosidases to form the EPS with little change in the pH of the environment. Thus, a hemostatic state is achieved in which neutrophilic organisms thrive.

A lot of interspecies interaction takes place among the microbes of the biofilms which can be symbiotic or antagonistic, and this results in the prevention of non-oral microbes from colonizing the biofilm. Thus, an oral biofilm associated with health has a dynamic balance between slow acid and alkali production, thereby maintaining the pH near neutrality. This restricts the growth of disease-causing microbes on the tooth surface.

Dietary sugar, especially sucrose, is considered as the most cariogenic not only because it is metabolized to acids but also because it produces intra- and extracellular polysaccharides. When dietary sugars are present, the biofilm shifts to a more acid producing, saccharolytic microbes which lower the pH by the breakdown of sugars and inhibit the neutrophilic microbes from growing which were associated with tooth health.

Biofilms associated with root canal infections are a major issue, and it has been dealt with in detail in another chapter. Figure 12.3 shows the SEM images of biofilms on root canal walls and dentinal tubules [25].

In the gingival sulcus, the microbes in the biofilm associated with periodontal disease use gingival crevicular fluid as the main source of nutrition and are mostly proteolytic and anaerobic in nature. Complex food webs with a lot of interspecies interactions lead to sequential breakdown of host proteins and glycoproteins while scavenging on oxygen, thereby creating an environment suitable for thriving of anaerobic microbes. Not only are some of these microbes capable of causing an alteration of the host environment but are also capable of taking advantage of the host defense mechanisms for which they have been termed "inflammophiles" [18].

Both dental caries and periodontal diseases-causing microbial biofilms have some common characteristics. Firstly, both are composed of highly organized, metabolically active multiple species of microbes. Secondly, both are controlled by a positive feedback loop mechanism though the stressors are different (altered salivary composition and flow along with dietary sugars for caries and alteration of the activities of key bacterial species and altered immune response for periodontal diseases). In periodontal diseases, dental plaque accumulation at the gingival margins causes inflammation of the gingiva, thereby activating the positive feedback loop and resulting in an increasing number of inflammophiles, while frequent exposure to sugar and its resultant fermentation to acids lowers the pH and promotes the growth of acidogenic bacteria in caries susceptible individuals. Another important property of disease-causing microbial biofilms are the emergent properties, which means that as a community, the properties are more than the sum of all the microbes put together. This has been demonstrated by the poor growth of the microbes in pure cultures on

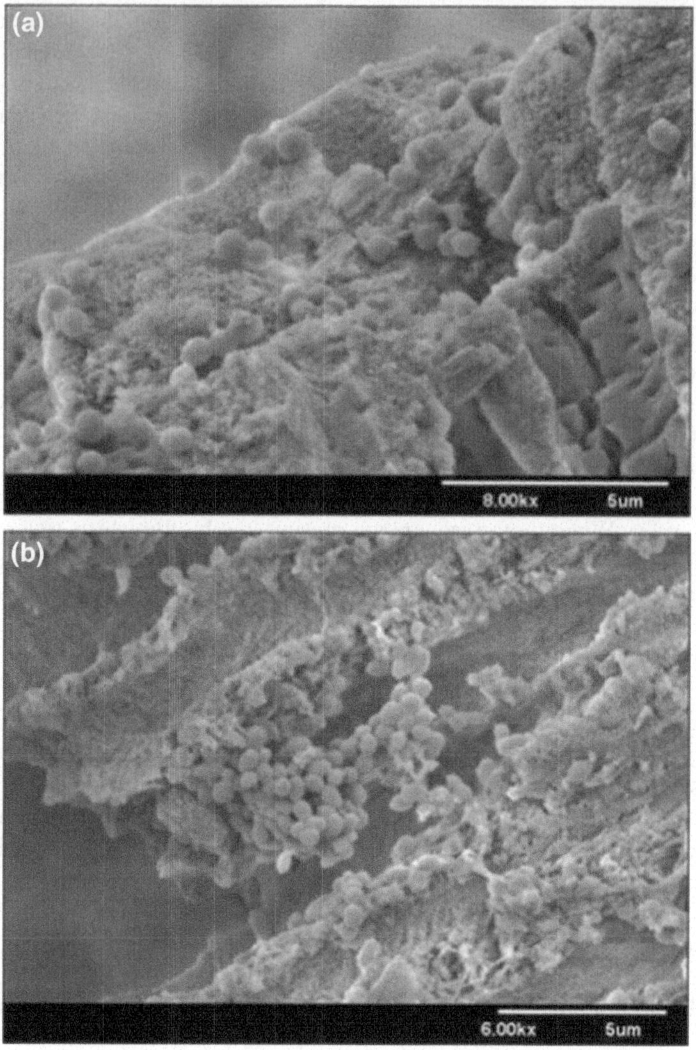

Fig. 12.3 Scanning electron micrographs showing biofilm (**a**) on the root canal wall (**b**) and within dentinal tubules, The Root Canal Biofilm. Editors: Luis E. Chávez de Paz, Christine M. Sedgley, Anil Kishen, Springers publications

glycoproteins in gingival crevicular fluid or saliva which is in contrast to the abundant growth as a microbial community [17].

Though the above-mentioned similarities exist between periodontal and caries-related biofilms, but some important differences need to be acknowledged. In caries-associated biofilm, the biodiversity has been observed to be less than in health, which is unlike biofilms associated with periodontal diseases where the diversity is more [17].

12.3 Controlling Biofilms

The etiology of pathogenesis in the oral cavity is complex, and a primary goal for dentistry is the prevention of disease. Oral hygiene instruction is the primary step in preventing both carious lesions and periodontal disease.

Control of oral biofilms is primarily accomplished through dentifrices containing abrasives, detergents, and antimicrobials, which are used in conjunction with mechanical toothbrushing [19]. Controlling biofilm formation, growth, and its re-aggregation is of importance for improvement and maintenance of oral health. Conversely, inability to control plaque leads to gingival inflammation and eventually to chronic periodontitis.

Anti-plaque agents usually function by disrupting or removing biofilms or preventing the formation of new biofilms. However, it is not necessary that they kill the microorganisms within the biofilm. The agents classified as antimicrobial act by inhibiting the growth (bacteriostatic) or killing (bactericidal) microorganisms are defined by minimum inhibitory concentration (MIC) and minimum bactericidal concentration (MBC), respectively. A key consideration in the administration of therapeutics is the uptake and penetration of these agents into biofilms [20]. This is of considerable importance, especially when these agents have to reach less accessible stagnation sites within the oral cavity. There is a huge interest and challenge in developing plaque control measures that require a minimum patient compliance and professional healthcare intervention [21].

Several alternatives to plaque control measures including the prevention of microbial adhesion surfaces, complementary approaches (electrical therapy, ultrasound, photodynamic therapy), and nanomaterials (liposomes, nanospheres, solid lipid nanoparticles, polymeric nanoparticles, dendrimers, polymeric micelles) have been used to control and eradicate the biofilms [22]. Considering this context, nanoparticles have completely different mechanisms of antibacterial activity than traditional antimicrobials, making nanotechnology a promising strategy to target and overcome biofilms within the oral cavity. Research has shown that certain nanostructures have the ability of controlled and sustained release, thereby improving stability and reducing drugs toxicity [23].

Some characteristics of nanoparticles which can be modified to facilitate the eradication of biofilms are its surface charges, the degree of hydrophobicity, and the ability of the nanoparticles to be taken up by the surface of biofilms [24].

12.4 In Vitro Biofilm Quantification

In order to evaluate the antibiofilm efficacy of the nanoparticles, it is important to understand the methods that can be used to quantify biofilms. Common methods used to assess the antibiofilm effect of nanoparticles are

(A) Bacterial growth rate

It is the simplest method in which the bacterial growth rate is measured by reading the optical density (OD) at 590 nm. This is based on the fact that higher the bacterial cell density more is the light scattering.

(B) Live/dead staining

This is a well-established method to evaluate bacterial cell viability. Bacterial cells with intact or damaged cytoplasmic membranes are stained with SYTO 9 (fluorescent green) and propidium iodide (fluorescent red) [26]. The drawback is that this method is time-consuming, and in general only a small portion of biofilm cultures can be examined using confocal laser microscopy.

(C) Metabolic assays

In order to assess the overall biofilm bacterial cell viability after treatments of nanoparticles with or without an alternating magnetic field, metabolic assays, such as the XTT assay and Alamar Blue assay, can be used to estimate bacterial viability as an indirect measurement [27].

(D) Staining

Crystal violet staining and measuring their absorbance in biofilms are used to assess the effect of nanoparticles on biofilm formation.

(E) Combination

The bacterial cell viability cannot be measured using crystal violet as it stains the entire biomass, including extracellular matrices as well as live and dead biofilm bacterial cells [28]. Usually in such cases a combination of live/dead staining and crystal violet staining is used for a thorough in vitro characterization of biofilms.

12.5 Nanoparticulate Metals as Antimicrobial Agents

Silver, zinc, and copper have been used as antimicrobial agents since time immemorial. These metals and their oxides have been integrated into a wide range of dental products, either alone, or in combination with other components, each having different properties and spectra of activity [29].

Powdered (micron-sized) zinc citrate or acetate has been incorporated into various oral care products to control the formation of dental plaque [30]. Powdered titanium dioxide has been used extensively as a whitener in toothpastes. In context to nanoparticulate metals, focus has been predominantly on the antimicrobial properties of silver [31] and copper [32]. Both these metals have been coated on to or incorporated into various base materials [33] and hydrogels [34]. An inverse relationship between the size of nanoparticles and antimicrobial activity has been clearly demonstrated, where particles in the size range of 1–10 nm have been shown to possess the greatest biocidal activity against bacteria [35]. Indeed, it has been shown that smaller silver nanoparticles are more toxic than larger particles, more so when oxidized [36]. At the nanoscale, Ag+ ions are known to be released (leached) from the surface [37].

12.5.1 Silver

Nanoparticulate silver is the most widely used antimicrobial agent among the metals. Ionic silver has the following advantages,

- It has broad-spectrum antimicrobial action.
- It is well tolerated by tissues.
- It is compatible with most materials used in making medical devices.
- It can be compounded into the submatrix or applied on the surface, and resistance to it is rare.

Silver can inhibit microbial growth and also even kill microorganisms. Silver works in a number of ways to disrupt critical functions in a microorganism. For example, it has a high affinity for negatively charged side groups on biological molecules such as sulfhydryl, carboxyl, phosphate, and other charged groups distributed throughout microbial cells. This binding reaction modifies the molecular structure of the macromolecule, thereby making them worthless to the cell.

Silver attacks the bacterial cell at multiple sites at the same time, thereby disrupting critical physiological functions of the cell such as cell wall synthesis, nucleic acid (RNA and DNA) synthesis and translation, membrane transport, electron transport and protein folding function, which are the key energy sources of the cell. Disruption of these functions either inhibits bacterial growth or results in the death of the microorganism. Resistance to the antimicrobial action of silver is highly unlikely as the organism would have to undergo simultaneous mutations in every critical function within a single generation to escape the influence of silver.

Various studies available in the literature on the nano-silver activities cannot be compared because of differences in the chemistry and physical properties of the particles employed. Furthermore, the bactericidal effect of AgNPs is dependent on the size and shape of the particles [38, 39]. The specific surface area of a dose of AgNPs increases as the particle size decreases, thereby felicitating greater material interaction with the surrounding environment. In addition to these triangular-shaped particles of silver displays more potent bactericidal activity than rods or spherical particles [39]. Other characteristics influencing the biological activity of nanoparticles are zeta potential and particle chemistry, with the latter likely to play a significant role in the ability of particles to penetrate into the cell [40].

A number of studies have demonstrated an inverse relationship between the size of silver nanoparticles and antimicrobial activity, where the highest biocidal activity has been shown by particles in the size range of 1–10 nm. Studies have also demonstrated higher toxicity from smaller silver nanoparticles when compared to larger particles, especially when oxidized [36]. At the nanoscale, Ag+ ions are known to be leached from the surface [37]. There is a high possibility that silver nanoparticles have multiple mechanisms of antibacterial action; however, due to the limited knowledge on this subject, the exact basis for the antibacterial activity of AgNPs is still undefined.

Some studies have reported that AgNPs release Ag+ ions in the presence of water [41–43]. According to Lok and coworkers, approximately 12% of the silver is present in the ionic form, tightly associated with the oxidation layer [36]. However, their experimental design made it difficult to differentiate between the mechanisms of action of AgNPs and dissolved Ag+ ions. Their study concluded that nano-silver affects bacterial membrane permeability by attaching to the cell membrane surface and modifying the cell potential.

Microbial responses to ionic silver and nanoparticles are different, and knowledge of both is required for a complete understanding of the antibacterial activity. It has been also shown that AgNPs interact with bacterial membrane proteins, intracellular proteins, phosphate residues in DNA and interfere with cell division, leading to bacterial cell death [31, 44]. DNA conglomeration defense mechanisms, which protect the cell from toxic effects, gets activated in the presence of biocidal Ag+ ions released from the nanoparticle surfaces, but at the same time compromises on its replication capabilities.

Few studies have reported that nano-silver causes oxidative damage, leading to the production of reactive oxygen species (ROS), i.e., free radicals [45, 46], and it has been suggested that the production of ROS is one of the primary mechanisms of nanoparticle toxicity [47]. Schematic representation of the effect of silver nanoparticles on microbial cell targets and biofilm formation is presented in Fig. 12.4 [48].

Many studies have reported the antibiofilm activity of silver nanoparticles. Small but significant decreases in the biomass of 24-h *P. putida* biofilms were observed by Fabrega and coworkers, and they were the first to discuss interactions between well-quantified and characterized bacterial biofilms and silver nanoparticles [49]. They reported small but significant decreases in the biomass of 24-h *P. putida* biofilms. The average diameter of the AgNPs employed in this study was 65 ± 30 nm. Kalishwaralal et al. [50] in their study found that AgNPs (mean diameter 50 nm) at a concentration of 100 nM almost completely prevented biofilm formation by *P. aeruginosa* and *S. epidermidis* by impeding the initial step: bacterial adhesion to the surface. Studies which have tended to employ smaller AgNPs have shown greater biological activity. Nano-silver (average diameter 25.2 ± 4 nm) was found to effectively prevent the formation of *P. aeruginosa* biofilms and kill bacteria in established biofilm structures [51]. There was 98.7% decrease in colonization and growth of *M. smegmatis* biofilms on membranes coated with nano-silver at a concentration of 100 μm. In addition,

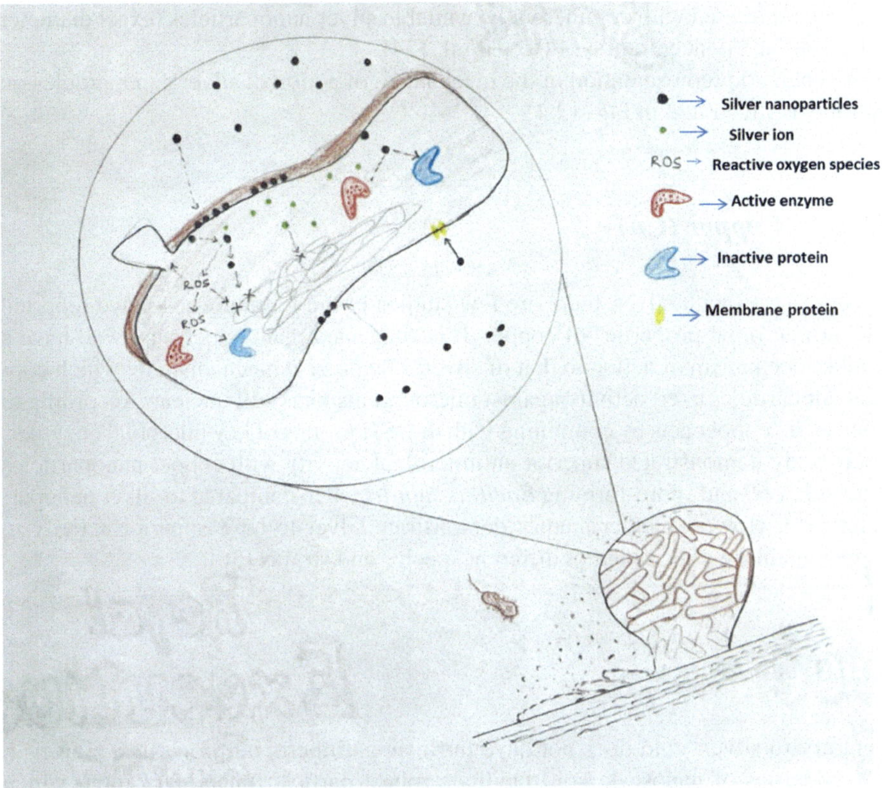

Fig. 12.4 Schematic representation of the silver nanoparticle mechanism of action on the biofilm forming microbial cell

this bacterium was reduced to only 0.03% in the presence of silver nanoparticles [52]. The variations observed in the antimicrobial and antibiofilm efficacy in the aforementioned studies may be attributed to the differences in the chemical and physical properties of nano-silver. Some researchers have proposed the usage of stabilized AgNPs to overcome the relatively low stability of colloidal solutions. This delays the aggregation of particles into larger forms that can significantly decrease their activity. Radzig and colleagues found that stabilizing AgNPs (about 8 nm in diameter) with hydrolyzed casein peptides strongly inhibited biofilm formation by some gram-negative bacteria. AgNPs concentration between 5 and 10 μg/mL showed decrease in the bacterial mass in biofilms of *E. coli*, *P. aeruginosa*, and *Serratia proteamaculans* [53].

Hartmann et al. introduced a chip-calorimetric approach to detect bacterial metabolic changes which allows non-invasive and real-time investigation of the effects of nanoparticles on bead-grown biofilms. With this system, their study revealed a complete inhibition of the cell metabolic activity in a bead-grown biofilm of *P. putida* treated with commercially available silver nanoparticles (exact diameter not given) at a concentration of 0.5 μg/mL [54].

A schematic representation of the mechanism of action of silver nanoparticles on biofilms is illustrated in Fig. 12.4.

12.5.2 Copper (Cu)

In comparison with silver, there are few studies in the literature that have reported the antimicrobial properties of copper. It is suggested that copper may well have a similar mechanism of action to that of silver. The precise mechanism by which copper nanoparticles exert activity against microorganisms is still unclear. According to Yoon et al., copper acts by combining with the -SH groups of key microbial enzymes. Their study demonstrated superior antimicrobial activity with copper nanoparticles against *E. coli* and spore-forming *Bacillus subtilis* when compared to silver nanoparticles [55]. However, other studies demonstrate silver to have superior activity to copper against a wide range of different species and strains [56].

12.5.3 Gold

Contrary to silver, gold does not have intrinsic antimicrobial properties. However, the properties of nanoscale gold facilitate robust particle functionalization, which has been explored by researchers for the possibility of using gold nanoparticles in biofilm treatment. Similar to silver, gold nanoparticles are most commonly synthesized by the reduction of gold salts [57–59]. Gold nanoparticles showed 97% *Staphylococcus aureus* biofilm inhibition and disrupted 24-h-old *S. aureus* biofilms up to 95% and *Acinetobacterbaumannii* biofilms up to 40%. However, silver and bimetallic nanoparticles showed better antimicrobial activity against biofilms in all strains tested in terms of inhibition and disruption [60]. Ramasamy et al. investigated the antibiofilm efficacy of gold-silver bimetallic nanoparticles synthesized from γ-proteobacterium *Shewanella oneidensis* MR-1. Gold-silver bimetallic nanoparticles inhibited all strains effectively at 250 μM. Particle concentration as low as 10 μM was enough to inhibit *E. coli* biofilm formation [61].

Gold exhibits a weak antimicrobial effect in comparison with silver and copper. Nevertheless, gold nanoparticles have been employed in multiple applications involving biological systems. The exceptional binding properties of gold make it particularly suitable for attaching ligands to enhance bio-molecular interactions. Gold nanoparticles also exhibit an intense color in the visible range and contrast strongly for imaging by electron microscopy. Despite all the current and potential applications for gold nanoparticles, there remains little information as to how these particles affect microorganisms.

12.6 Nanoparticulate Metal Oxide

12.6.1 Copper Oxide

CuO nanoparticles exhibit effective antimicrobial activity against various bacteria, however, higher concentrations are required to achieve desired antimicrobial effects, and at these concentrations CuO nanoparticles could be toxic to mammalian cells [62–64]. Although CuO NPs have excellent antibacterial effects, their antibiofilm effects are limited to a narrow antibacterial window [65]. CuO is cheaper than Ag, with the advantage of being a stable chemical, and with physical properties that allow it to be easily mixed with polymers [62]. However, further research is warranted to investigate CuO NPs as a potential antibacterial agent in dental materials.

12.6.2 Zinc Oxide

Similar to that of other nanoparticulate metals and metal oxides, the antimicrobial mechanisms of zinc are not completely understood. Nano-zinc oxide has received increasing attention, not only because it is stable under harsh processing conditions but also because it is generally regarded as safe and biocompatible [66]. Studies have shown that ZnO exhibits a degree of selective toxicity to bacteria with a minimal effect on human cells [67, 68]. The important factor that contributes in bacterial growth inhibition could be the release of zinc ions from ZnO NPs and the generation of reactive oxygen species [69, 70]. Liu et al. [71] in their study have shown that exposure to ZnO NPs causes leakage of the intracellular contents of bacteria and the disruption of cell walls and cell membranes. ZnO is said to be relatively more biocompatible than other NPs with known antimicrobial activity, such as silver NPs, as zinc is a necessary trace element in humans [72]. ZnO demonstrated greater sensitivity under anaerobic conditions for organisms implicated in oral infections, including *A. actinomycetemcomitans*, *P. gingivalis*, *Prev. intermedia,* and *F. nucleatum*, with growth inhibitory and killing concentrations of 0.25–2.5 and 0.25–2.5 mg/mL, respectively [73]. Khan et al. [74, 75] suggested the possible incorporation of ZnO NPs in oral pastes and mouthwashes at concentration of 100 lg/ml. Use of these nanoparticles in toothpastes is also reported by other authors [76, 77]. ZnO NPs with an average size of 35 nm have shown to exhibit good antimicrobial activity against oral bacteria in vitro [74, 78]. In addition to the antimicrobial and antibiofilm formation activities, ZnO NPs have also shown to inhibit dentine demineralization [79].

12.6.3 *Titanium Dioxide (TiO_2)*

TiO_2 NPs are one of the most widely produced NPs globally with some studies estimating a global production of 3000 tons per year [80, 81]. Its ability to act as a photocatalytic antimicrobial compound is well established [82]. TiO_2 is widely used in a number of applications in a nanoparticulate form and is generally considered to be nontoxic at the given concentrations. TiO_2 NPs have shown significant antimicrobial activity against a number of microorganisms including *E. coli, S. aureus, P. aeruginosa, E. faecium, B. subtilis,* and *Klebsiella pneumonia* [83, 84]. The MIC value of the TiO_2 NPs (62–74 nm) is reported in the range of 40–80 lg/ml against the above-given bacteria. ROS generation, DNA damage after internalization, peroxidation of membrane phospholipids, and inhibition of respiration are some of the possible modes of action of TiO_2 NPs against bacteria [85, 86]. Photoactivation of TiO_2 NPs remarkably increased its antimicrobial activity against *Bacteroidesfragilis, E. coli, E. hire, P. aeruginosa, S. typhimurium, and S. aureus* [87]. However, there have been concerns about the health hazards of nano-titanium oxide through inflammation as generated by release of interleukin-1α [88].

12.7 Anti-adhesive Nanoparticles

12.7.1 *Chitosan*

Chitosan is a biopolymer obtained by deacetylation of chitin, a natural polymer, and has been used extensively in biomedical applications due to its high biocompatibility and antimicrobial properties [89].

The exact mechanism of the antimicrobial action of chitin, chitosan, and their derivatives is still unknown, but various mechanisms have been proposed. Interaction between positively charged chitosan molecules and negatively charged microbial cell membranes leads to the leakage of proteinaceous and other intracellular constituents [90]. Chitosan acts mainly on the outer surface of the bacteria. At a lower concentration (<0.2 mg/mL), the polycationic chitosan probably binds to the negatively charged bacterial surface to cause agglutination, while at higher concentrations, the larger number of positive charges may have imparted a net positive charge to the bacterial surfaces to keep them in suspension [91]. Greater antibacterial activity was demonstrated against *Staphylococcus aureus* than against *E. coli*. A concentration of 500 ppm was needed to kill all *S. aureus* within 120 min [92].

An in vitro study conducted by Kishen et al. [93] demonstrated the effect of a zinc oxide and a combination of zinc oxide and chitosan nanoparticulates on root canal surfaces and concluded that cationic antibacterial nanoparticulates are able to significantly reduce *E. faecalis* adherence to dentin.

12.7.2 *Silica Nanoparticles*

Silica nanoparticles because of their high thermal and chemical stability, high surface area, and good biocompatibility have generated considerable interest. Colloidal silica is a highly versatile material, as it can be developed in different forms and sizes; its surface is easily modified and is relatively cheap. Silica particles and silica-modified nanoparticles loaded with different antimicrobial agents have been routinely used for the prevention and treatment of infections.

12.7.2.1 Nitric Oxide-Releasing Silica Nanoparticles

The antimicrobial activity of NO, a reactive free radical produced by inflammatory cells, has been demonstrated by a variety of approaches. NO production is part of a host defense mechanism in response to infection. It is stimulated by proinflammatory cytokines such as IL-1, IL-2, IFN-γ, and TNF-α, and microbial products such as lipopolysaccharide and lipoteichoic acid stimulate the release of NO synthase isoform (NOS2). NO-donor compounds when directly administered in vitro have shown to inhibit or kill microbes, demonstrating effectiveness against a remarkably broad range of pathogenic microorganisms including bacteria, fungi, viruses, and parasites [94]. Several strategies have been developed for delivering nitric oxide to bacteria. Nitric oxide-releasing silica nanoparticles have been used to successfully control biofilm formation. Although the mode of action of NO-releasing nanoparticles remains unclear, it may be attributed to the rapid diffusion properties of NO, which may result in enhanced penetration into the biofilm matrix and thus improve efficacy against biofilm-embedded bacteria [95].

12.7.2.2 Metal-Modified Silica Nanoparticles

The high cytotoxicity of copper and silver nanoparticles toward mammalian cells has been an area of concern. The incorporation of silver and copper into silica nanoparticles allows the reduction of concentrations of metal ions needed to achieve antimicrobial concentrations.

A. **Silver-SiNP**

There are a number of studies in the literature with the combination of silver and silica for the synthesis of nanoparticles. Spherical nanoparticles with a silver core and an amorphous silica shell were successfully fabricated by using tetraethoxysilane as silica precursor and reducing silver nitrate with ascorbic acid. These nanoparticles had excellent antibacterial effects against *E. coli* and *S. aureus* [96]. Wang et al. [97] in their study reported the antibacterial effects of Ag–SiO_2 hollow composite powders against *Escherichia coli*, *Staphylococcus aureus,* and *bacillus*. Hollow silica nanospheres and nanotubes were synthesized as hosts for the immobilization of silver. Though both composites had excellent antibacterial activity, Ag-supported tubular hollow structure demonstrated a stronger antibacterial ability compared to the spherical hollow structure due to their ability to retain higher silver contents as well as smaller and more dispersed silver nanoparticles.

B. **Copper-SiNP**

Copper is a heavy metal which is an effective antibacterial agent with relatively low toxicity toward mammalian cells. It inactivates bacteria by altering structure of proteins and enzymes so that they can no longer perform their normal functions [98].

Copper can be deposited on a supporting material like spherical silica nanoparticles so that the releasing time of Cu can be delayed over a long period of time. Kim et al. [99] in their study demonstrated excellent antibacterial properties of the Cu-SiO_2 nanocomposites when examined with disk diffusion assays.

12.8 Surface-Modified Silica Nanoparticles

Grafting methods constitute a functionalization technique that involves a wide range of modifications made on the surface of the nanoparticles resulting in an ordered secondary structure with new interesting properties including antibacterial and antifungal abilities. In this way, polypeptide polymer-grafted silica nanoparticles (NPs) were synthesized from poly-*L*-lysine covalently attached and their efficacy as antimicrobial agents was proved on both gram-negative *E. coli* and gram-positive *Bacillus subtilis* [100].

Silica nanoparticles could also be used to covalently link molecules with antimicrobial effect to their surface through the selection of proper linkers. In this context, hybrid-silica nanoparticles (NPs) containing the FDA-approved antimicrobial triclosan (Irgasan) were found to be superior in killing bacteria, as compared with the free biocide [101]. Botequim et al. [102] tested silica nanoparticles coated with a quaternary ammonium cationic surfactant, didodecyldimethylammonium bromide (DDAB) against bacteria (*S. aureus*) and fungi (*C. albicans*), resulting in lower minimal inhibitory concentrations (MIC) against bacteria and fungi compared to a soluble surfactant.

12.9 Bioglasses and Bioceramics

Bioactive glasses of the SiO_2-Na_2O-CaO-P_2O_5 composition have demonstrated antimicrobial activity in aqueous solutions due to the release of their ionic compounds over time [103]. This results in an alkaline environment [103], which is not tolerated by bacteria [104]. In addition, antibacterial bioactive glass effect has been attributed to the release of silica [105]. The bactericidal efficacy of bioactive 45S5 Bioglass®-derived glass-ceramic substrates against common gram-positive and gram-negative bacteria and also against yeast has been demonstrated [106]. Moreover, the shift from micro- to nano-sized bioglass materials caused a tenfold increase in silica release and also in solution pH by more than three units. As a result, the killing efficacy was substantially higher for the nanomaterial against all tested strains of *Enterococcus fecalis* [107]. However, their antibacterial efficacy in human teeth is still inferior to that of other materials. Efforts have been made to spike bioactive glass with antimicrobial agents to increase its antimicrobial efficacy when used on dental hard tissues.

12.10 Photodynamic Therapy and the Use of Nanoparticles to Control Oral Biofilms

Antimicrobial photodynamic therapy (aPDT) is increasingly being explored for treatment of oral diseases and infections. In aPDT, a photoactivated compound or photosensitizer (PS), which has been taken up by microorganisms, is activated by visible light of a specific wavelength to produce reactive oxygen species (ROS) , including highly cytotoxic singlet oxygen (Fig. 12.5) [108]. Elimination of resistant microorganisms and secreted virulence factors, local delivery of PS, and double selectivity (deleterious effect only on sites where both PS and light are delivered concomitantly) are some of the advantages of aPDT over conventional antimicrobial therapies [109]. The adjunctive use of aPDT as an alternative to chemical antimicrobial agents has been suggested in the treatment of periodontitis for elimination of subgingival bacterial species [110].

Several studies have also shown incomplete destruction of oral biofilms using methylene blue (MB)-mediated aPDT [111, 112]. Biofilms exhibit reduced susceptibility not only to PDT but to antimicrobial treatments in general, which is attributed to reduced penetration of the PS and other drugs deep in the biofilm matrix [113]. In addition, it has been shown that MB and other phenothiazine derivatives are substrates of multidrug resistance efflux pumps in bacteria, which decreases the effectiveness of the few PS molecules that are able to penetrate the biofilm matrix [114]. Such drawbacks can be overcome by nanoparticles (NPs), which can significantly improve the pharmacological characteristics of the PS, e.g., improved solubility and absorption increased local retention and protection against degradation and/or efflux [115–118].

The advantages of nanoparticles containing PS over free include [116, 117]: (1) a larger critical mass (concentrated package of PS) for the local production of ROS; (2) limit the target cell's ability to pump the drug back out, thus reducing the chances of multidrug resistance; (3) increase treatment selectivity by the localized delivery of agents, by passive or active targeting; and (4) the nanoparticle matrix is non-immunogenic.

In the context of PDT polymer-based NPs, they display a few advantages regarding drug delivery and have garnered considerable interests. Biocompatibility and low toxicity are two remarkable properties of polymeric NPs, besides their easy and

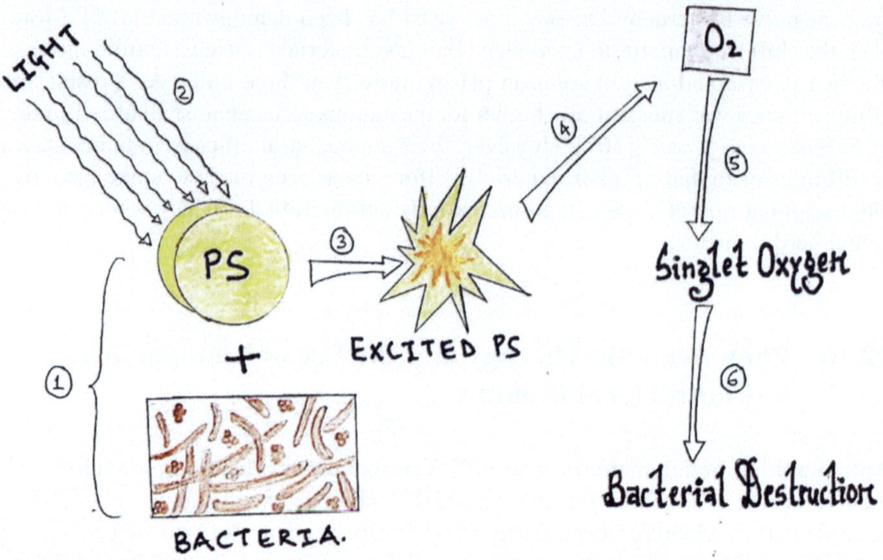

Fig. 12.5 Mechanism of photodynamic therapy action. A photosensitizer is taken up by microorganisms (1) and following exposure to light of the appropriate wavelength (2) becomes activated to an excited state (3). Then the photosensitizer transfers energy from light to molecular oxygen (4) to generate singlet oxygen and free radicals (5) that are cytotoxic to cells (6)

straightforward process to fabricate stable formulations [118]. The most widely used polymeric NPs to date are those composed of poly(lactic-co-glycolic acid) (PLGA). The physicochemical properties, biodegradation rate, and in vivo behavior of PLGAs can be easily modified by manipulating its molecular weight, lactic acid–glycolic acid ratio, and end group [119], resulting in a minimal systemic toxicity thereby giving way to its approval for use in humans by the US Food and Drug Administration (FDA) and the European Medicine Agency (EMA).

Nobel metallic nanoparticles such as colloidal SeNPs are attracting significant interest as they have some unique chemical and physical properties that make them efficient carriers and enhancers. Haris and Khan [120] in their study used selenium nanoparticle-enhanced photodynamic therapy against *S. mutan* biofilm. It is a new combination for enhancement of performance of antimicrobial photodynamic therapy processes in coupling nanotechnology to PDT. Antibiofilm assays and microscopic studies showed a significant reduction of biofilm presence of conjugate. A crystal violet assay revealed a maximum percent inhibition of *S. mutans* biofilm formation after 24-h incubation.

Colloidal gold nanoparticles, loaded with methylene blue and exposed to red light at 665 nm, have been tested against planktonic *E. faecalis* and in experimentally infected root canals [121]. The rationale behind this is that gold nanoparticle conjugates because of their improved binding and cell wall penetration properties should deliver a higher concentration of photoactive molecules. However, it remains to be fully established whether such conjugates will show superior antibacterial activity when compared to more conventional treatments.

12.11 Biofilm Removal in Dental Unit Water Lines

Dental waterline contamination is widespread in any type of dental setting having serious implications on clinicians and patients alike, especially elderly and immune-compromised. Hence, international bodies like center for disease and control (CDC) and American Dental Association (ADA) have come up with stringent measures for maintenance of water quality.

A wide variety of microbes has been isolated from the biofilms in DUWL. Among all microorganisms isolated from DUWL, the ones having a pathogenic potential are Legionella pneumophila, *Mycobacterium* spp., Pseudomonas aeruginosa, Acinetobacter baumannii, Staphylococcus spp., and Stenotrophomonas maltophilia [122, 123]. Most of the isolated organisms are gram-negative mesoheterotrophic water bacteria.

Yeast-like fungi such as the *Candida albicans* and *Candida curvata* have been found in the samples of water and biofilms from DUWL. Other species isolated from dental water units undergoing a continuous waterline treatment were *Aspergillus flavus, Penicilliumexpansum, Exophialamesophila* [123, 124]. Some of these fungi have also been isolated from immunocompromised patients with respiratory infections (Table 12.1) [125].

Evolution of multiple drug-resistant organisms (MDRO) and their spread into hospital and clinical setting have all accentuated the need to monitor these organisms [123]. To keep the biofilm and the contamination of DUWLs under control, or rather reduce the risk of infection both for the patients and health personal, water treatment methods in usage are biocidal agents such as chlorine and its derivatives, use of ultraviolet light [126], boiling of water, ultrasonic irradiation at a low frequency [126], water distillation, reverse osmosis, filtration with fiber and ceramic filters, activated carbon, solid block, pitcher and faucet-mount filters, bottled water, ion exchange, water softener, ozone systems, and activated alumina "Altered" Water.

12.11.1 Nanotechnology for Water Decontamination

The ever-increasing dearth of potable water has almost reached a crisis which could be abated utilizing advancements in nanoscience and engineering that have led to the development of tools such as nanoabsorbents, nanocatalysts, bioactive nanoparticles, nanostructured catalytic membranes, submicron nanopowder, nanotubes, magnetic nanoparticles, granules, flake, and high area metal particle supramolecular assemblies with characteristic length scales of 9–10 nm including clusters, micromolecules, nanoparticles, and colloids having a significant impact on water qual-

Table 12.1 List of isolated organisms from DUWL

Bacteria	
Gram-positive bacteria	Gram-negative bacteria
Rods: *Brevibacterium epidermidis* Cocci: *Micrococcus luteus* *Micrococcus lylae* *Staphylococcus cohnii* *Staphylococcus lentus* *Staphylococcus pulvereri/vitulus* *Staphylococcus* spp. *Streptococcus* spp.	*Brevundimonas vesicularis* *Moraxella lacunata* *Moraxella* spp. *Ralstonia (Pseudomonas)* *Sphingomonas paucimobilis* *Stenotrophomonasmaltophilia*
Fungi	
Molds	Yeast like
Aspergillus amstelodani *Aspergillus fumigates* *Aspergillus glaucus* *Aspergillus repens* *Citromuces* spp. *Penicilliumpusillum* *Penicilliumturolense* *Sclerotium sclerotiorum*	*Candida albicans* *Candida curvata* *Geotrichum candidum*

ity in the natural environment [127]. Carbon nanotubes (CNTs) such as dioxin [128], dichlorodiphenyltrichloroethane and its metabolites [128], polynuclear aromatic hydrocarbons [129–131], polybrominated diphenyl ethers [132], chlorobenzenes and chlorophenols [133, 134], trihalomethanes [135, 136], nonylphenol [137], bisphenol A, phthalate esters [138], dyes [139] have attracted a lot of attention as very powerful adsorbents, for a wide variety of organic compounds from water.

Among the wide variety of materials, four categories could be considered for water purification: (1) carbonaceous nanomaterials, (2) metal-containing nanoparticles, (3) zeolites, and (4) dendrimers. Their versatile physicochemical properties make them particularly attractive as separation and reactive media for water purification [140].

Carbon-based nano-adsorbents

CNTs are known to eliminate organic chemicals more efficiently than activated carbon on adsorption [141]. Oxidized CNTs due to their fast kinetics have higher adsorption capacity for metal ions. Electrostatic attraction and chemical bonding on the surface of functional groups (e.g., carboxyl, hydroxyl, and phenol) of CNTs assist in adsorption of metal ions [141].

Metal-based nano-adsorbents

One of the major advantages of metal oxides such as iron oxide, titanium dioxide, and alumina for the removal of metals and radionuclides is its cost-effectiveness. Metal-based nanomaterials have shown great potential to outcompete activated carbon in removing a variety of heavy metals such as arsenic, lead, mercury, copper, nickel, and cadmium [142].

Polymeric nano-adsorbents

Dendrimers are customized adsorbents that are capable removing both heavy and organic compounds. Their interior shells can be made hydrophobic for sorption of organic compounds, while the exterior branches can be tailored for adsorption of heavy metals [143].

12.11.2 Other Techniques

Pollution detection and sensing

Sensors for the detection of different compounds can be fabricated from nanostructured materials [144]. Silver nanoparticle array membranes can be used as flow-through Raman scattering sensors for water quality monitoring [145].

Nanofiltration

Water treatment using nanofiltration which is used for drinking water production is one of the easiest techniques that can be employed [146]. They consist of pressure-driven membranes with pore size in the range of 0.2 and 4 nm and properties in between those of reverse osmosis and ultrafiltration membranes. These nanofilter (NF) membranes have been shown to be effective in removing turbidity, microorganisms, and inorganic ions such as Na+ and Ca++. CNTs have been arranged to form hollow monolithic cylindrical membranes, [147] which were efficient for the removal of bacteria or hydrocarbons. Silver nanoparticles have been considered for drinking water treatment due to their strong and broad spectrum of antimicrobial activities [66, 127, 148]. Among the various silver nanofilters tested, Ag/cation resin filter is said to have the highest bacteria removal efficiency. These filters completely removed all targeted bacteria, namely *Escherichia coli, Salmonella typhimurium, Shigella dysenteriae,* and *Vibrio cholera* [148]. Figure 12.6 schematically represents a composite nanomaterial-packed bed reactor for the purification of water [127].

Adsorbents

Slurry reactors or adsorbers have included nano-adsorbents into their existing treatment procedures. These are considered to be most effective in powdered form since all surfaces of the adsorbents are utilized, and the mixing greatly facilitates the mass transfer. Nano-adsorbents can also be used in fixed or fluidized adsorbers in the form of pellets/beads or porous granules loaded with nano-adsorbents [149].

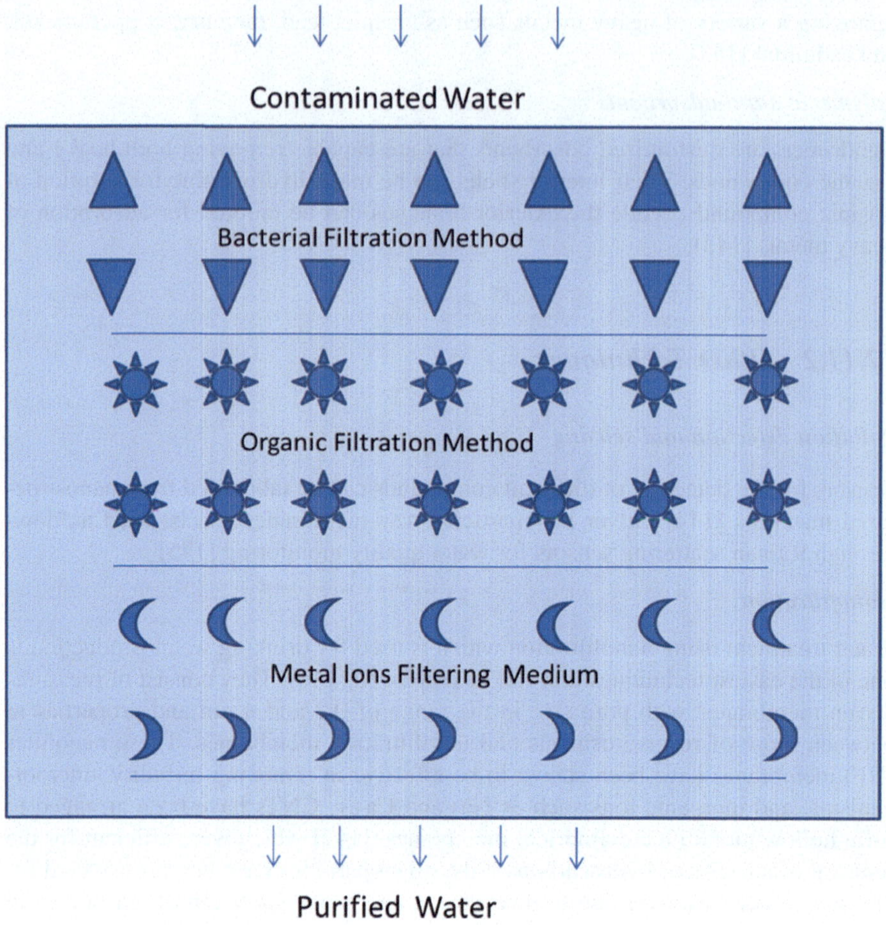

Fig. 12.6 Schematic representation of a composite nanomaterial-packed bed reactor for purification of water contaminated by mixtures of (i) metal ions, (ii) organic solutes, and (iii) bacteria

Nanoemulsions

Nanoemulsions are uniform population of droplets of high energy ranging in diameter from 100 to 300 nm [150]. These nanoemulsions exhibit a wide range of biocidal activity even against enveloped viruses, spores, fungi, and bacteria [151]. They act by nonspecific disruption of bacterial cell membranes, thereby preventing the development of resistant strains. Nanoemulsions can be diluted and stored for up to 2 years at a broad range of temperatures [152–154]. A significant reduction in live bacteria observed in nanoemulsion-treated DUWL tubing showed that nanoemulsion effectively reduced DUWL communities.

Mechanism of action

The kinetics of the nanoemulsion is accredited to their production, wherein nanoemulsions are under produced using high shear forces in a microfluidizer. The shear energy is stored in oil droplets, giving them the high energy, which is passed on to bacteria upon fusion of the droplet with the bacteria, disrupting the bacterial membrane. Nanoemulsions are usually supplemented with cetylpyridinium chloride and quaternary ammonium salt, by its incorporation as a co-surfactant to the nanodroplet. Nanoemulsion containing cetylpyridinium chloride is considered to be effective in DUWL decontamination as they exhibit multiple modes of antibacterial activity such as disruption of intermolecular interactions, cellular membrane, cellular permeability controls, and inducing leakage of cellular contents [153].

12.11.3 Disadvantages of Using Nanotechnology for Water Treatment

Ecotoxicity in water is a major concern due to engineered nanoparticle production. TiO_2 nanoparticles aggregate severely when added to water. Aqueous fullerene particles can get coagulated in the presence of salt. However, these problems can be addressed by the use of model wastewater treatment, by clearing sludge for the removal of oxide nanoparticles [154].

12.12 Conclusion

The high costs and complexities of the current antibiofilm agents press upon the urgent need for developing cost-effective alternatives. As is made clear by this chapter, recent developments in nanotechnology-based approaches aimed at preventing, controlling, and treating bacterial biofilm infections in the oral cavity and also in DUWL are worthy of serious consideration.

Biofilms have varied architecture and adaptive capabilities which pose unique challenges to conventional anti-infective agents. Nanoparticles by virtue of their adaptation to the environment via change of charge, hydrophobicity, etc., can be looked upon as an exciting avenue for combating biofilms and the microbes associated with it.

References

1. Donlan RM (2002) Biofilms: microbial life on surfaces. Emerg Infect Dis 8(9):881
2. Jones HC, Roth IL, Saunders WM III (1969) Electron microscopic study of a slime layer. J Bacteriol 99:316–325

3. Costerton JW, Geesey GG, Cheng K-J (1978) How bacteria stick. Sci Am 238:86–95
4. Characklis WG, McFeters GA, Marshall KC (1990) Physiological ecology in biofilm systems. In: Characklis WG, Marshall KC (eds) Biofilms. Wiley, New York, pp 341–394
5. Pringle JH, Fletcher M (1983) Influence of substratum wettability on attachment of freshwater bacteria to solid surfaces. Appl Environ Microbiol 45:811–817
6. Bendinger B, Rijnaarts HHM, Altendorf K, Zehnder AJB (1993) Physicochemical cell surface and adhesive properties of coryneform bacteria related to the presence and chain length of mycolic acids. Appl Environ Microbiol 59:3973–3977
7. Loeb GI, Neihof RA (1975) Marine conditioning films. Adv Chem 145:319–335
8. Rijnaarts HH, Norde W, Bouwer EJ, Lyklema J, Zehnder AJ (1993) Bacterial adhesion under static and dynamic conditions. Appl Environ Microbiol 59:3255–3265
9. Becker P, Hufnagle W, Peters G, Herrmann M (2001) Detection of different gene expression in biofilm-forming versus planktonic populations of Staphylococcus aureus using micro-representational-difference analysis. Appl Environ Microbiol 67:2958–2965
10. Lewandowski Z (2000) Structure and function of biofilms. In: Evans LV (ed) Biofilms: recent advances in their study and control. Harwood Academic Publishers, Amsterdam, pp 1–17
11. Flemming H-C, Wingender J, Griegbe, Mayer C (2000) Physico-chemical properties of biofilms. In: Evans LV (ed) Biofilms: recent advances in their study and control. Harwood Academic Publishers, Amsterdam, pp 19–34
12. Sutherland IW (2001) Biofilm exopolysaccharides: a strong and sticky framework. Microbiology 147:3–9
13. Hussain M, Wilcox MH, White PJ (1993) The slime of coagulase-negative staphylococci: biochemistry and relation to adherence. FEMS Microbiol Rev 104:191–208
14. Zhang L, Mah TF (2008) Involvement of a novel efflux system in biofilm-specific resistance to antibiotics. J Bacteriol 190(13):4447–4452
15. Jakubovics NS, Kolenbrander PE (2010) The road to ruin: the formation of disease-associated oral biofilms. Oral Dis 16(8):729–739
16. Marsh PD (1991) The significance of maintaining the stability of the natural microflora of the mouth. Br Dent J 171(6):174
17. Sanz M, Beighton D, Curtis MA, Cury JA, Dige I, Dommisch H, Ellwood R, Giacaman R, Herrera D, Herzberg MC, Könönen E (2017) Role of microbial biofilms in the maintenance of oral health and in the development of dental caries and periodontal diseases. Consensus report of group 1 of the Joint EFP/ORCA workshop on the boundaries between caries and periodontal disease. J Clin Periodontol 44(S18)
18. Hajishengallis G (2014) Immunomicrobial pathogenesis of periodontitis: keystones, pathobionts, and host response. Trends Immunol 35(1):3–11
19. Sanz M, Serrano J, Iniesta M, Cruz IS, Herrera D et al (2013) Monographs in oral science, vol 23. Karger, Basel (Toothpastes)
20. Stewart PS (2003) Diffusion in biofilms. J Bacteriol 185:1485–1491
21. Wilson M (1996) Susceptibility of oral bacterial biofilms to antimicrobial agents. J Med Microbiol 44:79–87
22. Kasimanickam RK, Ranjan A, Asokan GV, Kasimanickam VR, Kastelic JP (2013) Prevention and treatment of biofilms by hybrid- and nanotechnologies. Int J Nanomedicine. 18:2809–2819
23. Sun LM, Zhang CL, Li P (2012) Characterization, antibiofilm, and mechanism of action of novel PEG-stabilized lipid nanoparticles loaded with terpinen-4-ol. J Agric Food Chem 60:6150–6156
24. Nel AE, Madler L, Velegol D, Xia T, Hoek EMV, Somasundaran P et al (2009) Understanding bio physicochemical interactions at the nano-biointerface. Nat Mater 8:543–557
25. de Paz LE, Sedgley CM, Kishen A (eds) (2015) The root canal biofilm. Springer
26. Berney M, Hammes F, Bosshard F, Weilenmann H-U, Egli T (2007) Assessment and interpretation of bacterial viability by using the LIVE/DEAD BacLight Kit in combination with flow cytometry. Appl Environ Microbiol 73(10):3283–3290

27. Welch K, Cai Y, Strømme M (2012) A method for quantitative determination of biofilm viability. J Funct Biomater 3(2):418–431
28. Peeters E, Nelis HJ, Coenye T (2008) Comparison of multiple methods for quantification of microbial biofilms grown in microtiter plates. J Microbiol Methods 72(2):157–165
29. Melo MA, Guedes SF, Xu HH, Rodrigues LK (2013) Nanotechnology-based restorative materials for dental caries management. Trends Biotechnol 31:459–467
30. Giersten E (2004) Effects of mouth rinses with triclosan, zinc ions, copolymer, and sodium lauryl sulphate combined with fluoride on acid formation by dental plaque invivo. Caries Res 38:430–435
31. Sondi I, Salopek-Sondi B (2004) Silver nanoparticles as antimicrobial agent: a case study on E. Coli as a model for gram-negative bacteria. J Colloid Interface Sci 275:177–182
32. Cioffi N, Torsi L, Ditaranto N, Sabbatini L, Zambonin PG, Tantillo G et al (2005) Copper nanoparticle/ polymer composites with antifungal and bacteriostatic properties. Chem Mater 17:5255–5262
33. Li Z, Lee D, Sheng X, Cohen RE, Rubner MF (2006) Two-level antibacterial coating with both release-killing and contact-killing capabilities. Langmuir 22:9820–9823
34. Lee WF, Tsao KT (2006) Preparation and properties of nano composite hydrogels containing silver nano particles by ex situ polymerization. J Appl Poly Sci 1003653–1003661
35. Verran J, Sandoval G, Allen NS, Edge M, Stratton J (2007) Variables affecting the antibacterial properties of nano and pigmentary titania particles in suspension. Dyes Pigm 73:298–304
36. Lok CN, Ho CM, Chen R, He QY, Yu WY, Sun H et al (2007) Silver nano particles: partial oxidation and anti-bacterial activities. J Biol Inorg Chem 12:527–534
37. Benn TM, Westerhoff P (2008) Nanoparticle silver released into water from commercially available sock fabrics. Environ Sci Technol 42:4133–4139
38. Panacek A, Kvítek L, Prucek R, Kolar M, Vecerova R, Pizúrova N, Sharma VK, Nevecna T, Zboril R (2006) Silver colloid nanoparticles: synthesis, characterization, and their antibacterial activity. J Phys Chem B 110(33):16248–16253
39. Pal S, Tak YK, Song JM (2007) Does the antibacterial activity of silver nanoparticles depend on the shape of the nanoparticle? a study of the gram-negative bacterium Escherichia coli. Appl Environ Microbiol 73(6):1712–1720
40. Seil TS, Websters TJ (2012) Antimicrobial applications of nanotechnology: methods and literature. Inter J Nanomed 7:2767–2781
41. Santoro CM, Duchsherer NL, Grainger DW (2007) Minimal in vitro antimicrobial efficacy and ocular cell toxicity from silver nanoparticles. NanoBiotechnology 3:55–65
42. Asharani PV, Lian WuY, Gong Z, Valiyaveettil S (2008) Toxicity of silver nanoparticles in zebrafish models. Nanotechnology 19:102–255
43. Damm C, Münstedt H (2008) Kinetic aspects of the silver ion release from antimicrobial polyamide/silver nanocomposites. Appl Phys A Mater Sci Process 91:479–486
44. Xu X-HN, Brownlow WJ, Kyriacou SV, Wan Q, Viola JJ (2004) Real time probing of membrane transport in living microbial cells using single nanoparticle optics and living cell imaging. Biochemistry 43:10400–10413
45. Kim JS, Kuk E, Yu KN, Kim JH, Park SJ, Lee HJ, Kim SH, Park JK, Park YH, Hwang CY, Kim YK, Lee YS, Jeong DH, Cho MH (2007) Antimicrobial effects of silver nanoparticles. Nanomed-Nanotechnol 3:95–101
46. Hwang ET, Lee JH, Chae YJ, Kim YS, Kim BC, Sang BI, Gu MB (2008) Analysis of the toxic mode of action of silver nanoparticles using stress-specific bioluminescent bacteria. Small 4:746–750
47. Khan AU (2012) Medicine at nanoscale: a new horizon. Int J Nanomedicine 7:2997–2998
48. Katarzyna M, Anna MG, Krystyna IW (2013) Silver nanoparticles as an alternative strategy against bacterial Biofilms. Acta Biochim Pol 60(4):523–530
49. Fabrega J, Renshaw JC, Lead JR (2009) Interactions of silver nanoparticles with Pseudomonas putida biofilms. Environ Sci Technol 43:9004–9009
50. Kalishwaralal K, Barath Mani Kanth S, Pandian SRK, Deepak V, Gurunathan S (2010) Silver nanoparticles impede the biofilm formation by Pseudomonas aeruginosa and Staphylococcus epidermidis. Colloids Surf B 79:340–344

51. Martinez-Gutierrez F, Boegli L, Agostinho A, Sánchez EM, Bach H, Ruiz F, James G (2013) Anti-biofilm activity of silver nanoparticles against different microorganisms. Biofouling 29:651–660
52. Islam MS, Larimer C, Ojha A, Nettleship I (2013) Antimycobacterial efficacy of silver nanoparticles as deposited on porous membrane filters. Mater Sci Eng C Mater Biol Appl 33:4575–4581
53. Radzig MA, Nadtochenko VA, Koksharova OA, Kiwi J, Lipasova VA, Khmel IA (2013) Antibacterial effects of silver nanoparticles on gram-negative bacteria: Influence on the growth and biofilms formation, mechanisms of action. Colloids Surf B 102:300–306
54. Hartmann T, Mühling M, Wolf A, Mariana F, Maskow T, Mertens F, Neu TR, Lerchner J (2013) A chip-calorimetric approach to the analysis of Ag nanoparticle caused inhibition and inactivation of beads-grown bacterial biofilms. J Microbiol Methods 95:129–137
55. Yoon KY, Byeon JH, Park JH, Hwang J et al (2007) Susceptibility constants of Escherichia coli and Bacillus subtilis to silver and copper nanoparticles. Sci Tot Environ 373:572–575
56. Ren G, Hu D, Cheng EWC, Vargas-Reus MA, Reip P, Allaker RP (2009) Characterisation of copper oxide nanoparticles for antimicrobial applications. Int J Antimicrob Agents 33(6):587–590
57. De Jong WH, Borm PJ (2008) Drug delivery and nanoparticles: applications and hazards. Int J Nanomedicine 3(20):133–149
58. Taylor EN, Kummer KM, Durmus NG, Leuba K, Tarquinio KM, Webster TJ (2012) Superparamagnetic iron oxide nanoparticles (SPION) for the treatment of antibiotic-resistant biofilms. Small Weinh Bergstr Ger 8(19):3016–3027
59. Durmus NG, Taylor EN, Kummer KM, Webster TJ (2013) Enhanced efficacy of superparamagnetic iron oxide nanoparticles against antibiotic-resistant biofilms in the presence of metabolites. Adv Mater 25(40):5706–5713
60. Salunke GR, Ghosh S, Santosh Kumar R (2014) Rapid efficient synthesis and characterization of silver, gold, and bimetallic nanoparticles from the medicinal plant Plumbagozeylanica and their application in biofilm control. Int J Nanomed 9:2635–2653
61. Ramasamy M, Lee J-H, Lee J (2016) Potent antimicrobial and antibiofilm activities of bacteriogenically synthesized gold-silver nanoparticles against pathogenic bacteria and their physiochemical characterizations. J Biomater Appl 31(3):366–378
62. Ren G, Hu D, Cheng EWC, Vargas-Reus MA, Reip P, Allaker RP (2009) Characterisation of copper oxide nanoparticles for antimicrobial applications. Int J Antimicrob Agents 33(6):587–590
63. Esteban-Tejeda L, Malpartida F, Esteban-Cubillo A, Pecharromn C, Moya JS (2009) Antibacterial and antifungal activity of a soda-lime glass containing copper nanoparticles. Nanotechnology 20(50):505701
64. Ruparelia JP, Chatterjee AK, Duttagupta SP, Mukherji S (2008) Strain specificity in antimicrobial activity of silver and copper nanoparticles. Acta Biomater 4(3):707–716
65. Murthy PS, Venugopalan V, Das DA, Dhara S, Pandiyan R, Tyagi A (2011) Antibiofilm activity of nano sized CuO. In: Proceedings of the International Conference on Nanoscience, Engineering and Technology, pp 580–583
66. Stoimenov PK, Klinger RL, Marchin GL, Klabunde KJ (2002) Metal oxide nanoparticles as bactericidal agents. Langmuir 18:6679–6686
67. Brayner R, Ferrari-Iliou R, Brivois N, Djediat S, Benedetti MF, Fievet F (2006) Toxicological impact studies based on Escherichia coli bacteria in ultra fine ZnO nanoparticles colloidal medium. Nano Lett 6:866–870
68. Zhang LL, Jiang YH, Ding YL, Povey M, York D (2007) Investigation into the antibacterial behavior of suspensions of ZnO nanoparticles (ZnO nano fluids). J Nanopart Res 9:479–489
69. Fukui H, Horie M, Endoh S, Kato H, Fujita K, Nishio K, Komaba Lk, Maru J, Miyauhi A, Nakamura A, Kinugasa S, Yoshida Y, Hagihara Y, Iwahashi H (2012) Association of zinc ion release and oxidative stress induced by intratracheal instillation of ZnO nanoparticles to rat lung. Chem Biol Interact 198:29–37

70. Eshed M, Lellouche J, Matalon S, Gedanken A, Banin E (2012) Sonochemical coatings of ZnO and CuO nanoparticles inhibit Streptococcus mutans biofilm formation on teeth model. Langmuir 28:12288–12295
71. Liu Y, He L, Mustapha A, Li H, Hu Zq, Lin M (2009) Antibacterial activities of zinc oxide nanoparticles against Escherichia coli O157:H7. J Appl Microbiol 107:1193–1201
72. Sandstead HH (1995) Requirements and toxicity of essential trace elements, illustrated by zinc and copper. Am J Clin Nutr 61(Suppl):621S–624S
73. Vargas-Reus MA, Memarzadeh K, Huang J, Ren GG, Allaker RP (2012) Antimicrobial activity of nanoparticulate metal oxides against peri-implantitis pathogens. Int J Antimicrob Agents 40:135–139
74. Khan ST, Ahmad M, Al-Khedhairy AA, Musarrat J (2013) Biocidal effect of copper and zinc oxide nanoparticles on human oral microbiome and biofilm formation. Mater Lett 97:67–70
75. Khan ST, Ahamed M, Alhadlaq HA, Musarrat J, Al-Khedhairy AA (2013b) Comparative effectiveness of $NiCl_2$, Ni- and NiO-NPs in controlling oral bacterial growth and biofilm formation on oral surfaces. Arch Oral Biol 58:1804–1811
76. Yu JX, Li TH (2011) Distinct biological effects of different nanoparticles commonly used in cosmetics and medicine coatings. Cell Biosci 1:19
77. Vandebriel RJ, De Jong WH (2012) A review of mammalian toxicity of ZnO nanoparticles. Nanotechnol Sci Appl 5:61–71
78. Fang M, Chen JH, Xu XL, Yang PH, Hildebr HF (2006) Antibacterial activities of inorganic agents on six bacteria associated with oral infections by two susceptibility tests. Int J Antimicrob Agents 27:513–517
79. Takatsuka T, Tanaka K, Iijima Y (2005) Inhibition of dentine demineralization by zinc oxide: in vitro and in situ studies. Dent Mater 21:1170–1177
80. Piccinno F, Gottschalk F, Seeger S, Nowack B (2012) Industrial production quantities and uses of ten engineered nanomaterials for Europe and the world. J Nanopart Res 14:1109–1120
81. Keller AA, McFerran S, Lazareva A, Suh S (2013) Global life cycle releases of engineered nanomaterials. J Nanopart Res 15:1692
82. Blake DM, Maness PC, Huang Z, Wolfrum EJ, Jacoby WA, Huang J (1999) Application of the photocatalytic chemistry of titanium dioxide to disinfection and the killing of cancer cells. Sep Purif Methods 28:1–50
83. Rajakumar G, Rahuman AA, Roopan SM et al (2012) Fungus mediated biosynthesis and characterization of TiO_2 nanoparticles and their activity against pathogenic bacteria. Spectrochim Acta A 91:23–29
84. Kuhn KP, Cahberny IF, Massholder K et al (2003) Disinfection of surfaces by photocatalytic oxidation with titanium dioxide and UVA light. Chemosphere 53:71–77
85. Kumar A, Pandey AK, Singh SS, Shanker R, DhawanA (2011) Engineered ZnO and TiO_2 nanoparticles induce oxidative stress and DNA damage leading to reduced viability of Escherichia coli. Free Radic Biol Med 51:1872–1888
86. Tsuang YH, Sun JS, Huang YC et al (2008) Studies of photo killing of bacteria using titanium dioxide nanoparticles. Artif Organs 32:167–174
87. Maness PC, Smolinski S, Blake DM, Huang Z, Wolfrum EJ, Jacoby WA (1999) Bactericidal activity of photocatalytic TiO_2 reaction: toward an understanding of its killing mechanism. Appl Environ Microbiol 65:4094–4098
88. Yazdi AS, Guarda G, Riteau N, Drexler SK, Tardivel A, Couillin I et al (2010) Nanoparticles activate the NLR pyrindoma in containing3 (Nlrp3) inflammasome and cause pulmonary inflammation through release of IL-1α and IL-1β. Proc Natl Acad Sci USA 107:19449–19454
89. Rabea EI, Badawy ME, Stevens CV, Smagghe G, Steurbaut W (2003) Chitosan as antimicrobial agent: applications and mode of action. Biomacromol 4:1457–1465
90. Seo HJ, Mitsuhashi K, Tanibe H (1992) In: Brine CJ, Sandford PA, Zikakis JP (eds) Advances in Chitin and Chitosan. Elsevier Applied Science, New York, pp 34–40
91. Sudarshan NR, Hoover DG, Knorr D (1992) Antibacterial action of Chitosan. Food Biotechnol 6(3):257–272

92. Kim C (1997) Synthesis and antibacterial activity of water-soluble Chitin derivatives. Polym Adv Technol 8:319–325
93. Kishen A, Shi Z, Shrestha A, Neoh KG (2008) An investigation on the antibacterial and anti biofilm efficacy of cationic nanoparticulates for root canal infection. J Endod 34:1515–1520
94. Fang FC (1997) Perspectives series: host/pathogen interactions. Mechanisms of nitric oxide-related antimicrobial activity. J Clin Invest 99:2818–2825
95. Lancaster JR Jr (1997) A tutorial on the diffusibility and reactivity of free nitric oxide. Nitric Oxide 1:18–30
96. Xu K, Wang JX, Kang XL, Chen JF (2009) Fabrication of antibacterial monodispersed Ag-SiO_2 core-shell nanoparticles with high concentration. Mater Lett 63:31–33
97. Wang J-X, Wen L-X, Wang Z-H, Chen J-F (2006) Immobilization of silver on hollow silica nanospheres and nanotubes and their antibacterial effects. Mater Chem Phys 96:90–97
98. Michels H, Wilks S, Noyce J, Keevil C (2005) Copper alloys for human infectious disease control. Stainless Steel 77000(27):20
99. Kim YH, Lee DK, Cha HG, Kim CW, Kang YC, Kang YS (2006) Preparation and characterization of the antibacterial Cu nanoparticle formed on the surface of SiO_2 nanoparticles. J Phys Chem B 110:24923–24928
100. Kar M, Vijayakumar PS, Prasad BLV, Gupta SS (2010) Synthesis and characterization of poly- *L*-lysine-grafted silica nanoparticles synthesized via NCA polymerization and click chemistry. Langmuir 26:5772–5781
101. Makarovsky I, Boguslavsky Y, Alesker M, Lellouche J, Banin E, Lellouche J-P (2011) Novel triclosan-bound hybrid-silica nanoparticles and their enhanced antimicrobial properties. Adv Funct Mater 21:4295–4304
102. Botequim D, Maia J, Lino MMF, Lopes LMF, Simoes PN, Ilharco LM, Ferreira L (2012) Nanoparticles and surfaces presenting antifungal, antibacterial and antiviral properties. Langmuir 28:7646–7656
103. Sepulveda P, Jones JR, Hench LL (2002) In vitro dissolution of melt-derived 45S5 and sol-gel derived 58S bioactive glasses. J Biomed Mater Res 61:301–311
104. Allan I, Newman H, Wilson M (2001) Antibacterial activity of particulate Bioglass® against supra- and subgingival bacteria. Biomaterials 22:1683–1687
105. Zehnder M, Luder HU, Schätzle M, Kerosuo E, Waltimo T (2006) A comparative study on the disinfection potentials of bioactive glass S53P4 and calcium hydroxide in contra-lateral human premolars ex vivo. Int Endod J 39:952–958
106. Cabal B, Malpartida F, Torrecillas R, Hoppe A, Boccaccini AR, Moya JS (2011) The development of bioactive glass-ceramic substrates with biocide activity. Adv Eng Mater 13:B462–B466
107. Waltimo T, Brunner TJ, Vollenweider M, Stark WJ, Zehnder M (2007) Antimicrobial effect of nanometric bioactive glass 45S5. J Dent Res 86:754–757
108. Allison RR, Moghissi K (2013) Photodynamic therapy (PDT): PDT mechanisms. Clin Endosc 46:24–29
109. Dai T, Huang Y-Y, Hamblin MR (2009) Photodynamic therapy for localized infections—state of the art. Photodiagn Photodyn Ther 6:170–188
110. Haag PA, Steiger-Ronay V, Schmidlin PR (2015) The in vitro antimicrobial efficacy of PDT against periodontopathogenic bacteria. Int J Mol Sci 16:27327–27338
111. Ogura M, Abernethy AD, Blissett R, Ruggiero K, Som S, Goodson J, Kent R, Doukas A, Soukos NS (2007) Photomechanical wave-assisted molecular delivery in oral biofilms. World J Microbiol Biotechnol 23:1637–1646
112. Müller P, Guggenheim B, Schmidlin PR (2007) Efficacy of gasiform ozone and photodynamic therapy on a multispecies oral biofilm in vitro. Eur J Oral Sci 115:77–80
113. Taraszkiewicz A, Fila G, Grinholc M, Nakonieczna J (2013) Innovative strategies to overcome biofilm resistance. Biomed Res Int 2013:13
114. Shiha M-H, Huang F-C (2013) Repetitive methylene blue-mediated photoantimicrobial chemotherapy changes the susceptibility and expression of the outer membrane proteins of Pseudomonas aeruginosa. Photodiagn Photodyn Ther 10:664–671

115. Danhier F, Ansorena E, Silva JM, Coco R, Le Breton A, Préat V (2012) PLGA-based nanoparticles: an overview of biomedical applications. J Control Release 161:505–522
116. Koo YEL, Fan W, Hah H, Xu H, Orringer D, Ross B, Rehemtulla A, Philbert MA, Kopelman R (2007) Photonic explorers based on multifunctional nanoplatforms for biosensing and photodynamic therapy. Appl Opt 46:1924–1930
117. Pagonis TC, Chen J, Fontana CR, Devalapally H, Ruggiero K, Song X, Foschi F, Dunham J, Skobe Z, Yamazaki H et al (2010) Nanoparticle-based endodontic antimicrobial photodynamic therapy. J Endod 36:322–328
118. Kumari A, Yadav SK, Yadav SC (2010) Biodegradable polymeric nanoparticles based drug delivery systems. Colloid Surf B 75:1–18
119. Sah E, Sahis H (2015) Recent trends in preparation of poly(lactide-co-glycolide) nanoparticles by mixing polymeric organic solution with antisolvent. J Nanomater 22. https://doi.org/10.1155/2015/794601
120. Haris Z, Khan AU (2017) Selenium nanoparticle enhanced photodynamic therapy against biofilm forming streptococcus mutans. Int J Life Sci Sci Res 3(5):1287–1294
121. Pagonis TC, Chen J, Fontana CR, Devalapally H, Ruggiero K, Song X et al (2010) Nanoparticle-based endodontic antimicrobial photodynamic therapy. J Endod 36(2):322–328
122. Britto RP, Muralidharan NP (2013) Contamination of dental waterline and its control measures. Asian J Pharm Clin Res 6:19–23
123. Brooke JS (2012) Stenotrophomonas maltophilia: an emerging global opportunistic pathogen. ClinMicrobiol Rev 25:2–41
124. Porteous NB, Redding SW, Thompson EH, Grooters AM, De Hoog S, Sutton DA (2003) Isolation of an unusual fungus in treated dental unit waterlines. J Am Dent Assoc 134:853–858
125. Szymanska J, Sitkowska J, Dutkiewicz J (2008) Microbial contamination of dental unit waterlines. Ann Agric Environ Med 15:173–179
126. Gupta S, Behari J, Kesari K (2006) Low frequency ultrasonic treatment of sludge. Asian J Water Environ Pollut 3:101–105
127. Mamadou SD, Savage N (2005) Nanoparticles and water quality. J Nano Res 7:325–330
128. Zhou QX, Xiao JP, Wang WD (2006) Using multi-walled carbon nanotubes as solid phase extraction adsorbents to determine dichlorodiphenyltrichloroethane and its metabolites at trace level in water samples by high performance liquid chromatography with UV detection. J Chromatogr A 1125:152–158
129. Yang K, Zhu L, Xing B (2006) Adsorption of polycyclic aromatic hydrocarbons by carbon nanomaterials. Environ Sci Technol 40:1855–1861
130. Yang K, Wang X, Zhu L, Xing B (2006) Competitive sorption of pyrene, phenanthrene, and naphthalene on multiwalled carbon nanotubes. Environ Sci Technol 40:5804–5810
131. Gotovac S, Hattori Y, Noguchi D, Miyamoto J, Kanamaru M, Utsumi S et al (2006) Phenanthrene adsorption from solution on single wall carbon nanotubes. J Phys Chem B 110:16219–16224
132. Wang JX, Jiang DQ, Gu ZY, Yan XP (2006) Multiwalled carbon nanotubes coated fibers for solid-phase microextraction of polybrominated diphenyl ethers in water and milk samples before gas chromatography with electron-capture detection. J Chromatogr A 1137:8–14
133. Peng XJ, Li YH, Luan ZK, Di ZC, Wang HY, Tian BH et al (2003) Adsorption of 1, 2-dichlorobenzene from water to carbon nanotubes. Chem Phys Lett 376:761–766
134. Cai YQ, Cai YE, Mou SF, Lu YQ (2005) Multi-walled carbon nanotubes as a solid-phase extraction adsorbent for the determination of chlorophenols in environmental water samples. J Chromatogr A 1081:245–247
135. Chin CJ, Shih LC, Tsai HJ, Liu TK (2006) Adsorption of o-xylene and p-xylene from water by SWCNTs. J Hazard Mater B138:304–310
136. Lu C, Chung YL, Chang KF (2005) Adsorption of trihalomethanes from water with carbon nanotubes. Water Res 39:1183–1189
137. Cai Y, Jiang G, Liu J, Zhou Q (2003) Multiwalled carbon nanotubes as a solid-phase extraction adsorbent for the determination of bisphenol A, 4-*n*-nonylphenol, and 4-tert-octylphenol. Anal Chem 75:2517–2521

138. Cai YQ, Jiang QB, Liu JF, Zhou QX (2003) Multi walled nano carbon tubules packed cartridge for the solid phase extraction of several phthalate esters from water samples and their determination by high performance liquid chromatography. Anal Chim Acta 494:149–156
139. Fugetsu B, Satoh S, Shiba T, Mizutani T, Lin YB, Terui N et al (2004) Caged multiwalled carbon nanotubes as the adsorbents for affinity-based elimination of ionic dyes. Environ Sci Technol 38:6890–6896
140. Tiwari DK, Behari J, Sen P (2008) Application of nanoparticles in waste water treatment. World Appl Sci J 3:417–433
141. Rao GP, Lu C, Su F (2007) Sorption of divalent metal ions from aqueous solution by carbon nanotubes: a review. Sep Purif Technol 58:224–231
142. Sharma YC, Srivastava V, Singh VK, Kaul SN, Weng CH (2009) Nano-adsorbents for the removal of metallic pollutants from water and wastewater. Environ Technol 30:583–609
143. Crooks RM, Bradley K, Ishigami M, Zettl A (2001) Extreme oxygen sensitivity of electronic properties of carbon nanotubes. Acc Chem Res 34:181–190
144. Vaseashta A, Vaclavikova M, Vaseashta S, Gallios G, Roy P, Pummakarnchana O (2007) Nanostructures in environmental pollution detection, monitoring, and remediation. Sci Technol Adv Mater 8:47–59
145. Taurozzi JS, Tarabara VV (2007) Silver nanoparticle arrays on track etch membrane support as flow-through optical sensors for water quality control. Environ Eng Sci 24:122–137
146. Hilal N, Zoubi HL, Darwish NA, Mohammad AW, Arabi MA (2004) A comprehensive review of nanofiltration membranes: Treatment, pretreatment, modelling, and atomic force microscopy. Desalination 170:281–308
147. Srivastava A, Srivastava ON, Talapatra S, Vajtai R, Ajayan PM (2004) Carbon nanotube filters. Nat Mater 3:610–614
148. Mpenyana-Monyatsi L, Mthombeni NH, Onyango MS, Momba MN (2012) Cost-effective filter materials coated with silver nanoparticles for the removal of pathogenic bacteria in groundwater. Int J Environ Res Public Health 9:244–271
149. Aragon M, Kottenstette R, Dwyer B, Aragon A, Everett R, Holub W et al (2007) Arsenic pilot plant operation and results. Sandia National Laboratories, Anthony, New Mexico
150. Lawrence MJ, Rees GD (2000) Microemulsion-based media as novel drug delivery systems. Adv Drug Deliv Rev 45:89–121
151. Hamouda T, Myc A, Donovan B, Shih AY, Reuter JD, Baker JR Jr (2000) A novel surfactant nanoemulsion with a unique non-irritant topical anti-microbial activity against bacteria enveloped viruses and fungi. Microbiol Res 156:1–7
152. Lee VA, Karthikeyan R, Rawls HR, Amaechi BT (2010) Anti-cariogenic effect of a cetylpyridinium chloride-containing nanoemulsion. J Dent 38:742–749
153. Ramalingam K, Frohlich NC, Lee VA (2013) Effect of nanoemulsion on dental unit waterline contamination. J Dent Sci 8:333–336
154. Lyon DY, Brown DA, Alvarez PJ (2008) Implications and potential applications of bactericidal fullerene water suspensions: Effect of nC(60) concentration, exposure conditions and shelf life. Water Sci Technol 57:1533–1538

Chapter 13
Physical Properties and Biocompatibility of Nanostructural Biomaterials Based on Active Calcium Silicate Systems and Hydroxyapatite

Opačić-Galić Vanja, Petrović Violeta, Popović-Bajić Marijana, Jokanović Vukoman, Živković Slavoljub, Nikolić Biljana and Mitić-Ćulafić Dragana

13.1 Introduction

During the last two decades, a significant progress has been made in the area of development and application of new biomaterials with superior features comparing to the conventional ones [1]. At the same time, the progress in understanding the biology of stem cells has provided introduction of regenerative therapy in endodontics [2]. Researchers in the area of regenerative endodontics have shown that materials used in these procedures are crucial for induction of pulp stem cells and periodontal ligament, and interaction of materials with cells depends on the chemical composition of the materials [3]. Nowadays, the main focus is the application of biomaterials, that are biocompatible, bioactive, and bio-inductive, which results in the induction of cell differentiation and proliferation with tissue regeneration as a consequence.

Introduction of calcium silicate cements in clinical practice brought significant progress in the treatment of dental pulp and apical periodontium. Researchers have shown that these materials could induce formation of hard dental tissues and periodontal regeneration [4–6]. Also, they possess antimicrobial activity and adequate physical-chemical properties, good marginal sealing and setting ability in the wet conditions [7, 8]. However, the clinical use of calcium silicate is disturbed by grainy consistency of fresh cement, short working time, and long setting time. Long setting

---

Opačić-Galić Vanja and Petrović Violeta: equally contributed.

O.-G. Vanja · P. Violeta · P.-B. Marijana · Ž. Slavoljub
University of Belgrade-School of Dental Medicine, 11000 Belgrade, Serbia

J. Vukoman
University of Belgrade-Institute of Nuclear Sciences Vinča, Belgrade, Serbia

N. Biljana · M.-Ć. Dragana (✉)
University of Belgrade-Faculty of Biology, 11000 Belgrade, Serbia
e-mail: mdragana@bio.bg.ac.rs

R. S. Chaughule (ed.), *Dental Applications of Nanotechnology*,
https://doi.org/10.1007/978-3-319-97634-1_13

time carries a risk of leaching of materials from the site of application and delays the end of endodontic treatment [9]. Thus, researches aimed on the synthesis of materials with good physical and biological characteristics, with a shorter setting time still in progress [10–12].

Hydroxyapatite is known as material, with a good biocompatibility, bioactivity, and bioconductivity. However, due to the porous structure and inadequate mechanical properties, its use as endodontic cement is limited. Therefore, current research is focused on testing of new formulations of materials based on calcium phosphate with the calcium silicates. In these composite cements, calcium silicate hydrate phase has the role of filler, providing improved mechanical characteristics and prominent bioactivity to the cement, compared to pure calcium phosphate cements [13–15].

Recently, nanostructured material testing in the field of endodontic therapy has begun. The main advantage of the nanomaterials compared to conventional materials lies in the distinct activity of particles, as a consequence of the large reaction surface with respect to the total weight and volume of the particles. Large reaction surface and small size of particles provide their distinct activity in vivo environment. Smaller size particles enhance the hydration of the material and consequently exhibit a positive effect on the hardening and setting time [16] and also improve the physical and chemical characteristics of material [12].

13.2 New Nanostructured Biomaterials

The new materials were synthesized by a nanotechnology-innovative combination of hydrothermal sol-gel method and method of self-combustion waves (combination of sol-gel and high-temperature synthesis). The use of enhanced technology resulted with calcium silicate (CS) material whose structure consists of three hierarchical levels: agglomerates of few microns in size, which are built of ellipsoidal particles (the size of 117–477 nm), which are further composed of even smaller elements—crystallite (the size of 20 nm) [17, 18].

Another material (HA-CS) is a mixture of hydroxyapatite, obtained by the hydrothermal method and the calcium silicate. HA-CS particles have irregular, generally circular shape, with the diameter of 60–470 nm. Both of the materials contain a plasticizer. Nanostructure of newly synthesized material provides distinct particle activity that results in fast setting of the cement. Setting of the CS material starts after 3 ± 0.1 min, and setting of the HA-CS material 5 ± 0.1 min from the start of mixing the material with water, and completely ends after 10 ± 0.15 min (CS) or 15 ± 0.15 min (HA-CS) [18]. CS and HA-CS were compared with White MTA Angelus®.

Table 13.1 The weight of the samples during 28 days in artificial tissue fluid (mg)

	m_1	24 h	7 dan	14 dan	21 dan	28 dan
MTA	99.0±2.8	111.0±3.9	111.7±3.3	107.9±4.6	107.9±4.5	107.8±5.0
CS	99.2±2.9	113.3±3.3	108.8±4.3	108.4±4.0	109.2±4.3	109.1±4.2
HA-CS	73.8±2.2	96.3±1.8	95.2±2.3	93.7±2.1	93.1±2.1	92.9±1.7

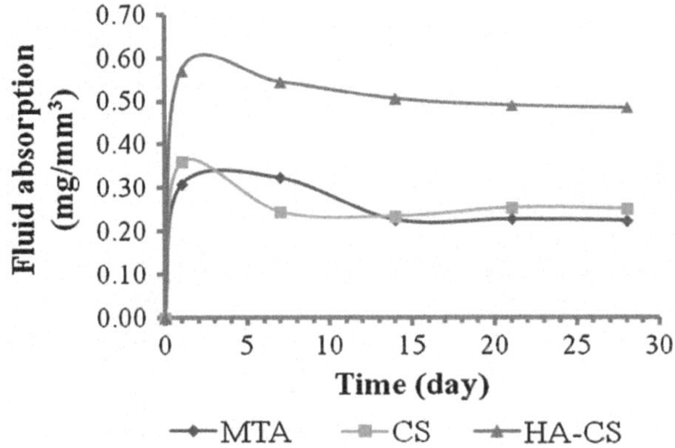

Fig. 13.1 Fluid absorption during 28 days

13.3 Physical Properties

13.3.1 Solubility

Solubility of a material is directly related to its porosity and according to ISO standard should not be larger than 3% of material weight. Higher porosity/solubility may result in disintegration of a material at the site of its application. Solubility and porosity of the new nanostructured biomaterials has been tested, after the storage of materials in artificial tissue fluid in the duration of 28 days.

The greatest absorption of fluid (swelling patterns) in all materials was observed after 24 h, and after this time there were no significant changes (Table 13.1). The least absorption of fluid was observed in MTA, and it was similar to CS (Fig. 13.1).

The absorption of the liquid indicates the initially porous structure of all the tested materials. Although porosity is a feature of all dental materials in the mixture with water, a specific amorphous structure of the calcium silicate cements, consisting of a number of pores and capillaries, contributes to its greater porosity [19]. Since the pores within set cement can be filled with water from the environment, number of pores decreases with cement solidification, which is manifested by lower absorption of fluid in function of time [11, 20]. The porosity of the HA-CS was significantly higher than the porosity of both CS and MTA materials ($p < 0.05$), and it could be attributed to the hydroxyapatite, which is one of the main elements of this material (Fig. 13.2). The obtained values for all materials tested were higher than the values prescribed with ISO standards.

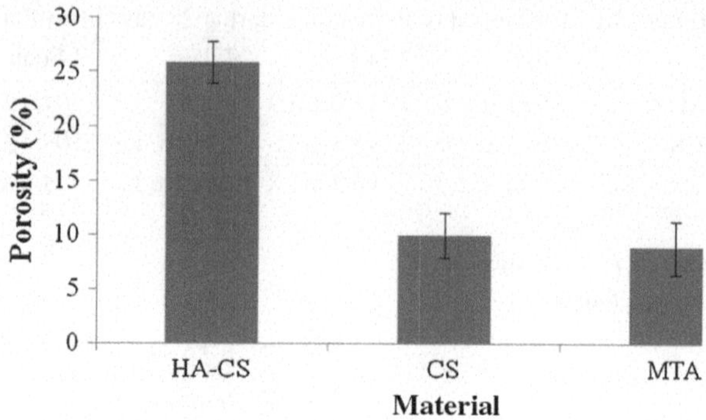

Fig. 13.2 Porosity of the materials

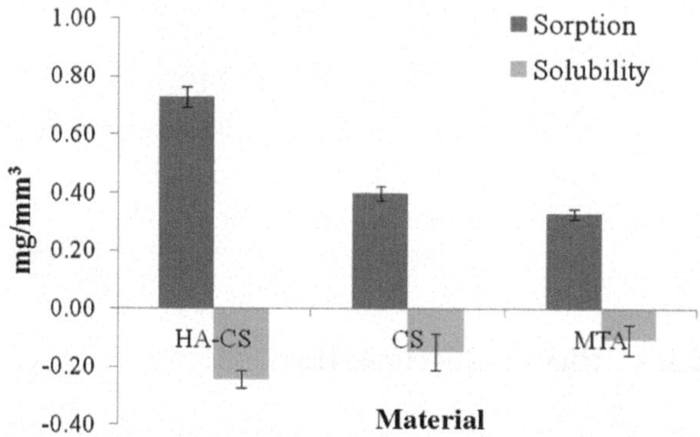

Fig. 13.3 Fluid sorption and solubility of the tested materials after 28 days in artificial tissue fluid

After 28 days, minimal sorption of liquid was measured in the MTA material, the values for CS was slightly higher, while the highest sorption of liquid was observed in the HA-CS. The minimum solubility was observed in the MTA material, similar as for CS. Solubility of the HA-CS was significantly higher compared to both CS and MTA materials ($p < 0.05$) (Fig. 13.3).

Although with a similar chemical composition, the CS showed a slightly higher fluid sorption compared to the MTA, which could be due to different particle sizes. Namely, the smaller particles of the CS material, and consequently, higher reactive surface, could enable stronger contact of the molecules of liquid with the particles, which could further result in a greater sorption of liquid and somewhat higher solubility in relation to the microstructure of the MTA. The highest sorption of fluid in the HA-CS, in addition to its nanoparticle size, could be attributed to the chemical composition of materials.

According to the results of considerable research, calcium silicates are insoluble or minimally soluble cements [11, 19]. Yet, in the literature the data on explicit solubility of these materials are available [21]. From the clinical point of view, the greatest risk of dissolution and subsequent leaching of calcium silicates exists in the first hours after application and is connected with the long setting time of the commercial cements (few hours). Also, the long setting time raises the question of possible

Fig. 13.4 Marginal microleakage of the materials

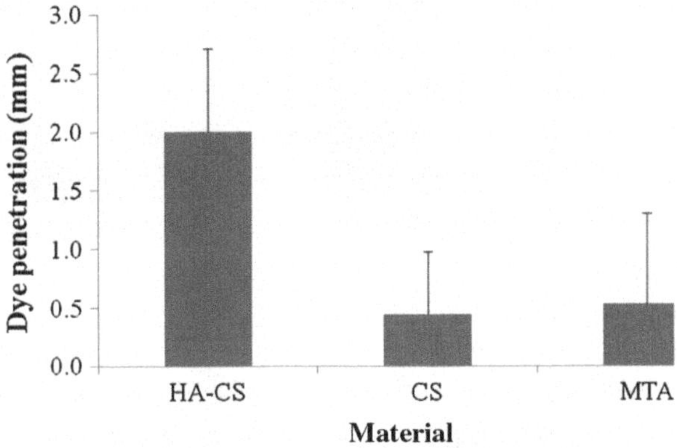

interaction of calcium silicate with the components of restorative dental materials [22]. In this regard, the application of new nanostructured materials, particularly the CS, with significantly shorter setting time if compared to conventional calcium silicates could further reduce the risk of disintegration and leaching of materials, creating the conditions for a definitive restoration at the same appointment.

13.3.2 Marginal Microleakage

A good marginal seal prevents bacterial microleakage, which is considered to be a significant factor for long-term success of endodontic treatment. MTA exerts good marginal sealing, even in the cavities that are contaminated with blood [8, 23]. According to the Hawley et al. 2010, good marginal seal of calcium silicate cements is the result of the expansion of materials during setting [24], while according to Sarkar et al. (2005) good marginal seal of the MTA is attributed to the creation of the hydroxyapatite on a surface of the material after contact with tissue fluids [25].

Marginal microleakage of CS, HA-CS, and MTA was analyzed on extracted human molars, after the experimentally prepared furcal perforations of the pulp chamber floor, and their closure with CS, HA-CS, and MTA. The lowest microleakage was obtained for CS and MTA materials, while in the HA-CS material microleakage was significantly higher ($p < 0.05$) (Fig. 13.4).

Herein, in the groups of samples filled with the CS and MTA, variable results were obtained. These results are in accordance with the results of De Deus et al. [26], who also found variable results within the groups [26]. With the HA-CS material, dye penetration was detected in all samples, and in most of the samples an explicit disintegration of the material was observed. Sanghavi et al. [27] and Tsatsas et al. [28] previously pointed out inadequate sealing ability of calcium phosphate ceramics as well [27, 28].

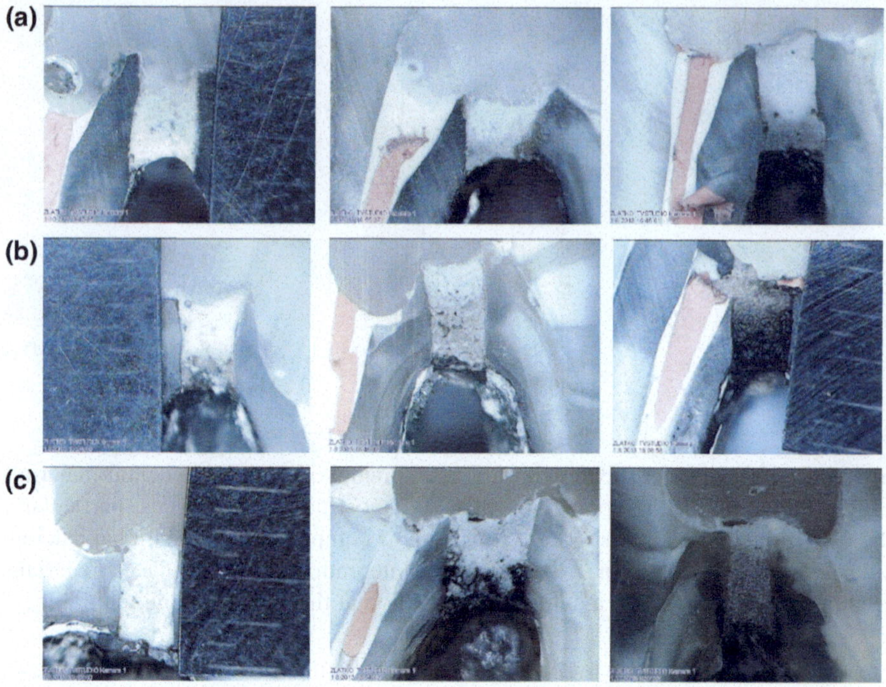

Fig. 13.5 Furcal perforations filled with the CS, HA-CS, and MTA after dye penetration A-CS, B-HA-CS, C-MTA (×30)

It is important to note the way how the dye penetrated (Fig. 13.5). A strictly marginal microleakage cannot be discussed, because, besides the color detected at the junction of the material and dental structures, discoloration of the material was observed. This is probably due to dye absorption by the material itself. Similar results were shown by Tobón-Arroyave et al. [29], who found that the method of dye penetration in the MTA is different compared to conventional cements [29]. Namely, the penetration in conventional cements was detected circularly around the material, or between the materials and the dentin of the root canal. In contrast, in all the samples with the MTA, absorption of the color was observed throughout the whole thickness of materials which might be result of the porous structure of the tested materials.

13.3.3 The Bond Strength to Dentin

One of the most important requirements of endodontic materials is the resistance to dislocation during operative procedures (condensation of definitive restorative materials) and the regular teeth function. The bond strength of MTA to dentine is

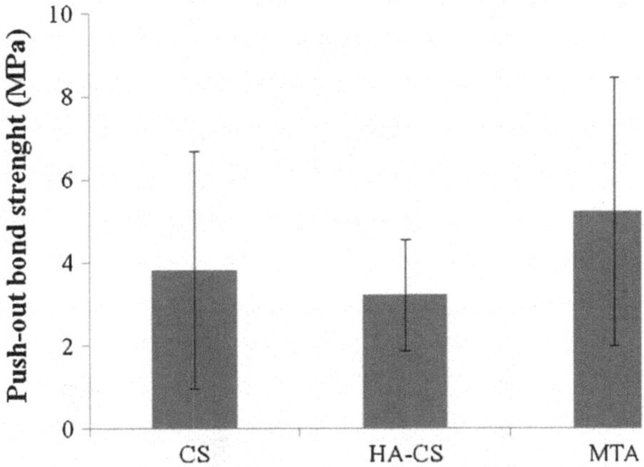

Fig. 13.6 Bond strength of the tested materials to dentin

weaker compared to Amalgam, compomer, and reinforced ZOE cement [30]. Bond strength of MTA to dentin increases after exposure to tissue fluids, but it is still weaker compared to resin-modified GIC and adhesives [31]. It has been found that the presence of moisture [32], a basic environment [30], and the presence of smear layer [33] positively affect on the bond strength. Atabek et al. [34] demonstrated that longer setting time of calcium silicate gives better results in bond strength to dentin [34].

Bonding of calcium silicates to dentin is achieved by deposition of hydroxyapatite on the surface of the material due to dissolution of materials and reactions of released calcium with phosphates from tissue fluid. Initially, the hydroxyapatite crystals fulfill the microspaces between the dentine and materials, providing a mechanical bond to dentin. Later, a chemical bond is achieved as a result of the reaction between apatite and dentin [25]. Also, in time, mineral crystals penetrated into dentinal tubules, contributing to the better adhesion of the material [35].

The bond strength of CS, HA-CS, and MTA to dentin was measured using a universal testing machine. Fractures of the materials were classified as: adhesive (at the junction of the materials and the dentin), cohesive (within the material or dentin), and mixed.

The highest values of bond strength to dentin of the root canal were found in MTA material (Fig. 13.6). In CS and HA-CS groups, the bond strength was slightly weaker, but it was not significantly different compared to MTA, which pointed out similar chemical composition of the materials and, consequently, a similar method of adhesion to the tooth structure.

In most of the samples of CS and HA-CS groups, the cohesive type of fracture was found. The mixed fracture was found in the most of the samples filled with MTA (Fig. 13.7). These results are in accordance with findings of Formosa et al. [36] and the EL Ma'aita et al. [33]. One could expect that over time the bond strength of all tested materials to dentin would grow, taking into account their bioactivity.

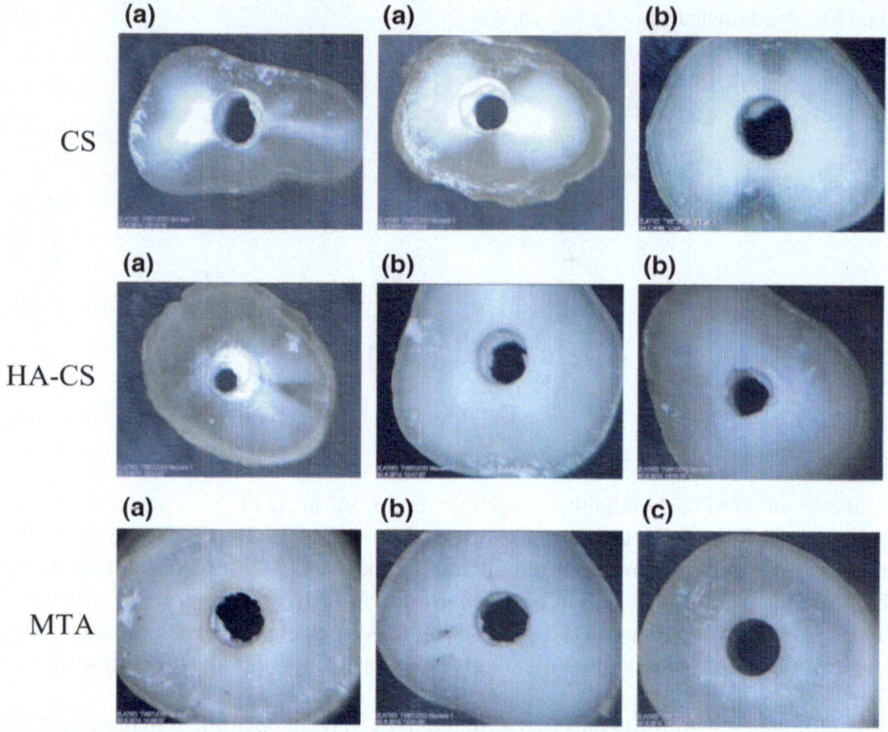

Fig. 13.7 Fractures of the materials after push out test **a** cohesive fracture, **b** mixed fracture, **c** adhesive fracture (×30)

13.4 Biocompatibility

Some of the most important initial steps in assessing biocompatibility are in vitro tests on tissues and cell cultures even though they do not show a complete mechanism of interaction between materials and tissues. Tests that can provide significantly more information about both inflammatory and immune response of the tested material are in vivo tests [37]. The reduction of toxic effect of the material caused by proteins, tissue fluids, and other factors is a fundamental difference between in vivo and in vitro studies [38].

Biomaterials demonstrate intention to react with biological systems, with a potential to restore damaged human tissues and organs, as well as their functions. Any material that evokes a host's positive response could be considered as a bioactive one [39, 40].

There are a few facts about the toxicity of nanostructured calcium silicate and calcium phosphate cements in endodontic therapy, because the application of these materials has recently been made available. In general, their toxicity is associated with particle size, which can lead to the absorption of the nanoparticles by the cells,

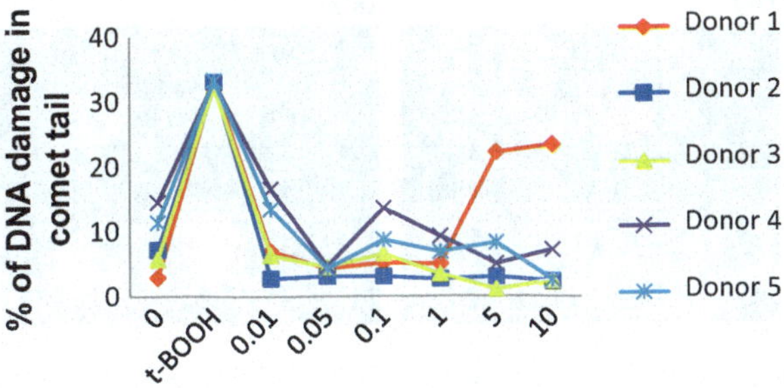

Fig. 13.8 Genotoxic effect of CS in human lymphocytes

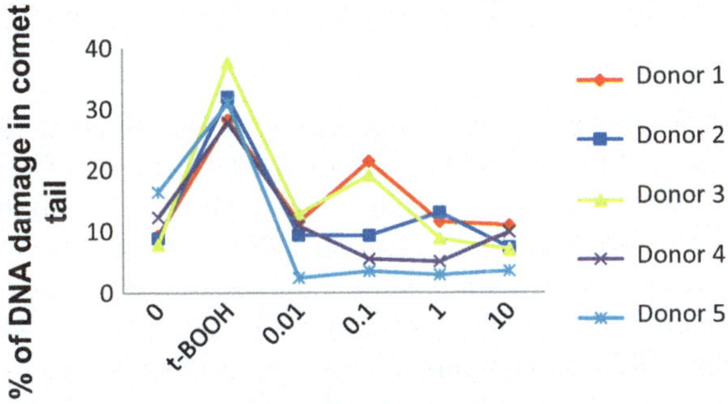

Fig. 13.9 Genotoxic effect of HA-CS in human lymphocytes

or dissolution of nanoparticles inside the cells. A higher concentration of released ions can affect cell survival [41]. Endodontic materials are mainly in a close contact with the dentin and periodontal tissue during a long period of time. The tissue response to these materials is very important and can significantly affect the outcome of endodontic treatment.

13.4.1 Genotoxicity

One of the most common assays for the genotoxicity testing is the single cell gel electrophoresis, known as the comet assay. Comet test is a sensitive and fast technique for quantification and analysis of DNA damaged in the cells [42].

Therefore, genotoxic potential of the nanostructural materials CS and HA-CS was tested in human lymphocytes, and both of the materials exhibited low genotoxic potential (Figs. 13.8, 13.9, and 13.10). However, HA-CS showed lower genotoxic potential compared to CS.

CS and HA-CS materials are nanostructured, and the number of harmful effects on nuclear structure and function (inhibition of replication, transcription, and cell proliferation) is associated with the penetration of nanoparticles into the cell nucleus. Additional attention during pre-clinical researches using results from more tests is highly recommended [43].

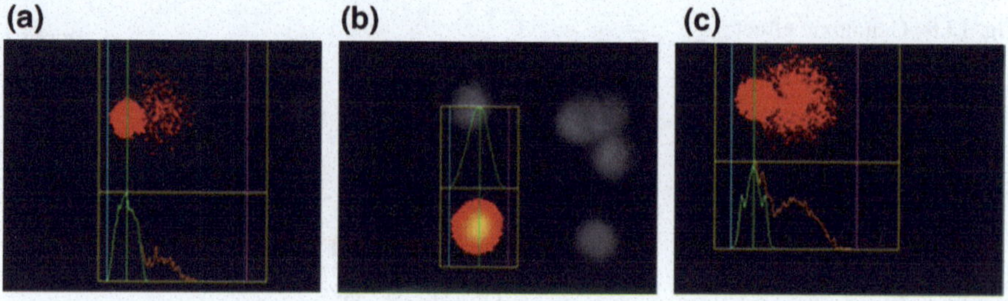

Fig. 13.10 The genotoxic effect of different compounds. **a** Damaged cell as a consequence of genotoxic potential of CS (conc 5 mg/ml) in donor 4. **b** Undamaged lymphocyte (negative control). **c** Fully damaged cell which has been treated with 0.5 mM *t*-BOOH, as positive control

A number of other studies have shown different cell response to genotoxic substances [42, 44–46]. Herein, different sensitivity of the cells obtained from different donors lymphocytes could be the result of the different genetic backgrounds and consequently differences in their DNA repair capacities. To underline such genetic differences between donors, a comparison of negative control (untreated cells) by themselves and the same comparison after the *t*-BOOH treatment were conducted. *t*-BOOH causes significant structural DNA damages due to synthesis of reactive oxygen species (ROS) which leads to oxygen stress and hydroxyl radical (OH) production. The obtained differences indicated various levels of induction of DNA repair enzymes in different donors, which is in line with findings reported by Hazra et al. [47]. Also, those studies showed high variability of the genes involved in base excision repair of DNA (BER) and nucleotide excision repair of DNA (NER) in lymphocytes from healthy subjects. As ROS is induced by *t*-BOOH that leads to induction of DNA damages mainly repaired via BER pathway (except double-strand breaks), it can be assumed that there is a significant variability of the genes involved in BER and NER repair enzymes into lymphocytes from normal healthy subjects.

It is well known that cells in suspensions were more exposed to the influence of the tested substance than cells growing in monolayer [48]. According to this, it can be assumed that lymphocytes in suspension were more exposed to testing material than cells in dental and periodontal tissues in real condition would ever be and, furthermore, that tested concentrations significantly exceeded the dose that would be absorbed regularly in vivo.

13.4.2 Cytotoxicity

13.4.2.1 Cytotoxicity-in Vitro

Investigation of the nanohydroxyapatite cytotoxicity indicates its biocompatibility, irrespective of the shape of particles [15], the concentration, and the duration of the

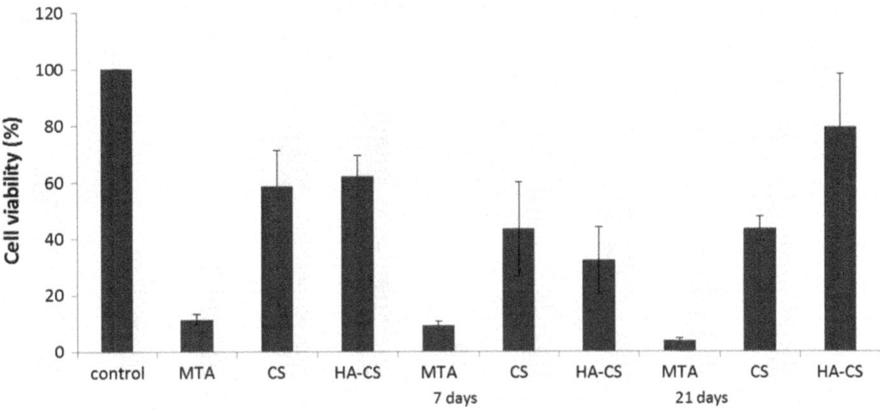

Fig. 13.11 Effect of the 100% eluates of the materials on the cell viability

leaching material [49]. Toxicity of nanocommercial endodontic cement (BioAggregate) based on calcium silicate and hydroxyapatite in the cell culture is similar [50] or less (in the system) compared to MTA [51]. Biocompatibility of nanostructured bioceramic materials based on calcium silicate and calcium phosphate was confirmed in a study by Ma et al. (2011) [52], but the authors noted that a shorter setting time with longer leaching of material may result in a higher amount of released ions with a negative effect on biocompatibility.

The cytotoxicity of new nanomaterials CS and HA-CS was evaluated according to ISO 10993-5 specifications and compared with commercial microstructured MTA Angelus. After elution of the samples, 100 and 50% eluates were tested on human lung fibroblasts. Additionally, the pH value of the 100% eluates was monitored (within 24 h, 4, 7, 21 days).

Exposure of the cells to undiluted eluates (100%) of all materials significantly reduced cell viability at all time periods. Interestingly, 100% concentration of MTA eluate resulted in significantly lower cell viability compared to CS and HA-CS eluates at all test periods; moreover, cytotoxicity of undiluted MTA increased over time (Fig. 13.11) [53].

Diluted eluates of the all tested materials showed significantly lower toxicity in comparison with undiluted eluates (Fig. 13.12).

Previous studies have shown variable results regarding cytotoxicity of MTA which spanned from the lack of any cytotoxicity [54], to measurable cytotoxic effects with increased cytotoxicity over time [50]. Also, for different MTA-based commercial materials have been reported variable cytotoxic effects [55]. Variability among results is associated with different cell lines, protocols for material preparation and setting, surface area/medium ratio, duration of exposure, as well as different parameters for measuring cell viability. The present findings indicate dose-dependent cytotoxicity of the tested materials.

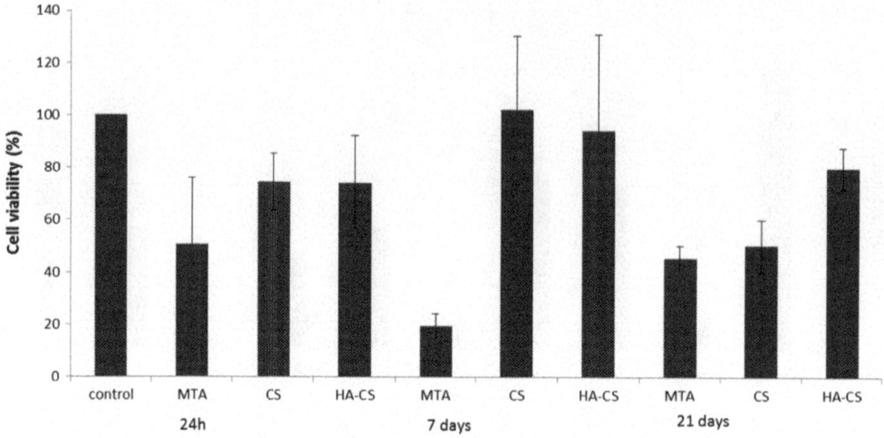

Fig. 13.12 Effect of the 50% eluates of the materials on the cell viability

Table 13.2 pH values of the undiluted eluates of the materials

	24 h	7 days	21 days
MTA	10.75	11.70	11.73
CS	9.73	11.49	11.49
HA-CS	9.62	10.59	9.95

Since pH data indicate different ion elution kinetics from the test materials (Table 13.2), cytotoxicity of the materials also could be associated with pH values obtained at the corresponding time-point. The most rapid ion release and the highest pH values were found for MTA. The observed cytotoxicity of MTA corresponded well with high pH, and the same correlation between cytotoxicity and pH values was recorded for CS and HA-CS.

Different cytotoxicity of MTA and CS in spite of similar chemical structure could be related to differences in radiopacifying agents. MTA contains bismuth oxide which interferes with hydration processes, becomes a constituent of calcium silicate hydrate, and may elute from set material [56]. Lower cytotoxicity of HA-CS compared to CS and MTA could be related to the chemical structure of this material and lower pH. The main ingredient in HA-CS is hydroxyapatite with calcium silicate added to improve mechanical properties [14]. HA-CS has previously been also reported to have lower genotoxic potential compared to CS [17].

13.4.2.2 Cytotoxicity-in Vivo

In vitro tests are an initial step in the evaluation of the biocompatibility, but to determine the type of interaction between materials and tissues in vivo tests are necessary. Subcutaneous tissues are common choice for biocompatibility evaluation, and local tissue reaction on implant could be quantitatively and qualitatively evaluated (his-

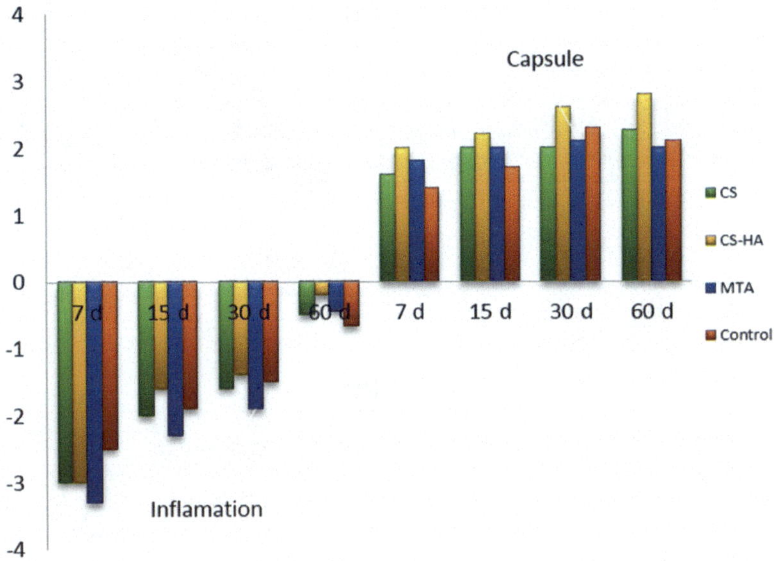

Fig. 13.13 The parameters of inflammation response and thickness of formed fibrous capsule

tology, histochemistry, SEM, TEM). Tissue reactions on materials are a cumulative pathophysiological consequence of healing after the surgery and material presence, chronic inflammation, and at the end, tissue renewal adjustment on material. Also, the dynamics of fibrous capsule formation are good marker of tissue recovery, and it is in reverse proportion with the level of tissue inflammation [57–59].

The tissue around CS and HA-CS showed the most notable inflammatory reaction at the beginning of the observation period (Fig. 13.13), with moderate disturbance of connective tissue. During the further tissue recovery, a significant decrease of the inflammation intensity level was observed with both materials with disappearance of inflammation infiltrate's cells and recovery and tissue remodeling as well. The most developed capsule was detected for HA-CS, indicating well tolerance of tissue to this material, pointed out a good interaction between materials and surrounding tissue's cells. However, the difference between tissue response obtained from negative controls (empty part of tubes), and tested materials, indicated that CS, HA-CS, MTA did not significantly contribute to the inflammation reaction (Figs. 13.14 and 13.15).

13.5 Bioinductivity

Bioinductivity of the CS and HA-CS was monitored after direct pulp capping (DPP) in experimental animals. DPP is a therapeutic procedure which is often used to preserve

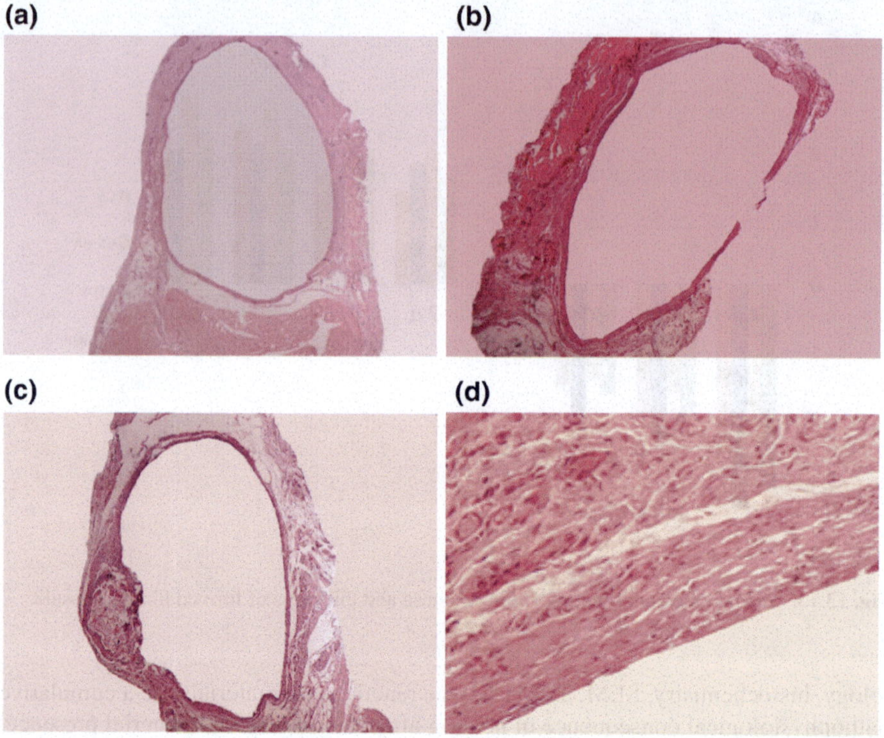

Fig. 13.14 The effect of nanomaterials on the induction of inflammation (30 days of exposure). **a** CS: well-developed fibrous capsule can be observed (HE, ×40). **b** HA-CS: the integrity of the connective tissue around the very well-developed capsule was maintained with no signs of vascular congestion; well-developed fibrous capsule (HE, ×40). **c** MTA: mild inflammation of connective tissue, very thin capsule around the implant (HE, ×40). **d** Negative control (empty tube): the integrity of connective was preserved; well-formed capsule (HE, ×400)

the dental pulp vitality. Calcium silicate- and hydroxyapatite-based materials are crucial for better and more efficient regenerative endodontic treatment [3, 60].

13.5.1 Direct Pulp Capping in Rabbits

First, the efficacy of the materials in DPP was evaluated in rabbits (*Oryctolagus cuniculus*). Class V cavities were prepared in gingival thirds of vestibular surfaces of incisors, and exposed pulps were closed with CS, HA-CS, and MTA. The animals were killed after 14 days, and tissue samples were analyzed according to the following parameters [61]:

(i) the inflammatory response of the pulp (intensity, scope [1–4])

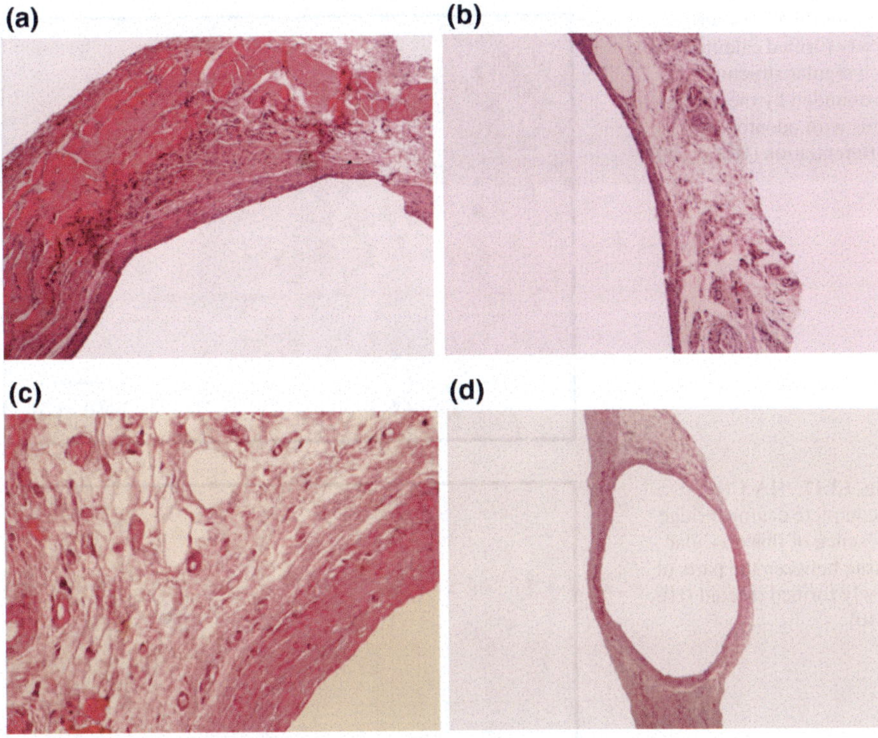

Fig. 13.15 The effect of nanomaterials on the induction of inflammation (60 days of exposure). **a** CS: the integrity of connective tissue was well preserved; capsule was well formed (HE, ×100). **b** HA-CS: the integrity of connective tissue was well preserved; a well-developed fibrous capsule around the implant (HE, ×100). **c** MTA: the connective tissue was with the maintained integrity; capsule was well developed (HE, ×400). **d** Negative control (empty tube), connective tissue was with well-preserved integrity; a thin capsule is formed (HE, ×40)

(ii) other findings in the pulp (giant cells, microorganisms, bulk material parts [1–4])
(iii) the formation of dentin bridge (continuity, morphology, thickness [1–4]).

The interaction between the capping material and the injured pulp tissue as well as the processes of healing and regeneration is not yet completely understood. Biomaterials that are in contact with the tooth/bone tissues should encourage tissue response in terms of adhesion, proliferation, and differentiation of cells. Topography and chemistry of the materials used play an important role in adhesion of cells [62]. Nanostructure of CS and HA-CS materials could probably increase adhesion, proliferation, and differentiation of the cells, since the bone is nanostructured by itself. The size and geometry of the crystals can modify the reaction of surrounding tissue. Inflammatory reaction on newly synthesized CS and CS-HA was observed near the material with no further expansion of the coronary pulp. Thin layers of newly formed calcified tissues (up to 149 μm) were found in samples with CS and HA-CS

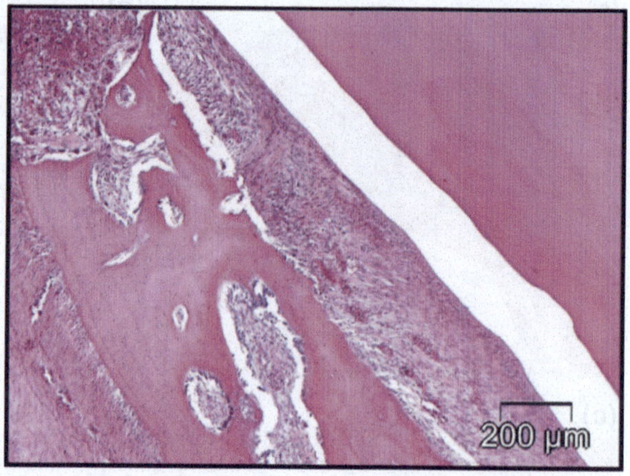

Fig. 13.16 CS material newly formed calcified tissue of a regular structure, surrounded by mesenchymal cells with odontoblasts differentiation (HE, ×100)

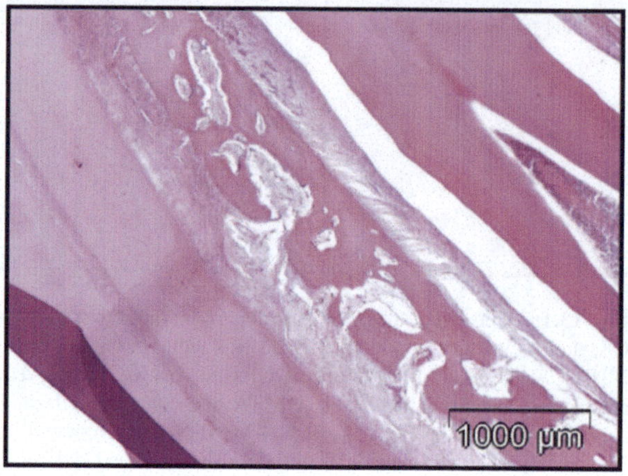

Fig. 13.17 HA-CS-incomplete dentine bridge. Presence of fibrovascular tissue between the parts of newly formed osteoid (HE, ×40)

(Figs. 13.16, 13.17, and 13.18), with regular structure similar to reparative dentin. This finding correlates with mild and moderate inflammation in samples with CS and CS-HA, and this finding is a confirmation of their biocompatibility.

Pulp–dentin complex has the regenerative potential that induces the formation of tertiary dentin. Dentin matrix is not just a scaffold for the development of mineralized tissue, but also a reservoir of growth factors secreted by odontoblasts and pulp fibroblasts [63]. These growth factors hypothetically give the signal for the proliferation, differentiation, and accumulation of pulp cells at the site of the damaged pulp. They initiate tissue regeneration. It is a very encouraging result that calcified tissue was formed in a relatively short period of time. Thicker layer of newly formed calcified tissue was observed in the samples of the experimental cements (CS and CS-HA).This finding correlates with mild and moderate inflammation in samples with CS and CS-HA, and this finding is a confirmation of their biocompatibility. These bioactive ceramic materials (CS, CS-HA, MTA) form a hydroxyapatite layer

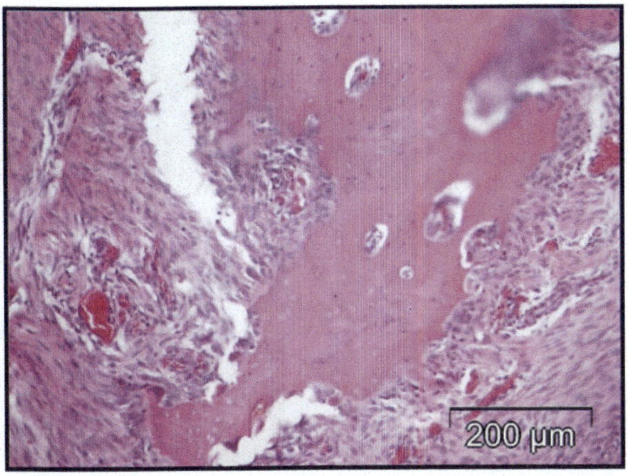

Fig. 13.18 MTA-hypercellular newly calcified tissue. (HE, K200)

after contact with biological fluids rich with phosphate ions. This layer can stimulate the deposition of mineralized tissue on its surface and later integration with the surrounding tissue in vivo [64]. The biocompatibility of these new cements can be explained by these processes. All three materials have shown the biocompatibility and the ability to induce calcification on the surface of the pulp tissue within 14 days.

13.5.2 Direct Pulp Capping in Vietnamese Pigs

Because of the differences between rabbit and human teeth and a relatively short observation period (14 days) in rabbits, it is interesting to describe the results of direct pulp capping with new nanostructured materials (CS and HA-CS) in Vietnamese pigs. The experimental study was completed on permanent teeth of Vietnamese pigs (*Susscrofadomesticus*) which morphologically resemble human teeth, with a longer observation period (30 days).

The criteria for histological evaluation of pulp reactions on applied materials are: dentin bridge formation (thickness, continuity with surrounding dentin, morphological aspects, and localization), reorganization of pulp tissue (morphological integrity of odontoblasts and odontoblast-like cells, as well as the integrity of deeper pulp tissue), inflammatory response of pulp (acute or chronic), and existence of bacterial cells [65].

The results of histological analysis showed that new dentine was formed in approximately all samples in both groups (CS and HA-CS). Newly formed dentin had characteristics of reparative dentin with a small number of irregular dentinal tubules. These were in continuity with the surrounding dentin [66].

Dentin bridge which completely enclosed pulp perforation was detected in most samples with HA-CS (Fig. 13.19).

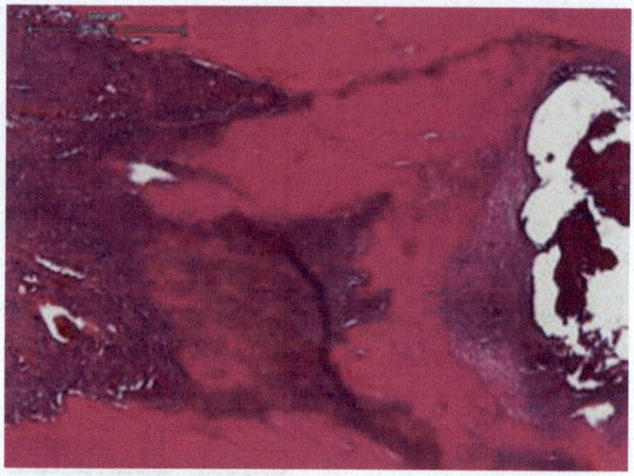

Fig. 13.19 Complete dentin bridge after direct pulp capping with HA-CS. Newly formed dentin completely closes preparation region of the pulp chamber, and it is very similar to original dentin. Particles of HA-CS in the area of perforation are visible (HE, 40×)

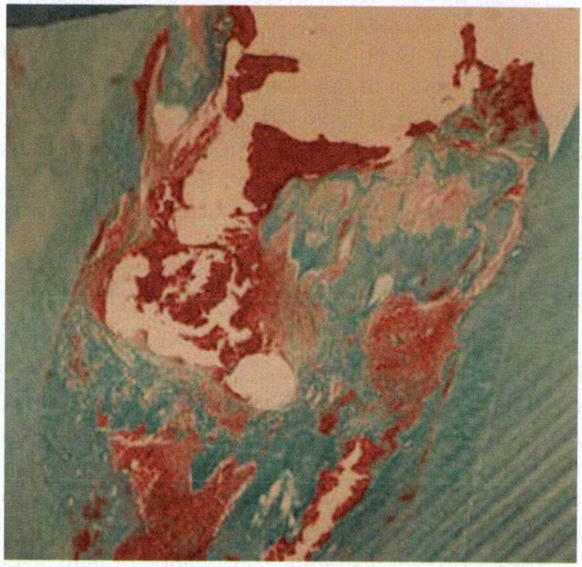

Fig. 13.20 Complete dentin bridge after direct pulp capping with CS. Newly formed dentin completely closes the preparation region of the pulp chamber. Particles of HA-CS are visible in the area of perforation (Goldner trichrome, 40×)

In CS group complete dentine bridge was detected in a similar number of samples as in MTA group (Fig. 13.20).

Incomplete dentin bridge in the form of dentin islands was observed in few samples of the CS and MTA groups (Fig. 13.21).

Reparative dentin in the form of a complete or incomplete dentin bridge (dentin islands) is formed after mineralization of the extracellular matrix. These islands tend to establish contact with dentin walls and seal and conserve exposed dental pulp (Fig. 13.22).

Continuous reparative dentin that extends along the lateral walls of dentin was noted in one case in both CS and HA-CS groups, while this form of dentin was not

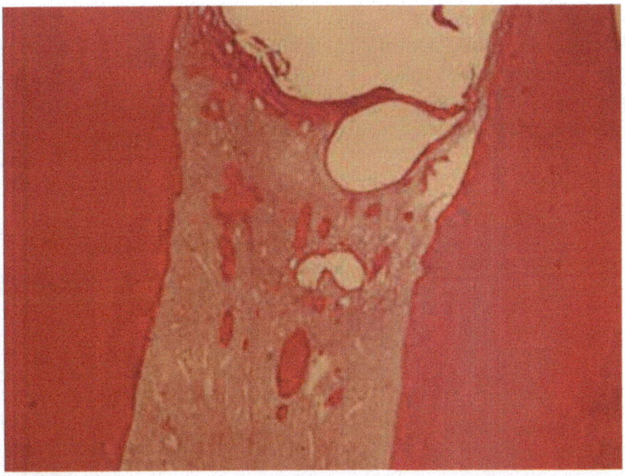

Fig. 13.21 Dentin islets after direct pulp capping with CS. Newly formed dentin is in the form of dentin islands and tends to close perforation of cavum dentis. Completely preserved pulp tissue without inflammatory reaction can be seen (HE, 40×)

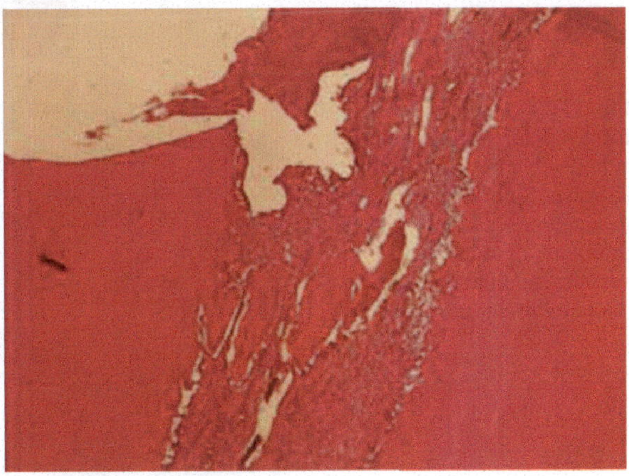

Fig. 13.22 Dentin islets after direct pulp capping with HA-CS. Three dentin islets that almost completely closed cavum dentis and signs of neoangiogenesis in the form of newly created blood vessels can be seen (HE, 40×)

registered in the samples with MTA (Fig. 13.23). Underneath the newly formed dentin bridge, odontoblasts with minor or major structural differences were registered in the most of the teeth from both experimental groups [66].

Tissue reorganization below the perforation was registered in the most of the samples of CS, HA-CS, and MTA groups in the form of odontoblast-like cells' hyperactivity. There was a correlation between the number of odontoblast-like cells, thickness of the bridge, and the preservation of deeper areas of the pulp similar to the report of Tziafas et al. 2002 [67]. Greater number of these cells increases the thickness of dentin bridge, while radicular pulp retains its physiological morphology [68]. Reorganization of subjacent area was in correlation with the number of odontoblast-like cells and showed prominence in the HA-CS group (Fig. 13.24) [66].

Fully preserved pulp tissue was a rarity in all applied materials after 30 days. Signs of neoangiogenesis and proliferation of blood vessels were recorded in most of the

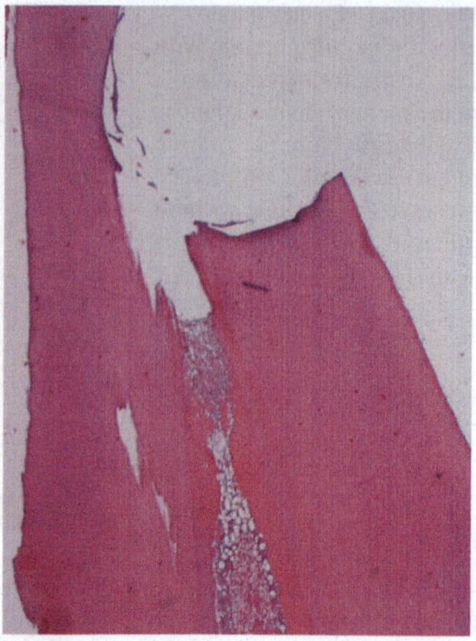

Fig. 13.23 CS-newly laterally deposited dentin that narrows pulpal space and mild inflammatory reaction (HE, 40×)

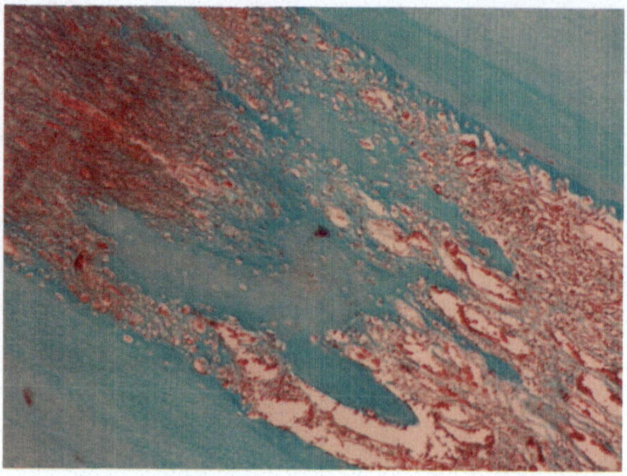

Fig. 13.24 HA-CS-newly formed dentin bridge which completely closed the pulp chamber. Signs of neoangiogenesis and proliferation of blood vessels. Mild inflammation (Goldner trichrome 40×)

cases. This was a clear indication of a healing process and complete revascularization (Fig. 13.25).

In CS group there were few samples with the presence of venous stasis, hemorrhage, and inflammation in the central part of the pulp. Complete disorganization of the pulp tissue was not registered in the samples of either experimental or control groups [66].

Histological analysis after 30 days indicates that the experimental pulp capping materials CS and HA-CS caused mild to moderate chronic inflammation of pulp tissue. Severe inflammation or abscesses were not observed in any sample. Mild inflammation was present in the most samples in the group with CS and HA-CS. Moderate inflammation with cellular proliferation in coronal and radicular pulp was

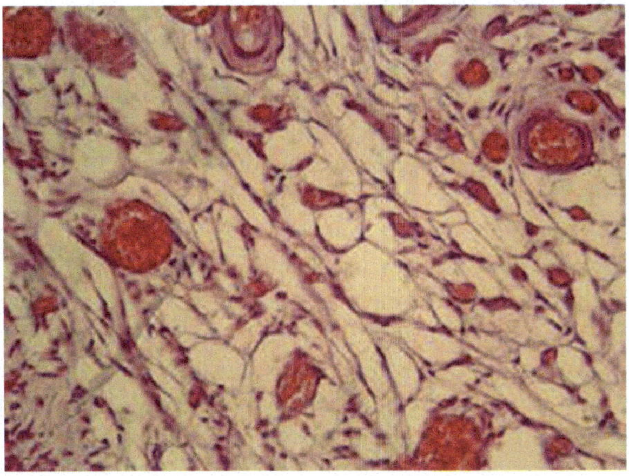

Fig. 13.25 Signs of neoangiogenesis and stasis in dental pulp. This finding was probably related to the chemical properties of the capping material and mechanical trauma during cavity preparation, rather than the presence of bacterial infection (HE, 400×)

observed in some cases with MTA, but in the CS and HA-CS groups there were fewer inflammatory cells. Acute inflammation and pulp necrosis were not observed in any sample. The findings of the study demonstrated the existence of inflammatory cells in the coronal and radicular pulp areas for all applied materials. Mononuclear cells were registered in the pulp tissue in all samples, suggesting the existence of chronic inflammation scoring from slight to average, according to earlier determined criteria. Unexpectedly, inflammatory cells were found in greater numbers in the central than in the peripheral parts of the pulp and were not associated with the presence of bacterial cells. In the control MTA group, the existence of lymphocytes, plasma cells, and macrophages was detected only in a few samples [66]. This is in agreement with the results of other authors [69].

Gram staining did not detect any gram-positive bacteria in the pulp of all samples. Small number of bacteria was observed just in dentinal tubules in some samples with all tested materials. It could be difficult to determine whether these bacteria invaded dentin during the capping procedure or they entered the dentinal tubules through the cavity margins during the healing process. This could be due to microleakage of the restorative material [66].

Therapeutic effects of CS and HA-CS were similar to MTA, and that indicates that new nanostructured materials based on calcium silicate cements and hydroxyapatite have favorable effects on reparative activities of the pulp of Vietnamese pigs primarily due to their physical and chemical properties. Different particle sizes of CS and HA-CS in regard to MTA considerably enhance the performance of these materials in terms of shorter setting time and workability in moist conditions. These properties affect the bonding quality of these materials to exposed pulp and their favorable effects on odontoblast activation and dentin bridge formation.

13.6 Conclusion

Nanotechnology is a relatively new field that involves substance manipulation at the molecular level, including individual molecules and interactions between them. Research in this area is focused on achieving high specificity to get desired physical and chemical properties of nanoparticles.

The authors agree that the use of nanomaterials in dentistry is very important and will be useful for improving the quality of life of patients. Advantages involving the use of new materials for preventive health care by nanoparticles are either antimicrobial or restorative properties for enamel.

Application of nanoparticles must overcome the challenges for their application, which contributes to more sophisticated diagnostic capabilities and provide more effective therapies and preventative properties.

Acknowledgements This work was supported by the Ministry of Education, Science and Technological Development of Republic of Serbia, Project No. 172058 and Project No. 172007.

References

1. Koch K, Brave D (2009) Bioceramic technology-the game changer in endodontics. Endod Practice 2:17–21
2. Murray PE, Garcia-Godoy F, Hargreaves KM (2007) Regenerative endodontics: a review of current status and a call for action. J Endod 33(4):377–390
3. Güven EP, Taşlı PN, Yalvac ME, Sofiev N, Kayahan MB, Sahin F (2013) In vitro comparison of induction capacity and biomineralization ability of mineral trioxide aggregate and a bioceramic root canal sealer. Int Endod J 46(12):1173–1182
4. Baek SH, Plenk H, Kim S (2005) Periapical tissue responses and cementum regeneration with amalgam, SuperEBA, and MTA as root-end filling materials. J Endod. 31(6):444–449
5. Da Silva GF, Guerreiro-Tanomaru JM, Sasso-Cerri E, Tanomaru-Filho M, Cerri PS (2011) Histological and histomorphometrical evaluation of furcation perforations filled with MTA, CPM and ZOE. Int Endod J 44(2):100–110
6. Danesh F, Vahid A, Jahanbani J, Mashhadiabbas F, Arman E (2012) Effect of white mineral trioxide aggregate compared with biomimetic carbonated apatite on dentine bridge formation and inflammatory response in a dental pulp model. Int Endod J 45(1):26–34
7. Parirokh M, Torabinejad M (2010) Mineral trioxide aggregate: a comprehensive literature review—part I: chemical, physical, and antibacterial properties. J Endod 36(1):16–27
8. Torabinejad M, Parirokh M (2010) Mineral trioxide aggregate: a comprehensive literature review—part II: leakage and biocompatibility investigations. J Endod 36(2):190–202
9. Okiji T, Yoshiba K (2009) Reparative dentinogenesis induced by mineral trioxide aggregate: a review from the biological and physicochemical points of view. Int J of Dentistry
10. Asgary S, Shahabi S, Jafarzadeh T, Amini S, Kheirieh S (2008) The properties of a new endodontic material. J Endod 34(8):990–993
11. Formosa LM, Mallia B, Camilleri J (2013) Mineral trioxide aggregate with anti-washout gel–properties and microstructure. Dent Mater 29(3):294–306
12. Saghiri MA, Garcia-Godoy F, Gutmann JL, Lotfi M, Asatourian A, Ahmadi H (2013) Push-out bond strength of a nano-modified mineral trioxide aggregate. Dent Traumatol 29(4):323–327
13. Pietak AM, Reid JW, Stott MJ, Sayer M (2007) Silicon substitution in the calcium phosphate bioceramics. Biomaterials 28(28):4023–4032

14. Huan Z, Chang J (2009) Calcium–phosphate–silicate composite bone cement: self-setting properties and in vitro bioactivity. J Mater Sci Mater Med 20(4):833–841
15. Zhao Y, Zhang Y, Ning F, Guo D, Xu Z (2007) Synthesis and cellular biocompatibility of two kinds of HAP with different nanocrystal morphology. J Biomed Mater Res B Appl Biomater 83(1):121–126
16. Asgary S, Eghbal MJ, Parirokh M, Ghoddusi J, Kheirieh S, Brink F (2009) Comparison of mineral trioxide aggregate's composition with Portland cements and a new endodontic cement. Journal of Endodontics. 35(2):243–250
17. Opačić-Galić V, Petrović V, Živković S, Jokanović V, Nikolić B, Knežević-Vukčević J, Mitić-Ćulafić D (2013) New nanostructural biomaterials based on active silicate systems and hydroxyapatite: characterization and genotoxicity in human peripheral blood lymphocytes. Int Endod J 46(6):506–516
18. Jokanović V, Čolović B, Jokanović B, Živković S (2015) Superplastic, Quick-Bonding Endodontic Mixtures and Their Hydration. International Journal of Applied Ceramic Technology 12(S2)
19. Fridland M, Rosado R (2003) Mineral trioxide aggregate (MTA) solubility and porosity with different water-to-powder ratios. Journal of Endodontics. 29(12):814–817
20. Bodanezi A, Carvalho N, Silva D, Bernardineli N, Bramante CM, Garcia RB, Moraes IG (2008) Immediate and delayed solubility of mineral trioxide aggregate and Portland cement. J Appl Oral Sci 16(2):127–131
21. Gandolfi MG, Siboni F, Prati C (2012) Chemical–physical properties of TheraCal, a novel light-curable MTA-like material for pulp capping. Int Endod J 45(6):571–579
22. Camilleri J (2011) Scanning electron microscopic evaluation of the material interface of adjacent layers of dental materials. Dent Mater 27(9):870–878
23. Montellano AM, Schwartz SA, Beeson TJ (2006) Contamination of tooth-colored mineral trioxide aggregate used as a root-end filling material: a bacterial leakage study. J Endod 32(5):452–455
24. Hawley M, Webb TD, Goodell GG (2010) Effect of varying water-to-powder ratios on the setting expansion of white and gray mineral trioxide aggregate. J Endod 36(8):1377–1379
25. Sarkar NK, Caicedo R, Ritwik P, Moiseyeva R, Kawashima I (2005) Physicochemical basis of the biologic properties of mineral trioxide aggregate. J Endod 31(2):97–100
26. De-Deus G, Reis C, Brandão C, Fidel S, Fidel RA (2007) The ability of Portland cement, MTA, and MTA Bio to prevent through-and-through fluid movement in repaired furcal perforations. J Endod 33(11):1374–1377
27. Sanghavi T, Shah N, Shah RR (2013) Comparative analysis of sealing ability of biodentin and calcium phosphate cement against mineral trioxide aggregate (MTA) as a furcal perforation repair material (an in vitro study). NJIRM 4:56–60
28. Tsatsas DV, Meliou HA, Kerezoudis NP (2005) Sealing effectiveness of materials used in furcation perforation in vitro. Int Dent J 55(3):133–141
29. Tobón-Arroyave SI, Restrepo-Pérez MM, Arismendi-Echavarría JA, Velásquez-Restrepo Z, Marín-Botero ML, García-Dorado EC (2007) Ex vivo microscopic assessment of factors affecting the quality of apical seal created by root-end fillings. Int Endod J 40(8):590–602
30. Guneser MB, Akbulut MB, Eldeniz AU (2013) Effect of various endodontic irrigants on the push-out bond strength of biodentine and conventional root perforation repair materials. J Endod 39(3):380–384
31. Iacono F, Gandolfi MG, Huffman B, Sword J, Agee K, Siboni F, Tay F, Prati C, Pashley D (2010) Push-out strength of modified Portland cements and resins. Am J Dent 23(1):43
32. Gancedo-Caravia L, Garcia-Barbero E (2006) Influence of humidity and setting time on the push-out strength of mineral trioxide aggregate obturations. J Endod 32(9):894–896
33. EL-Ma'aita AM, Qualtrough AJ, Watts DC (2013) The effect of smear layer on the push-out bond strength of root canal calcium silicate cements. Dental Materials 29(7): 797–803
34. Atabek D, Sillelioğlu H, Ölmez A (2012) Bond strength of adhesive systems to mineral trioxide aggregate with different time intervals. J Endod 38(9):1288–1292

35. Han L, Okiji T (2013) Bioactivity evaluation of three calcium silicate-based endodontic materials. Int Endod J 46(9):808–814
36. Formosa LM, Mallia B, Camilleri J (2014) Push-out bond strength of MTA with antiwashout gel or resins. Int Endod J 47(5):454–462
37. Martínez Lalis R, Esaín ML, Kokubu GA, Willis J, Chaves C, Grana DR (2009) Rat subcutaneous tissue response to modified Portland cement, a new mineral trioxide aggregate. Braz Dent J 20(2):112–117
38. Yavari HR, Shahi S, Rahimi S, Shakouie S, Roshangar L, Abassi MM, Khavas SS (2009) Connective tissue reaction to white and gray MTA mixed with distilled water or chlorhexidine in rats. Iranian Endodontic Journal. 4(1):25
39. Theiszova M, Jantova S, Dragunova J, Grznarova P, Palou M (2005) Comparison the cytotoxicity of hydroxiapatite measured by direct cell counting and MTT test in murine fibroblast NIH-3T3 cells. Biomedical Papers-Palacky University in Olomouc 149(2):393
40. Gandolfi MG, Taddei P, Tinti A, Prati C (2010) Apatite-forming ability (bioactivity) of ProRoot MTA. Int Endod J 43(10):917–929
41. Chen L, Mccrate JM, Lee JC, Li H (2011) The role of surface charge on the uptake and biocompatibility of hydroxyapatite nanoparticles with osteoblast cells. Nanotechnology 22(10):105708
42. Zeferino EG, Bueno CE, Oyama LM, Ribeiro DA (2010) Ex vivo assessment of genotoxicity and cytotoxicity in murine fibroblasts exposed to white MTA or white Portland cement with 15% bismuth oxide. Int Endod J 43(10):843–848
43. Chen M, von Mikecz A (2005) Formation of nucleoplasmic protein aggregates impairs nuclear function in response to SiO_2 nanoparticles. Exp Cell Res 305(1)
44. Brzovic V, Miletic I, Zeljezic D, Mladinic M, Kasuba V, Ramic S, Anic I (2009) In vitro genotoxicity of root canal sealers. Int Endod J 42(3):253–263
45. Camargo SE, Camargo CH, Hiller KA, Rode SM, Schweikl H, Schmalz G (2009) Cytotoxicity and genotoxicity of pulp capping materials in two cell lines. Int Endod J 42(3):227–237
46. Zhang W, Li Z, Peng B (2010) Ex vivo cytotoxicity of a new calcium silicate–based canal filling material. Int Endod J 43(9):769–774
47. Hazra TK, Das A, Das S, Choudhury S, Kow YW, Roy R (2007) Oxidative DNA damage repair in mammalian cells: a new perspective. DNA Repair 6(4):470–480
48. Mitić-Ćulafić D, Žegura B, Nikolić B, Vuković-Gačić B, Knežević-Vukčević J, Filipič M (2009) Protective effect of linalool, myrcene and eucalyptol against t-butyl hydroperoxide induced genotoxicity in bacteria and cultured human cells. Food Chem Toxicol 47(1):260–266
49. Shahoon H, Hamedi R, Yadegari Z, Majd Hosseiny VA, Golgounnia P (2013) The comparison of silver and hydroxyapatite nanoparticles biocompatibility on L929 fibroblast cells: an in vitro study. J Nanomed Nanotechnol 4:1–4
50. De-Deus G, Canabarro A, Alves G, Linhares A, Senne MI, Granjeiro JM (2009) Optimal cytocompatibility of a bioceramic nanoparticulate cement in primary human mesenchymal cells. J Endod 35(10):1387–1390
51. Khalil WA, Eid NF (2013) Biocompatibility of BioAggregate and mineral trioxide aggregate on the liver and kidney. Int Endod J 46(8):730–737
52. Ma J, Shen Y, Stojicic S, Haapasalo M (2011) Biocompatibility of two novel root repair materials. J Endod 37(6):793–798
53. Petrović V, Opačić-Galić V, Živković S, Nikolić B, Danilović V, Miletić V, Jokanović V, Mitić-Ćulafić D (2015) Biocompatibility of new nanostructural materials based on active silicate systems and hydroxyapatite: in vitro and in vivo study. Int Endod J 48(10):966–975
54. Modareszadeh MR, Di Fiore PM, Tipton DA, Salamat N (2012) Cytotoxicity and alkaline phosphatase activity evaluation of endosequence root repair material. J Endod 38(8):1101–1105
55. Lee BN, Son HJ, Noh HJ, Koh JT, Chang HS, Hwang IN, Hwang YC, Oh WM (2012) Cytotoxicity of newly developed ortho MTA root-end filling materials. J Endod 38(12):1627–1630
56. Camilleri J (2008) Characterization of hydration products of mineral trioxide aggregate. Int Endod J 41(5):408–417
57. Silva-Herzog D, Ramírez T, Mora J, Pozos AJ, Silva LA, Silva RA, Nelson-Filho P (2011) Preliminary study of the inflammatory response to subcutaneous implantation of three root canal sealers. Int Endod J 44(5):440–446

58. Parirokh M, Mirsoltani B, Raoof M, Tabrizchi H, Haghdoost AA (2011) Comparative study of subcutaneous tissue responses to a novel root-end filling material and white and grey mineral trioxide aggregate. Int Endod J 44(4):283–289
59. Gomes-Filho JE, Watanabe S, Lodi CS, Cintra LT, Nery MJ, Dezan E, Bernabé PF (2012) Rat tissue reaction to MTA FILLAPEX®. Dent Traumatol 28(6):452–456
60. Zhang S, Yang X, Fan M (2013) BioAggregate and iRoot BP Plus optimize the proliferation and mineralization ability of human dental pulp cells. Int Endod J 46(10):923–929
61. Maria de Lourdes RA, Holland R, Reis A, Bortoluzzi MC, Murata SS, Dezan E, Souza V, Alessandro LD (2008) Evaluation of mineral trioxide aggregate and calcium hydroxide cement as pulp-capping agents in human teeth. J Endod 34(1):1–6
62. Gandolfi MG, Ciapetti G, Perut F, Taddei P, Modena E, Rossi PL, Prati C (2009) Biomimetic calcium-silicate cements aged in simulated body solutions. Osteoblast response and analyses of apatite coating. J Appl Biomater Biomech 7(3)
63. Chen CL, Huang TH, Ding SJ, Shie MY, Kao CT (2009) Comparison of calcium and silicate cement and mineral trioxide aggregate biologic effects and bone markers expression in MG63 cells. J Endod 35(5):682–685
64. Liu WN, Chang J, Zhu YQ, Zhang M (2011) Effect of tricalcium aluminate on the properties of tricalcium silicate–tricalcium aluminate mixtures: setting time, mechanical strength and biocompatibility. Int Endod J 44(1):41–50
65. Shayegan A, Petein M, Vanden Abbeele A (2009) The use of beta-tricalcium phosphate, white MTA, white Portland cement and calcium hydroxide for direct pulp capping of primary pig teeth. Dent Traumatol 25(4):413–419
66. Popović-Bajić M, Petrović V, Galić VO, Danilović V, Jokanović V, Prokić B, Prokić BB, Živković S (2016) Direct pulp capping with novel nanostructural materials based on calcium silicate systems and hydroxyapatite. Serbian Dent J 63(4):183–192
67. Tziafas D, Pantelidou O, Alvanou A, Belibasakis G, Papadimitriou S (2002) The dentinogenic effect of mineral trioxide aggregate (MTA) in short-term capping experiments. Int Endod J 35(3):245–254
68. Orhan EO, Maden M, Senguüven B (2012) Odontoblast-like cell numbers and reparative dentine thickness after direct pulp capping with platelet-rich plasma and enamel matrix derivative: a histomorphometric evaluation. Int Endod J 45(4):317–325
69. Tabarsi B, Parirokh M, Eghbal MJ, Haghdoost AA, Torabzadeh H, Asgary S (2010) A comparative study of dental pulp response to several pulpotomy agents. Int Endod J 43(7):565–571

Correction to: Advanced Nanomaterials and Their Functionalization in Clinical Endodontics

Hend Mahmoud Abou El Nasr and Makbule Bilge Akbulut

Correction to:
Chapter 3 in: R. S. Chaughule (ed.), *Dental Applications of Nanotechnology*, https://doi.org/10.1007/978-3-319-97634-1_3

The original version of the book was inadvertently published with incorrect co-author name, which should be corrected to read as "Makbule Bilge Akbulut" in Chapter 3. The correction chapter and the book have been now updated with the change.

The updated online version of this chapter can be found at
https://doi.org/10.1007/978-3-319-97634-1_3

R. S. Chaughule (ed.), *Dental Applications of Nanotechnology*,
https://doi.org/10.1007/978-3-319-97634-1_14

Index

A
Acidophilic and aciduric *Actinomyces*, 3
Acrylic resins, 92, 125, 184
Adper single bond plus, 190
Adsorbents, 236, 237
Amalgam, 2, 59, 64, 74, 191, 253
American Physical Society, 82
Antiadhesive nanoparticles
chitosan, 230
Anti-biofilm approaches, 207
Antimicrobial activity, 52, 89, 91, 108, 111, 112, 115, 130, 161, 162, 211, 225, 226, 228–232, 247
Antimicrobial agent, 39, 47, 91, 108, 109, 112, 124, 162, 209, 220, 224, 225, 231–233
Auto-phosphorylation, 3
Azarin red staining, 13

B
Bioactive fixation, 7
Bioactivity, 6, 7, 9–11, 25–27, 49–51, 75, 97, 248, 253
Bioactivity index (BI), 10
Biocompatibility, 9, 10, 25, 27, 43, 49, 51, 73, 95, 113, 115, 125, 150, 165, 166, 183, 194, 198, 230, 231, 233, 248, 254, 256–258, 262, 263
Biofilm, 3, 32, 42–46, 86, 89, 109, 110, 115, 124, 143, 161, 162, 206, 207, 211, 217, 219–224, 226–228, 233, 235, 239
Biofilm architecture, 44, 217, 220
Biofilm removal
in dental unit water lines, 235
Biofilms in the oral cavity, 220
Bioglasses and bioceramics, 232
Bioinductivity, 259
Biomedical microclectromechanical systems, 166
Bionanotechnology, 82, 139, 147
Bioresorbable polymers, 198
Bis-GMA, 62, 63
Bond strength to dentin, The, 252
β-tricalcium phosphate (β-TCP), 28

C
Calcium phosphate biomaterials, 25
Calcium phosphate nanoparticles (CP-NPs), 48, 49
Calcium silicate, 6, 48–50, 52, 247–251, 253, 254, 257, 258, 260, 267
Camphorquinone (CQ), 72
Candida albicans, 47, 109–112, 115, 124, 235, 236
Candidiasis, 109, 112
Carbon nanomaterials, 38, 83, 85
Carbon nanotubes (CNTs)
reinforced polymethylmethacrylate (PMMA), 183
Caries, 2–5, 59, 62, 81, 91, 92, 152, 161, 163, 204, 221, 222
Carious attack, 2
Cells migration and proliferation, 8, 12
CHA layer formation, 12
Chitosan, 12, 13, 17, 43, 44, 46, 47, 230
Chitosan nanoparticles (CSNPs), 43, 44, 47
Cleaning of endodontic instruments, 41
Clearfil majesty nanocomposites, 189
Collagen-polycaprolactone, 9

R. S. Chaughule (ed.), *Dental Applications of Nanotechnology*,
https://doi.org/10.1007/978-3-319-97634-1

Controlling biofilms, 223
Crystalline quartz, 73
Cytotoxicity, 2, 13, 51, 92–94, 96, 99, 105, 108, 111, 151, 163, 206, 256–258
Cytotoxicity-in vivo, 258
Cytotoxicity on cell lines, 92

D
de Broglie wavelength, 87
Dendrimers, 38, 86, 142, 143, 149, 150, 156, 223, 236, 237
Dendrimers and dendritic copolymers
  liposomes, 144
  nanobelt, 143
  nanofibers, 144
  nanorods, 144
  nanospheres, 144
Dental adhesives, 108, 114, 116, 144, 190
Dental biomaterials, 7
Dental drug delivery systems, 147, 148
Dental hypersensitivity, xii
Dental nanocomposites, 72
Dental resin matrix, 61
Dental resin nanocomposites, 61
Dental unit water lines, 235
Dentifrices, 39, 204, 223
Dentin-enamel junction (DEJ), 1
Dentin hypersensitivity, 41, 204
Dentistry
  nanostructures and its types
    nanopores, 142
    nanoscale cantilevers, 143
    nanoshells, 143
    nanotubes, 142
    quantum dots, 143
Dentistry and green chemistry, 76
Denture base resin, 109, 110, 124, 183
Denture *stomatitis/Candidiasis*, 109
Dimethacrylate-based resins, 61
Direct pulp capping in rabbits, 260
Direct pulp capping in Vietnamese pigs, 263
Drug delivery, 34, 38, 39, 51, 72, 84, 86, 87, 89, 99, 136, 144, 147, 148, 150, 168, 205, 207, 209, 233
Drug delivery by liposomes, 148

E
EDTA, *see* Ethylenediamine tetraacetic acid
E. faecalis biofilm, 42, 52
Efficiency of coatings, 93
Electro spinning technique, 11
Endodontic filling materials, 42, 47
Endodontics, 39, 50, 52, 135, 247
Endodontic surgery
  bone replacement materials, 52
  mineral trioxide aggregate (MTA), 51
  root end filling materials, 51
Eric Drexler, 107, 133, 179
Ethylenediamine tetraacetic acid, 9
Extracellular matrices, 31, 33, 224

F
Fabrication
  of hollow wires, 94
  of orthodontic brackets, 95
Feynman, Richard, 37, 73, 107, 133
Fibroblast growth factor, 3, 167
Filtek supreme universal restorative, 188, 189
Fluorohydroxyapatite, 1, 91

G
GC G-Bond, 190, 191
Gene therapy, 33, 150
Genotoxicity, 110, 255
Glass-ionomer cements (GICs), 2, 17
Gutta percha, 47–49

H
HA nanoparticles, 25, 26, 91
Hexametaphosphate associated with chlorhexidine, 211
Hollow NiTi wires, 94, 158
HPA's solubility, 4
Human microbiome project, 3
Hybrid fillers, 64
Hydrophobic-to-hydrophilic transition, 17
Hydroxiapatite-calcium silicate, 258
Hydroxyapatite (HPA), 1, 9
Hydroxyapatite nanoparticles, 27
Hypersensitivity management, 204

I
Implant coatings, 31, 135, 165, 196
Implant dentistry, 32
Implant-supported fixed prosthodontics, 196
Indirect pulp treatment, 5
Indocyanine green (ICG), 46
Inorganic nanoparticles, 87, 107, 122, 123, 141, 156
Insolubility of NP, 92
Insulin-like growth factor, 1
International System of Units (SI), 24
Intracellular method, 88
In vitro biofilm quantification, 224
IPT procedure, 5
Irrigation, 42

L

L. acidophilus, 6, 163
Low-intensity pulsed US (LIPUS), 167, 170
Luting cements and nanoparticles, 114

M

Management of peri-implantitis, 32
Marginal microleakage, 251, 252
Matrix metalloproteinases, 46
Metallic nanoparticles, 86–89, 98
Metalloproteinase, 1, 46
Metal modified silica nanoparticles
  Copper-SiNP, 232
  Silver-SiNP, 231
Metal oxide nanoparticles, 141, 156, 209
Micro filled composites, 66, 67
Microimplants, 91, 95, 96, 164, 165
Microleakage, 91, 92, 251, 267
Mineral trioxide aggregate, 2, 51
Minimally invasive dentistry (MID), 4, 18

N

Nano-bioglass (nano-BAG), 10, 11, 17, 18
Nanobiomaterials
  in oral cancer therapy, 149
Nanobiotechnology, 82
Nanoceramics, 113, 126, 127, 130, 178, 194, 195, 198
Nanocoatings in archwires, 157
Nanocomposite denture teeth, 185
Nanocomposites, 34, 39, 64, 65, 67, 68, 73, 76, 86, 97, 159, 160, 179, 186–188, 198, 199, 232
Nanocrystalline bone grafts, 24
Nanocrystalline hydroxyapatite, 25, 28, 50
Nanodentistry in prosthodontics
  application of, 32, 123
  in the management of periimplantitis, 32
Nanodiagnostics
  advantages of using nanodiagnostics, 145
  diagnosis of oral cancer and other diseases, 145
  techniques used in nanodiagnosis
    bio barcode assay, 147
    iodinated nanoparticles, 147
    lab-on-a-chip methods, 146
    nano electromechanical systems (NEMS), 146
    nanopores, 145
    nanoscale cantilevers, 146
    nanotexturing, 147
    nanotubes, 145
    nanowires, 147
    optical nanobiosensor, 147
    oral fluid nanosensor test (OFNASET), 147
    quantum dots, 145
Nanoemulsions, 238, 239
Nanoenamel remineralization, 161
Nanoencapsulation, 148, 149
Nanoencapsulation and nanotherapeutics
  nanoparticles as dental drug delivery systems, 147, 148
Nanofilled resin-modified glass ionomer, 192
Nanofillers, 73, 112, 159, 160, 162, 185, 188, 191–193, 195
Nanofiltration, 237
Nanohybrid, 68, 75–77
Nano-hydroxyapatite modified surfaces, 211
Nano-hydroxyapatite (nano-HPA), 9, 10, 13, 17, 18, 48, 50, 52, 91, 97, 204
Nano LIPUS devices
  for bone growth, 167
  to prevent root resorption in orthodontics, 167
Nanomaterials
  biocompatibility of, 150
  biological consequences of, 151
  for brachytherapy, 150
  in prosthodontics, 122, 123
Nanomedicine, 34, 39, 72, 135, 179, 182
Nanometal material and dental implants, 128
Nano-metal oxides to silicone elastomer, 185
Nano-optimized moldable ceramics, 194
Nanoparticles
  applications of, 81, 89, 90, 98
  as enamel remineralizing agents, 97
  in bioimaging, 98
  in denture base resins (Poly Methyl Methacrylate)–PMMA, 109
  in evolving the cell matrix interface, 95
  in hybrid resin luting cement, 128
  in impression material, 112, 127
  in maxillofacial materials, 127
  in resin polymethyl methacrylate, 124, 125
  in tissue conditioner, 111, 130
  shapes and types of, 83
  synthesis of, 88, 89, 107, 231
  to control oral biofilms, 233
Nanoparticulate metal oxide
  Copper oxide, 229
  Titanium dioxide (TiO_2), 230
  Zinc oxide, 229
Nanoparticulate metals as antimicrobial agents, 224
Nanophase crystals of HA, 75

Nanophase materials, 136, 141
Nano-proresolving medicines in management of periodontal diseases, 212
Nanorobotic dentifrices, 203
Nanorobotics coupled with smart brackets-futuristic approach, 168
Nanorobots, 41, 72, 136, 137, 140, 150, 151, 168, 182, 205
Nano sealers, 49
Nanosensors, 136
Nanoshells, 142, 143, 149, 156
Nanosized calcium phosphate particles (NCaPP), 33
Nano-sized fluoroapatite (NFA), 91
Nanosized hydroxyapatite (NHA), 25, 96
Nanostructured hydroxyapatite coatings, 32
Nanostructured metallo-ceramic coatings, 32
Nano-stenciled rgd-gold patterns, 95
Nanotech Elite H-D+, 191
Nanotechnology
  advantages of using nanotechnology in dentistry, 134
  and elastomeric ligatures, 160
  and orthodontic miniscrews, 164
  applications of nanotechnology in dentistry, 135
  approaches in
    bottom up approach, 122
    functional approach, 122
    speculative approach, 122
    top-down approach, 122
  for bone growth and prevention of root resorption in orthodontics, 167
  in endodontics, 39
  in periodontics, 23
  prevention of root resorption in orthodontics, 167
  principle of nanotechnology, 134
  probable risks of nanotechnology, 169
  social implications of, 199
Nanotechnology and orthodontic brackets, 168
Nanotechnology in dentistry
  future perspective, 152
Nanotherapeutics, 138, 148
Nanotopographic patterning, 33
Natural silica, 73
Newkome's "arborol" systems, 86
Nickel-phosphorous and tungsten disulfide (WS_2), 92, 95
Nitric oxide-releasing silica nanoparticles, 231
Nitrogen doped titanium dioxide, 91
Non-collagenous proteins, 1, 2

O

Odontoblasts, 2, 18, 262, 263, 265
Organic–inorganic nanocomposites, 87
Orthodontic archwires, 156
Orthodontic brackets, 91, 95, 156, 159, 160, 163, 168
Ossiointegration, 116
Osteoblast proliferation, 13
Osteoconductive property, 25

P

Peri-implantitis lesions, 28
Periodontal ligament cells, 17, 25, 166
Periodontal regeneration, 23, 33, 152, 247
Periodontal therapy, 23, 33, 203
Periodontal tissue engineering, 13, 33
Periodontics, 135
Phosphorylated and non-phosphorylated matrix proteins, 1
Photodynamic therapy (PDT)/photosensitizers, 45, 81, 223, 234
Plaque accumulation, 91, 206, 221
Platelet-derived growth factor (PDGF), 1, 33
Pollution detection and sensing, 237
Polycationic chitosan nanoparticles, 47
Polyhedralsilsesquioxanes (POSSs) reinforced silicone elastomers, 184
Polyisoprene or polycaprolactone based root canal, 48
Poly(lactic-co-glycolic acid), 9
Polymerization, 34, 60–62, 64, 72, 83, 91, 98, 124, 125, 183
Polymerization shrinkage, 61, 62, 69, 73, 75, 97, 159, 186, 188
Polymethyl methacrylate (PMMA), 92, 124, 125
Polyvinylsiloxane impression materials, 112
Potentiodynamic polarization test, 93
Premise (Kerr Corporation), 189
Primary stability and partial osseointegration of TADs, 95
Primary stability of TADs, 95
Prime & bond NT adhesive, 190
Probing pocket depth (PPD), 28
Proliferation enhancement ability, 13
Prophylaxis programs, 2
Prosthodontics, 107, 113, 116, 121, 123, 130, 177, 182
Prosthodontics material, 113, 116
Proteoglycans, 1
Pseudomonas aeruginosa, 116, 235
Pulp regeneration, 7, 16, 18

PureNano™, 189
Pyrolysis, 26, 88, 97

Q
Quaternary ammonium polyethyleneimine (QPEI), 50

R
Radiography, 41
Rapid tooth movement in orthodontics, 166
Reactive oxygen species (ROS), 38, 43, 226, 233, 256
Regenerative endodontics, 7, 51, 247
Regenerative medicine and tissue engineering, 23
Remaining dentin thickness (RDT), 3
Remineralization, 1, 4–6, 8, 11, 18, 39, 41, 49, 59, 97, 152, 160, 204
Remineralizing potential of bioglass, 11
Reosseointegration, 32
Resin modified glass ionomer cement (GIC), 162
Resin nanocomposites, 110, 186
Root canal disinfection
  nanoparticle-based disinfection, 42

S
Self-healing materials, 97
Shape-memory esthetic polymer wires, 157
Shear bond strength, 91, 92, 159, 162
Silica nanoparticles, 141, 189, 231, 232
Silver nanoparticles
  in dental implants, 115
  in maxillofacial prosthesis, 115
Silver nanoparticles (AgNPs), 32, 40, 42, 99, 108, 110–112, 114–116, 124, 130, 162–164, 209, 225–228, 231, 232, 237
Single-walled carbon nanotubes (SWNTs), 85, 142
Sol–gel method, 26
S. sobrinus, 3
Streptococcus, 3, 92, 112, 160, 163, 164, 211, 221, 236
Surface-modified silica nanoparticles, 232
Synthesized nanoceramic particles, 13

T
TEGDMA, 62, 63
Tetracalcium aluminoferrite, 6
Tetric EvoCeram, 69
Therapeutic nanostructures in dentistry
  fullerenes, 141
TiO_2 nanoparticles, 75, 76, 91, 239
Tissue engineering together with the regenerative medicine (TERM), 7
Tissue regeneration scaffolds, 196, 197
Tomalia's PAMAM dendrimers, 86
Top-down and bottom up approaches, 138
Toxicity, 46, 59, 98, 99, 108, 151, 163, 183, 226, 229, 232, 233, 254, 257
Transforming growth factor-beta 1 (TGF-β 1), 1

U
Ultrafine grain sized titanium (UFG Ti), 95
Ultrasonic spray pyrolysis, 94

V
Variables related to cell adhesion and biofilm formation, 217, 218
Vinyl poly siloxanes, 112
Vital pulp therapy, 50

X
X-ray photoelectron spectroscopy (XPS), 93

Y
Young's modulus, 12, 114

Z
Zeolites, 111, 236
Zinc oxide/eugenol, 2
Zirconia, 7, 73, 74, 97, 113, 114, 125, 129, 187, 194
ZnO NPs, 44, 45

The manufacturer's authorised representative in the EU is Springer Nature Customer Service Centre GmbH, Europaplatz 3, 69115 Heidelberg, Germany. If you have any concerns regarding our products, please contact ProductSafety@springernature.com

Printed and bound by CPI Group (UK) Ltd, Croydon, CR0 4YY
22/09/2026
02221068-0001